J. ERNEST RENAN

AF873607

THE FUTURE OF SCIENCE

IDEAS OF 1848

Elibron Classics
www.elibron.com

Elibron Classics series.

© 2005 Adamant Media Corporation.

ISBN 1-4212-0306-5 (paperback)
ISBN 1-4212-0305-7 (hardcover)

This Elibron Classics Replica Edition is an unabridged facsimile of the edition published in 1891 by Chapman & Hall, Ltd., London.

Elibron and Elibron Classics are trademarks of Adamant Media Corporation. All rights reserved.

This book is an accurate reproduction of the original. Any marks, names, colophons, imprints, logos or other symbols or identifiers that appear on or in this book, except for those of Adamant Media Corporation and BookSurge, LLC, are used only for historical reference and accuracy and are not meant to designate origin or imply any sponsorship by or license from any third party.

THE FUTURE OF SCIENCE.

WORKS BY THE SAME AUTHOR.

RECOLLECTIONS OF MY YOUTH. Crown 8vo, 8*s*.

HISTORY OF THE PEOPLE OF ISRAEL.

VOL. I.—TILL THE TIME OF KING DAVID. Demy 8vo, 14*s*.

VOL. II.—FROM THE REIGN OF DAVID UP TO THE CAPTURE OF SAMARIA. Demy 8vo, 14*s*.

VOL. III.—*In the press.*

THE FUTURE OF SCIENCE

IDEAS OF 1848.

BY

ERNEST RENAN.

"Hoc nunc os ex ossibus meis et caro de carne mea."

TRANSLATED FROM THE FRENCH.

LONDON: CHAPMAN AND HALL, LIMITED.
1891.

Translated by Albert D. Vandam *to page* 284, C. B. Pitman *from page* 284, *and Notes corresponding.*

PREFACE.

The year 1848 made an exceedingly keen impression upon me. Until then I had never given a thought to socialistic problems. Those problems, starting from the earth, as it were, and frightening people, got hold of my mind and became an integral part of my philosophy. A paper on the study of Greek in the Middle Ages which I had begun in answer to a question of l'Académie des Inscriptions et Belles Lettres had engrossed all my thoughts. After that I passed my examination as Doctor of Philosophy in September. Towards October I felt myself again. I felt the need of summing up in a volume the new faith which with me had replaced shattered Catholicism. This took me the last two months of 1848 and the first four or five months of 1849. The beginner's naïve but ambitious dream was to publish that big volume there and then. On the 15th July, 1849, I gave an extract from it to *La Liberté de Penser* with a note to the effect that the volume would appear in a few weeks.

It was a great piece of presumption indeed. About the time I wrote those lines, M. Victor Le Clerc bethought himself to have me, in conjunction with my friend Charles Daremberg, entrusted with various researches in the public libraries of Italy in connection with the literary history of France and a thesis I had begun on Averroism. This journey which lasted eight months influenced my mind very materially. The artistic side of life which had till then been almost closed to me revealed itself resplendent and comforting. A fairy wielding an enchanting power seemed to say to me what the Church in her hymn says to the wood of the Cross.

Flecte ramos, arbor alta,
Tensa laxa viscera,
Et rigor lentescat ille
Quem dedit nativitas.

A sort of soothing breeze made me unbend, nearly all my illusions of 1848 vanished as utterly impossible of realization. I became aware of the fatal necessities of human society, resigned to a state of things in which a great deal of evil is the necessary condition of a small amount of good, in which an imperceptible quantity of aroma is extracted from an enormous *caput mortuum* of spoiled matter.

I became reconciled in certain respects to the reality, and when on my return I took up the book written a twelvemonth before I found it to be harsh, dogmatic, sectarian and hard.

My thought in its primary shape lay on my back like a load sticking out on all sides and getting entangled everywhere. My ideas too autocratic for conversation were still less fit for publication as a whole. Germany, whose pupil I had been for some years, had made me too much in her own image and that in a kind of production in which she does not shine, in "book-making." I felt convinced that French readers would find all this insufferably clumsy.

I consulted several friends, especially Augustin Thierry, who treated me like a son. This worthy man finally persuaded me not to make my *entrée* in the literary world with this enormous bundle on my head. He predicted a complete failure with the public and advised me to proceed piecemeal, to contribute to the *Revue des Deux Mondes* and to the *Journal des Débats* articles on various subjects in which I would get rid retail of the stock of ideas which presented in compact bulk would inevitably frighten the reader. The boldness of the theories would in that way be less likely to shock people's notions. Men of the world often accept in small doses what they refuse to swallow as a whole.

A little while after M. de Sacy encouraged me to do the same thing. The old Jansenist was fully alive to my heresies. When I read my articles to him I noticed his smile at every respectful or cajoling sentence.

There is no doubt that the huge volume whence all this came with its heavy style and very indifferent literary form would have simply horrified him. It was plain enough that if I meant to appeal to cultivated people, I should be bound to leave a great deal of my baggage at the door. My ideas dawn upon me in an involved way, they only become clear after a labour similar to that of a gardener who trims a tree, lops away its dead branches and trains it against the wall.

In that way I retailed the huge volume which thanks to sound advice and friendly counsel had been consigned to the bottom of one of my drawers.

The Coup d'État which happened shortly afterwards had the effect of binding me more firmly to the *Revue Deux Mondes* and the *Journal des Débats*, in my disgust at the people who on the 2nd December greeted the signs of grief of the honest citizen with ironical smiles. I was engrossed with special studies, travel, my "Origins of Christianity" left me no time to think of anything else for twenty-five years. I made up my mind that the old MSS. should be published after my death, that it would then afford pleasure to a select section of enlightened minds, that it might succeed perhaps in fixing on me once more that attention of the world of which the dead stand in so much need in the unequal competition the living thrust upon them in that respect.

My life having been prolonged beyond my expectations I lately made up my mind to be my own publisher. I flattered myself that perhaps some people would read these ancient, honest pages not without profit to themselves and that the rising generation especially, which seems to be somewhat uncertain about its road, would be pleased to find out how a young man, very frank and very sincere thought forty years ago, face to face with himself only. Young people like the work of young people.

In my writings intended for men and women of the world I have been compelled to make many sacrifices. In the following pages which have undergone no process of boiling down the reader will meet with the young, conscientious Breton lad who one day ran away frightened from Saint-Sulpice because he fancied that part of what his masters had told him was perhaps not strictly true. A day may come when the critics will maintain that the *Revue des Deux Mondes* and the *Journal des Débats* spoilt me insomuch that they taught me to write, that is, constantly to condense and prune my ideas. That day they may perhaps like these pages for which I only claim one merit, that of showing in his natural and as yet uncorrupted state and suffering from violent inflammation of the brain a young man living solely with his own thoughts and believing frantically in the truth.

In fact, the blemishes of this first work are enormous, and if I had the slightest literary pride, I should have suppressed it altogether. The way I introduced my ideas lacks the commonest skill. It is a dinner in which the primary materials are good but which has been cooked and served up anyhow, the parings of the various ingredients not even having been removed. I was too anxious not to lose anything. Lest I should not be understood I insisted too much, in order to drive home the nail I fancied myself bound to knock with all my might. The art of composing implying the cutting away of the tangled growths that might obstruct the light in "the forest of thought" was unknown to me. No one is brief at his first start.

The clearness and tact exacted by the French which, I am bound to confess, compel one to say only part of what one thinks and are damaging to depth of thought seemed to me so much tyranny. The French only care to express that which is clear, as it happens the most important truths, those that relate to the transformations of life are not clear; one only perceives them in a kind of half-light. That is why, after having been the first to perceive the truth of what is called Darwinism nowadays, France has been the last to rally to it. They saw it well enough, but it was out of the beaten track of the language, it did not fit the mould of well-constructed phrases. In that way France passed by the side of precious truths not without seeing them, but simply flinging them among the waste paper as useless or impossible to express. At the start I wanted to say everything and I often said it badly. At the risk of tumbling into the realm of the unintelligible I endeavoured to fix the fleeting essence, hitherto considered as not worthy of consideration.

However much and wisely or the reverse I may have modified my habits of style as regards exposition as little have I changed my fundamental ideas from the moment I began to think for myself. My religion is now as ever the progress of reason, in other words the progress of science. But in looking over these pages of my youth, I often found a certain confusion which distorted certain deductions. Intensive culture constantly adding to the sum total of human knowledge is not the same thing as extensive culture disseminating that knowledge more and more for the welfare of the countless human beings in existence. The sheet of water in expanding continues to lose in depth. Towards 1700 Newton had acquired views on the system by which the earth was governed infinitely superior to everything that had been thought out before him without his matchless discoveries having in the least affected the education of the people. On the other hand one might reasonably conceive a state of exceedingly perfect elementary education without the higher sciences deriving much benefit therefrom, Our real motive for advocating elementary education is that a nation without education is fanatical and that fanatical nations are dangerous to science, governments being in the habit of shackling freedom of thought in the names of popular beliefs and the so-called sanctity of family institutions.

Hence, the idea of a state of civilization with a levelling mission, such as it is presented in a few pages of this book is nothing more than a dream. A school in which the pupils would make the laws would be a sorry school indeed. Enlightenment, morality, art will always be represented among mankind by a magistracy, by a minority, preserving the traditions of the true, the good and the beautiful. But we must be on our guard against this magistracy disposing of the public forces and appealing to superstition and imposture in order to maintain its power.

There were also a great many illusions in my acceptance of the socialistic ideas of 1848 in bygone days. While still believing that science alone is capable of improving the unhappy lot of man here below, I have ceased to believe in the solution of the problem being as near as I believed it to be then. Inequality is one of nature's written laws, it is the consequence of liberty and the liberty of the individual is a necessary postulate of human progress. This progress implies a great sacrifice of individual happiness. The actual condition of humanity, for instance, demands the maintenance of separate nations which are establishments exceedingly heavy to bear. A condition which would afford the greatest possible happiness to individuals would probably, from the point of view of the ennobling pursuits of mankind, be a condition of profound abasement.

The main error with which these old pages teem is an exaggerated optimism which fails or is determined not to see that evil still exists and that we have to pay dearly, that is in privileges, the power that protects us against this evil. The reader will also notice an old leaven of Catholicism; the idea that we shall behold once more the age of belief when a compulsory and universal religion will prevail as it prevailed in the first half of the Middle Ages. Heaven preserve us from being saved in that fashion. Uniformity of belief, that is, fanaticism, can only come back in this world of ours in company with the ignorance and credulity of bygone centuries. I would by far have an immoral people than a fanatical people for immoral masses are by no means difficult to deal with while fanatical masses reduce the world to a state of imbecility, and a world condemned to a state of imbecility has no longer any claim on my interest; I would as lief see it perish. Let us suppose every orange tree to be smitten with a disease impossible of cure except on the condition of its no longer producing oranges. It would be so much time wasted seeing that the orange tree which does not produce oranges is worthless.

In order not to make this publication utterly devoid of all interest I had to submit to one condition, namely; to reproduce my essay in its simple, matted, often abrupt form. I might just as well have written a new book as have attempted to correct numberless inaccuracies, to modify a great number of thoughts which at present appear to me either expressed in an exaggerated manner or which are no longer just, and moreover the framework of my old essay is by no means such as I would choose to-day.* Hence, I confined myself to the striking out of mistakes resulting from carelessness, those big blunders which one only notices in proof and which would assuredly have been corrected if I had published the book in times gone by. I have left the notes as a whole at the end of the volume. Many a passage will provoke a smile on the reader's part. It will make no difference to me as long as he, the reader, acknowledges that these pages contain the expression of great intellectual rectitude and perfect sincerity.

A great difficulty resulting from my decision of printing my *purana* as it stands was the resemblance between certain pages of the present work and many of works published before, a resemblance which cannot fail to strike the reader. Besides the fragment published in *La Liberté de Penser* which has been reproduced in my "Contemporary Essays," there are many other passages that have found their way either as regards the mere idea or as regards both the idea and its expression in several of my writings, notably in those belonging to my first period. I tried at first to excise this dualism, but it soon became patent to me that the book would not stand on its legs at all in that way. The parts that had been repeated were the most important, the whole structure, like a wall from which the most necessary stones have been abstracted, was toppling over. The simplest way out of the difficulty, I thought, was to appeal to the indulgence of the reader. Those who

* I have left all the passages in which I presented German culture as being synonymous with aspiration towards the idealistic. They were true when I penned them. It is not I who have changed.

do me the honour of reading my writings in the order they were written will, I trust, pardon those repetitions, if the present volume should succeed in showing them my ideas arranged and combined in a way that may present something novel and interesting to them.

In attempting to strike a balance between what has remained merely so much vision and what has been realized in those dreams of half a century ago I must confess to a feeling of appreciable moral satisfaction. After all, I was right. Excepting a few disappointments progress has travelled on the lines laid down in my imagination. At that period I did not see sufficiently clearly what man had left behind in the purely animal kingdom, I had not a sufficiently clear perception of the inequality of races but I had a just conception of what I may call the origin of life.

I perceived well enough that everything is accomplished in humanity and nature, that creation has no part nor parcel in the series of effects and causes. Too little of a naturalist to track the paths of life in the labyrinth which we see without seeing it, I was a determined evolutionist in all that appertains to the productions of humanity, language, literature, social forms, writings. I began to perceive that the morphological draughtboard of the vegetal and animal species was indeed the indication of a genesis; that everything is born in accordance with a design of which we can only see the obscure canvas. The aim of science is an immense development of which the cosmological sciences give us the first perceptible links, of which history proper shows us the last expansions. Like Hegel I made the mistake of being too confident in attributing to mankind a central part in the universe.

The whole of human development may be of no more consequence than the moss or lichen with which every moist surface is covered. To us, though, the history of man stands first and foremost, seeing that humanity alone creates the conscience of the universe. A plant's only worth lies in its producing flowers, fruit, aroma, nourishing tubercules, which are of no account as a mass, if they are compared to the mass of the plant itself, but which possess the character of finality in a much greater degree than the leaves, branches and trunk.

Historical science and its auxiliaries, philological sciences, have made immense conquests since I took to them so fondly forty years ago. But the end can already be foreseen. In another century mankind will pretty well know everything that can be known about its past; and then it will be time to stop, because the tendency of these studies is to begin their own destruction the moment they have reached comparative perfection. The danger of a revival of superstition will alone keep up the habit of critical disquisition at first hand.

The history of religion has been cleared up in its most important branches. It has become patent, not from *a priori* arguments, but from the very discussion of evidence that in the centuries open to men's researches there has been neither revelation nor supernatural fact. The onward course of civilization has been made manifest in its general laws. The inferiority of certain races to others is proved. The claims of each human family to a more or less honourable mention in the history of progress are pretty well decided.

With regard to the political and social sciences, one may safely say that progress during the last forty years has been slow. The old political economy whose pretensions were so noisily shouted forth in 1848, has been wrecked. Socialism which has been taken up again by the Germans so earnestly and with so much study continues to trouble the world without arriving at a clear solution. Prince Bismarck who was to have stopped its progress in five years by means of repressive legislation has evidently been mistaken, at any rate this time. What appears very probable indeed is that there will be no end of socialism. But assuredly the socialism that will gain the victory will be different from the Utopism of 1848. A keen observer might have seen in the year 300 of our era that Christianity will not end, but ought also to have seen that the world will not end, that the latter will adapt the former to its needs and out of a belief destructive of all society, will make a sedative, a political machine, conservative to a degree.

In politics the situation is by no means more clear. The national principle has since 1848 been developed to an extraordinary extent. Representative government is established nearly everywhere.

But evident signs of the fatigue caused by national burdens are looming on the horizon. Patriotism is becoming local, national enthusiasm decreases. Modern nations resemble the heroes borne down by their armour, on the tomb of Maximilian at Innsbruck, ricketty bodies with iron masters over them. France who was the pioneer on that road will, following ordinary laws be the first to react against the movement she started. In fifty years the national principle will be on the decline. The terrible harshness of the proceedings by which the ancient monarchial States obtained the sacrifice of the individual will have become impossible in free States; scarcely any one nowadays cares to provide the materials for those towers of Tamerlane, built up with corpses. In fact, it has become too clear that the happiness of the individual is not in direct proportion to the grandeur of the nation to which he belongs, and as a rule one generation cares very little about the why or wherefore a preceding generation has sacrificed its life.

These variations spring from the uncertainty of our ideas with regard to the object to be attained and the higher end of humanity. Between the two objects held out by political life, the grandeur of nations, and welfare of individuals the choice is prompted by interest or passion. There is no hint afforded to us either as to nature's will or the aim of the universe. For us, idealists, there exists but one true doctrine, the transcendental doctrine according to which the aim of humanity consists in constituting a loftier consciousness of the universe, or as we used to say, the highest glory of God; but it is very clear that this doctrine will afford no basis for a practical policy. Such an aim must, on the contrary, be carefully dissimulated. Men would revolt if they knew they were being thus exploited.

How long will national spirit be able to hold out against individual egotism? Who, in centuries to come, will have served humanity most, the patriot, the liberal, the reactionary, the *savant?* No one knows and still it would be a capital thing to know, for what is good in one of these hypotheses is bad in the other. One works the switches without knowing whither one wants to go. According to the goal to be reached France is doing either detestable or excellent work. Other nations are more enlightened. Politics are like a desert in which one marches at random towards the north or towards the south; for we must keep on marching. No one knows where the good lies in the social order. There is one comfort, one is sure to land somewhere. In the kind of rifle competition with which humanity is amusing itself the mark hit is supposed to be the mark aimed at. In that way the good and true men always have a clear conscience. For the rest, in the existing state of general doubt, liberty in any case, has its value, since it is a means of allowing free play to the secret spring which moves humanity, and carries it along with or against its will.

To sum up; if through the constant labour of the nineteenth century the knowledge of facts has considerably increased, the destiny of mankind has on the other hand become more obscure than ever. The serious thing is that we fail to perceive a means of providing humanity in the future with a catechism that will be acceptable henceforth, except on the condition of returning to a state of credulity. Hence, it is possible that the ruin of idealistic beliefs may be fated to follow hard upon the ruin of supernatural beliefs and that the real abasement of the morality of humanity will date from the day it has seen the reality of things. Chimeras have succeeded in obtaining from the good gorilla an astonishing moral effort; do away with the chimeras and part of the factitious energy they aroused will disappear. Even glory, as a motive-power implies in some respects immortality, the fruit of it generally coming only after death. Suppress the alcohol on which the workman has hitherto relied for his strength, but you must not ask him for the same amount of work.

Candidly speaking, I fail to see how, without the ancient dreams, the foundations of a happy and noble life are to be relaid. The hypothesis that the true sage would be he who, barring to himself all distant horizons, would confine himself to the perspective of mere vulgar gratification, this perspective, I say, is absolutely repugnant to us. However, man's happiness and noble aims have rested before now on false foundations. The wisest thing to do, then, is to go on enjoying the supreme gifts vouchsafed to us, life and the faculty of seeing the reality. Science will always remain the gratification of the noblest craving of our nature; curiosity; it will always supply man with the sole means of improving his lot. It protects him against error, though it may not reveal the truth to him, but there is an advantage in being certain of not being duped. Man fashioned according to this discipline is on the whole a better man than the instinctive man of the ages of faith. He is not subject to the errors to which the uncultured fatally yield, he is more enlightened, he commits fewer crimes, he is less sublime, but he is also less ridiculous. All this, it will be said, is not worth the heaven science takes away from us. First of all, who knows whether it does take it away; secondly people are none the poorer for being robbed of bogus shares and false banknotes. A little true science is better than a great deal of bad science. One is less liable to error by confessing one's ignorance than by fancying that one knows a great many things one knows not.

Consequently I was right at the outset of my intellectual career firmly to believe in science and to make it the object of my life. If I had to begin again I should do exactly as I have done, and during the little time that remains to me I shall go on as I began.

Immortality means to labour at a lasting work. According to the primitive Christian idea, the true one, only those shall rise again who have contributed to the divine work; furthering God's kingdom on earth. The punishment of the wicked and frivolous will be utter annihilation. Here a formidable objection starts up against us. Can science be more everlasting than humanity whose end is written down from the very fact of its having had a beginning? It matters not; human reason has not been engaged consecutively for more than a hundred years on the problem of matters mundane. It has already made some wonderful discoveries that have increased man's power a hundred, nay a thousandfold. What then will it be a hundred thousand years hence? And pray remember that no truth is ever lost, that no error ever strikes root. All this makes us feel secure. We are really afraid of nothing except of the falling-in of the sky, and even if the sky came crashing down we should still go to sleep quietly with the thought, "The Being of whom we were the transitory blossom has always been, always will be."

To Monsieur Eugène Burnouf, Member of the Academy, Professor at the Collége de France.

MONSIEUR,

During the last twelvemonth my thoughts have frequently gone back to that memorable 25th February 1848, when after having scaled the barricades to get to the Collége de France we found our modest room transformed into a guardroom, to which we were welcomed like so many suspected individuals. That day I asked myself more seriously than ever whether a man could do better than devote every moment of his life to study and to thought, and after having consulted my conscience and strengthened my faith in human intellect, I resolutely answered; "No." If science were nothing more than an agreeable pastime, a kind of diversion for the idle, a mere costly ornament, a hobby for the amateur, in short *the least vain of all vanities*, there might be times when the *savant* would have occasion to say with the poet;

"Shame to him who sings, the while Rome burns."

But if science be a serious matter, if the destinies of mankind and the perfection of the individual be bound up with it, if it be a religion, it has like matters of religion its value every day, every moment of our lives. To devote to study and intellectual culture only our moments of peace and leisure is an insult to the human intellect, it is tacitly supposing that there is something more important than the pursuit of truth. If such were indeed the case, if science were only a matter of second rate importance how could the man who has resolved to devote his life to the attainment of the perfect, who wishes to be able to say at his last moments; "I have accomplished my task," how could that man devote so much as a single hour to it, when he knows that higher duties claim him?

That revolutions and the dread of the future offer a temptation to science ignorant of its aim, to science that has never endeavoured to ascertain its own value and true significance, is easily understood. As for serious and philosophical science which responds to a want of human nature, no social upheavals will succeed in affecting it, they may, on the contrary, be of use to it, by causing it to take itself to task, to verify its titles, to be no longer satisfied with the mere perfunctory judgment on which it was wont to base itself formerly.

These are the reflections, Monsieur, I have made for myself, while remaining isolated and calm amidst the universally prevailing agitation; and which I have embodied in these pages. Thanks to the sentiments with which they have inspired me, I have gone through many a sad day without cursing any one, full of trust in the natural rectitude of human reason and its necessary tendency towards a more enlightened, consequently a more moral and happier state. I hesitated a good deal before making up my mind to disclose in this way the thoughts of my youth, the critic of which I may become perhaps when older, and which no doubt will have but scant value to those further advanced on the road of science. Still, I fancied that some young people enamoured of the beautiful and true might derive comfort and strength from this confidence of mine amidst the struggles which every distinguished mind is bound to wage at a certain age in order to find out and to shape for himself the ideal of his life. I also wanted, at the outset of my scientific career to proclaim my deeply rooted belief in human reason and modern intelligence, at a moment when so many faint-hearted brethren drop tired and spent into the arms of those who profess to regret ignorance while they anathematize criticism. I would fain warn those who take advantage of our weakness, who discount beforehand our misfortunes, who found their hopes on the intellectual fatigue and depression resulting from great suffering, I would fain warn those against thinking that the generation just entering upon "the life of thought" is theirs. We shall be able to uphold the tradition of the modern spirit both against those who wish to bring back the past and against who those aim at substituting for our living multiple civilization a kind of architectural and petrified condition of society like that in which the pyramids were built.

It is not a mere commonplace sentiment which prompts the dedication of this essay to you. It was really thought out before you. Whenever, in my inward hesitations, my scientific ideal seemed to become obscured the thought of you was sufficient to dispel the clouds, you were the answer to all my doubts. It is your image I have had constantly before me when trying to express the lofty ideal which conceives life not as a part to be played or as an intrigue to be accomplished successfully, but as something earnest and true. In listening to your lectures on the most beautiful of languages and literatures of the primitive world I realized what until then had been only a dream; science becoming philosophy and the highest results springing from the most scrupulous analysis of details.

It is to this living proof that I would invite all those whom I may not be able to convince of my favourite thesis; that the science of the human intellect must be above all the history of the human intellect; and this history only becomes possible by the patient and philological study of the works it has produced at its different epochs.

Believe me, Monsieur,

Your respectful and admiring pupil,

ERNEST RENAN.

PARIS, *March*, 1849.

THE FUTURE OF SCIENCE.

CHAPTER I.

"BUT ONE THING IS NEEDFUL." I grant in its widest philosophical bearing this precept of the Great Teacher of morality. I look upon it as the principle of every life striving to be noble, as the expressive formula—though dangerous in its briefness—of human nature from the point of view of morality and duty. The first step of him who aspires to wisdom, as respectable antiquity expressed it, is to divide his life into two parts; the first, commonplace and having nothing sacred in it, consisting of wants and indulgences of an inferior order (material existence, pleasure, wealth, etc.); the other, which may be termed the ideal one, heavenly, divine, disinterested, taking as its aim the purer forms of truth, beauty, moral goodness. In other words, to employ the most comprehensive expression hallowed by reverence in the past, God Himself; God Himself, ever felt, ever perceived, ever touched in His thousand forms by the intelligent perception of all that is true, by the love of all that is beautiful. This is the great opposition of the *body* and the *soul* recognized by all religions, by all lofty philosophical systems; an opposition superficial indeed if it be meant to denote a dual substance in the human being, but perfectly true, if judiciously enlarging the sense of these two words and applying them to two orders of phenomena, we take them as signifying the two roads of life open to man. To admit the distinction between these two roads, is tantamount to admitting that the higher life, the ideal one is everything, and that the lower, the life spent in pursuit of pleasure and interest is nothing, that the latter disappears before the former as the finite before the infinite, and that if practical wisdom commands us to think of it, it is only in view and as a condition of the first-named.

I am aware that, by leading off with such ponderous truths I have virtually written myself down a slow-coach. But I am utterly without shame on this point. For many years already I have elected to take my stand among the simple and dull-witted who take things conscientiously. I am sufficiently weak-minded to look upon that pretence at delicacy which refuses to take life as a serious and sacred business as unbecoming, and very easy to imitate; and if there were no alternative I should prefer the most narrow dogmatism to this flippancy which one honours too much by calling it scepticism, the more appropriate names for it being sheer folly and trash. If it were true that human life is nothing more than a profitless succession of vulgar facts without any higher worth than that derived from the senses, the first serious reflection would lead man to make away with himself; there would be no choice between intoxication, a tyrannical occupation of one's every moment and suicide. To live the intellectual life, to inhale the infinite through every pore, to endeavour to realize the beautiful, to attain the perfect, each according to his ability, that is the only thing needful; all the rest is vanity and vexation of spirit.

Christian asceticism in proclaiming this grand simplification of life understood the one thing needful in so narrow a spirit that its principle became in the course of time a galling yoke to the human intellect. Not only did it wholly neglect the true and the beautiful (philosophy, science, poetry being mere "vanities"), but in clinging exclusively to the good, it conceived it in the meanest spirit. The good according to it meant the realization of the will of a superior being, a kind of *subjection* humiliating to human dignity; because the realization of moral good no more means obedience to imposed laws than the realization of the beautiful in a work of art means the carrying out of certain rules. Consequently human nature was mangled in its most noble members. In matters intellectual, which are all equally holy, a distinction was made between the sacred and the profane. The profane, thanks to the instincts of nature which happened to be stronger than the principles of an artificial asceticism, was not altogether banished; though "vanity," it was tolerated; sometimes Christian asceticism went even so far as to call it the least vain of vanities, but if it had been consistent the profane would have been pitilessly proscribed; it was considered a mere weakness of which the ascetically perfect would have none. It was a fatal distinction which poisoned the existence of many free and beautiful natures born to relish the ideal in all its infinity and whose lives were spent in sadness, crushed in the grip of the fatal vice. Oh, the struggles it has cost me! The first philosophical victory of my youth was to proclaim from the depth of my conscience that, "Everything appertaining to the soul is sacred."

Hence, it is no narrow limit which I am laying down for human nature in suggesting to its activity one thing only as being worthy of it, for this sole thing contains the infinite. It only excludes the commonplace, which has no value except in so far as it is felt and at the moment it is felt; and this inferior sphere is far more circumscribed than is generally believed. There are very few things in human life altogether profane. Our moral and intellectual progress will disclose new standpoints that will invest acts in appearance most homely with an ideal value. Has not Christianity, aided by the instincts of the Celtic and Germanic races, raised to the dignity of an æsthetic and moral sentiment a fact viewed by the whole of antiquity—with the exception of Plato perhaps—as a mere gratification of the senses? Has not the act of nutrition, the most material of life, received a most admirable mystic significance at the hands of the first Christians? Manual labour which nowadays is scarcely anything else but an irksome and brutalizing penalty to those condemned to it, was not such to the craftsmen of the Middle Ages who built cathedrals singing as they built. Who knows but what one day the sight of the general welfare of humanity for whom the work is constructed may soften and sanctify the sweat on the workman's brow? For from the point of view of humanity the most lowly works have an ideal worth seeing that they are the means or at any rate the implied condition of mental conquest. The sanctifying of the material lower life by outward ceremonies and practices is a trait common to all religions. The progress of rationalism at once—and without much merit—proclaimed these ceremonies so many acts of pure superstition. What has been the result? Baulked in its attempts at idealizing, life has become something profane, vulgar, prosaic, so much so that for certain acts in which the need of religious significance was more deeply felt, such as for instance in birth, marriage, and death, the world has preserved the ancient ceremonies, though it may no longer have faith in their efficacy. I am inclined to think that eventual progress will reconcile those two tendencies by substituting moral sentiment in all its purity for sacramental acts, whose only value lies in their signification, and which viewed in their material execution are utterly inefficacious.

Hence, everything connected with the higher life of man, the life by which he is distinguished from the animal, is sacred and deserving of the passionate devotion of high-minded natures. A beautiful sentiment is worth a beautiful thought; a beautiful thought is worth a beautiful action. A system of philosophy is worth a poem, a poem is worth a scientific discovery; a life spent in the pursuit of science is as good as a life spent in the practice of virtue. The perfect man will be he who is a poet, a philosopher, a *savant*, and a virtuous man in one and that not at intervals (periodically) and at distinct moments—for in that case he would only be such in a restricted sense—but by an intimate simultaneous interpenetration at every moment of his life, who will be a poet at the same time that he is a philosopher, a philosopher at the same time that he is a *savant*, in whom, in one word, all the elements of humanity will be blended in a superior harmony, as in collective humanity itself. The weakness of our age of analysis does not allow of such an elevated degree of unity; life has become a trade, a profession, a man is compelled to advertise his title as a poet, artist or *savant*, create for himself a little world of his own in which he lives apart without understanding anything beyond, nay, often denying that anything beyond exists. That this is a necessity of the actual condition of the human intellect, it is impossible to deny; nevertheless we are bound to admit that such a system of life, though warranted by the necessity of it, is contrary to human dignity and the perfection of the individual. Tested as man a Newton, a Cuvier, a Heyne sounds less beautiful than an antique sage, a Solon or a Pythagoras for instance. The final aim of man is not to know, to feel, to imagine, but to be perfect, that is, to be man in every acceptation of the word; to represent in an individual type a condensed picture of complete humanity, and to show blended in one powerful unit all the aspects of life which humanity has sketched at different epochs and places. Man too frequently fancies that morality and morality alone constitutes perfection, that the pursuit of the true and the beautiful is nothing more than a mere enjoyment, that the upright man is the perfect man, such, for instance, as the Moravian brother. The model of perfection is afforded to us by humanity itself, the most perfect life is that which best represents the whole of humanity. And cultured humanity is not only moral, it is also learned, inquiring, poetical, impassioned.

To think that the individual man may one day embrace the whole field of intellectual culture would no doubt be carrying one's hopes for the future of humanity beyond the limits observed by the boldest Utopian. But there are in the various branches of art and science two totally distinct elements which though equally necessary to the production of work scientific or artistic, contribute in a very unequal measure to the perfection of the individual; there is on one side the technical process, the practical skill, indispensable to the discovery of the true, the realization of the beautiful; on the other the mind that creates and animates, the soul that breathes life into the work of art, the great law that lends significance and value to such and such a scientific discovery. It will always be impossible for the same man to handle with the same skill the painter's brush, the musician's instrument, the chemist's apparatus. In all this there is a special education and a practical skill which, to become spontaneous habit calling for no previous consideration demands a life's practice. But that which may become possible in a more advanced form of intellectual culture is the sentiment that endows the composition of the poet or artist with life, the penetration of the philosopher or the *savant*, the moral sense of a lofty nature being united in order to make but one soul, sympathetic with everything that is good, true and beautiful, to constitute a moral type of humanity in the aggregate, an ideal which without being realized in this man or that may be to the future generations what

Christ has been to the past eighteen centuries, namely—a Christ who would no longer represent only the moral side in its highest power, but furthermore the æsthetic and scientific side of humanity.

For after all, all these categories of pure forms perceptible to the intellect constitute but the facets of a same unity. Divergence only begins at a lower level. There is a great central focus in which poetry, science and morality are identical, in which to know, to admire, and to love are one and the same thing, in which all opposing sentiments drop away, in which human nature recognizes the high harmony of all its faculties and that grand act of adoration which sums up the tendency of its whole being towards the everlasting infinite, in the identity of its aim. The saint is he who devotes his life to this grand ideal and votes all the rest useless.

Pascal has shown the necessarily pernicious circle of the positive life in a masterly way. Man labours to obtain rest and then rest becomes unbearable. He does not enjoy life, but only expects to enjoy it. The fact is that worldly people have no well defined system of life—at any rate, as far as I can see. They cannot exactly say what is essential, what is accessory, they are not sure what is the end and what the means. Wealth cannot be the final aim seeing that it has no value except in the enjoyments it procures. Nevertheless, we see the most serious faculties frittered away in the acquisition of wealth and pleasure is looked upon as a relaxation only for lost moments and useless years. The philosopher and the religious man only can take their fill of rest at any moment, seize upon and profit by the fleeting hour without postponing anything to the future.

A man said once to a philosopher of antiquity that he did not think he was born to be a philosopher. "You poor, unfortunate mortal," replied the sage; "for what then do you think you were born?" No doubt if philosophy were a specialty, a profession like any other, if to philosophize meant to study or to seek the solution of a certain number of more or less important questions, then the reply of the sage would be singularly nonsensical. And yet if we understand philosophy in its proper sense, the man who is not a philosopher, that is, who has not succeeded in grasping the loftier meaning of life is indeed a wretched being. A great many people equally willingly give up the title of poet. If to be a poet meant the skilful use of the mechanism of language they would be excusable. But if we understand by poetry the soul's faculty of being touched in a certain way, of yielding a response of a particular and undefinable nature when face to face with the beauty of things, he who is not a poet is not a man, and to give up the title is tantamount to abdicating voluntarily the dignity of his nature.

If needs were illustrious examples could be found to prove that this lofty harmony of the powers of human nature is not an idle fancy. The lives of men of genius nearly always present the delightful sight of vast intellectual capacity allied to very lofty poetical sentiment and charming good nature, to a degree such as to make in most cases their lives in their serene and sweet tranquillity their most beautiful work of all and an essential part of their complete works. Really and truly the words poetry, philosophy, art, science do not signify so much diverse objects offered to the intellectual activity of man as different ways of looking at the same object which is simply existence itself in all its manifestations. That is why there is no great philosopher who is not at the same time a poet; the great artist is often much more of a philosopher than those who bear the name. All these are merely so many different forms, which like those of literature are capable of expressing everything. Béranger found means to say everything in the guise of songs, another haply in the guise of novels, a third in the guise of history. All genius is universal with regard to the object of its efforts and the small minds are just as wrong in trying to establish the exclusive pre-eminence of their art as the great men are right in maintaining that their art is the whole of man, seeing that it enables them, in fact to express that which cannot be divided; namely —the soul, God.

Nevertheless one is bound to admit that the secret to blend those diverse elements is as yet, not found. In the actual condition of human intelligence a nature too richly endowed suffers constant martyrdom. The man born with one eminent faculty which absorbs all the others is far happier than the one who is always discovering within himself new wants which he cannot satisfy. He would need one life to acquire knowledge, another to feel and to love, a third to act, or to speak correctly he would like to lead abreast a series of parallel existences, while still possessing in one superior unity the simultaneous consciousness of each of these. Limited by time and extraneous necessities, his concentrated activity burns itself out inwardly. He requires so much time to live for himself that he finds none to live for the outer world. He does not wish to lose an atom of this all-devouring, multiple existence which escapes him and which he himself devours hurriedly and greedily. He rolls from one world on to another or rather worlds badly harmonized jostle one another in his breast. He envies in turns—for he is capable of understanding in turns—the simple soul that lives by love and faith, the virile nature that takes life like a muscular athlete, the critical and penetrating intellect which enjoys the handling of its exact and certain instrument at leisure. Then, when he finds out the impossibility of realizing this multiple ideal, when he sees how short, how fatally incomplete, how necessarily divided is life, when he reflects that whole sides of his rich and fruitful nature will never emerge from semi-obscurity a reaction sets in full of unparalleled bitterness. He anathematizes this superabundance of life which only leads to his wearing himself out without result, or if he throws his energy on some extraneous work, he still suffers in being unable to throw more than a part of himself into it. No sooner has he realized one side of life than a thousand others just as beautiful flash upon, deceive and lead him on in their turn until the day comes when he is bound to give up and when casting a glance behind he can say at last with some comfort to himself; "I have lived a great deal." It is the first time he has found his reward.

CHAPTER II.

To know is the keyword of the creed of natural religion; for to know is the first condition of the commerce of mankind with the things that are, of the penetrating study of the universe, which is the intellectual life of the individual; to know is self-initiation to God. By ignorance man is as it were sequestrated from nature, shut up within himself, and reduced to make himself a fanciful *non-ego* on the model of his personality. Hence arises the strange world in which infancy lives, in which primitive man lived. Man is only capable of communing with things by knowledge and love; without science he only loves so many chimeras. Science only can supply the foundation of reality necessary to life. If like Leibnitz we conceive the individual soul as a mirror in which the universe is reflected, it is by science that it will be able to reflect a smaller or greater portion of what is, and travel towards its final aim; namely, towards its perfect harmony with the universality of things.

To know is of all acts of life the least profane, for it is the most disinterested, the most independent of gratifications, the most *objective*, to employ the language of the schools. It is a waste of time to prove its sanctity, for those only for whom there is nothing sacred would dream of denying it. Those who go no further than the mere facts of human nature without venturing upon a qualification on the value of things, even those will not deny that science at any rate is the first and foremost necessity of mankind. Man face to face with things is necessarily impelled to seek their secret. The problem suggests itself, and that by virtue of man's faculty of penetrating beyond the phenomenon he perceives. It is first of all nature itself which whets this craving to know, and he attacks the latter with the impatience bred of a naïve presumption, which fancies itself able to draw up a system of the universe at the first attempt and in a few pages. Then his curiosity is tempted by the wish to know all about himself, and much later on by the desire to solve the problem of his species, of humanity at large, of its history. Then comes the final problem, the great cause, the supreme law. The problem gets varied, grows larger and larger, according to the horizons appertaining to each age, but it never ceases suggesting itself; face to face with the unknown man always experiences a dual sentiment; a reverence for the mysterious, a noble recklessness that prompts him to rend the veil in order to know what is beyond it.

To remain indifferent face to face with the universe is utterly impossible to man. As soon as he begins to think, he begins to seek, he puts problems to himself and solves them; he must needs have a system on the world, on himself, on the primary cause, on his origin, on his end. He lacks the necessary data to answer the questions he puts to himself, but no matter. He supplies them himself. Hence, primitive religions, improvised solutions of a problem that required long centuries of research, but to which an immediate answer was necessary. The scientific method is capable of resigning itself to no-knowledge, or it does at any rate submit to delay; primitive science wanted there and then to grasp the meaning of things. In fact, to ask man to adjourn certain problems, to postpone to future centuries the knowledge of what he is, what kind of place he occupies in the world, what is the cause of the world and of himself, to ask him to do this is to ask him to achieve the impossible. Even if he did get to know the enigma insoluble, one could not prevent him from worrying and wearing himself out about it.

I am aware that there is something irreverent, something unlawful, something savouring of high treason against the divine in this bold act of man by which he endeavours to penetrate the mystery of things. At any rate that is how all ancient peoples looked at it. According to them science was a robbery committed to the prejudice of God, an act of defiance and disobedience. In the beautiful myth with which the Pentateuch opens, it is the genius of evil that prompts man to emerge from his state of innocent ignorance in order to become like God by the knowledge distinct and antithetic of good and evil. The fable of Prometheus has no other meaning than that; the conquests of civilization presented as an attempt against, an illicit rape upon, a jealous divinity who wished to keep them to himself. Hence, the proud character of daring against the gods borne by the first inventors, hence, the theme developed in so many mythological legends: that the wish for a better state is the source of all evil in the world. It will be easily understood that, antiquity not having the "key-word" of the enigma, progress was, as it were, bound to feel a respectful dread in shattering the barriers erected, according to it by a superior power, that not daring to rely upon the future for a state of happiness, it conceived it as having existed in a primitive golden age (1), that it should have said, *Audax Iapeti genus*, that it called the conquest of the perfect a *vetitum nefas*. Humanity, in those days, had the sentiment of the obstacle, not that of victory, but though calling itself all the while audacious and daring, it kept marching onward and onward. As for us who have reached the grand moment of our consciousness, it is no longer a question of saying, "*Cœlum ipsum petimus stultitia!*" and to go on committing sacrileges as it were. We must proceed with proudly uplifted head and fearlessly towards that which is ours and when we do violence to things in order to drag their secret from them, feel perfectly convinced that we are acting for ourselves, for them and for God.

Man does not at once become fully conscious of his strength and creative power. Among primitive peoples, all the marvellous exploits of the human intellect are attributed to the Divinity; the wise men believe themselves inspired and thoroughly convinced of their mysterious relations with higher beings and boast of them. Often the supernatural agents themselves are credited with the authorship of works that seemingly exceed the powers of man. In Homer it is Hephæstos who creates all the ingenious mechanisms. The credulous centuries of mediævalism attribute all eminent science or all skill above the common level to secret faculties, to commerce with the Evil One. As a rule the non-reflecting or "little-reflecting" centuries are given to substitute theological for psychological explanations. It seems natural to believe that *grace* comes from on high; it is only later on discovered to emanate from the inmost conscience. The untutored fancies that the dew drops from the sky; he scarcely believes the *savant* who assures him that it emanates from the plants themselves.

When I wish to picture to myself fact as of the progenitor of science in all its primitive simplicity and disinterested impulse, I revert with a feeling of inexpressible charm to the first rational philosophers of Greece. To the psychologist there is a priceless ingenuousness and truth in this spontaneous ardour of a few men who without traditional precedent or official motive but from mere inward impulse of their nature take to grappling with the eternal problem in its true form. Aristotle is already a deep-thinking *savant* conscious of his process who produces science and philosophy as Virgil produces verse. Those first thinkers, on the contrary, are moved by their spontaneous curiosity in a totally different manner. The object is before them, whetting their appetite; they attack it like the child who, growing impatient when confronted with a complicated piece of machinery, tries it in every way in order to get at its secret, and does not stop until he has found the to him sufficiently satisfactory explanation. This primitive science is nothing more than the constantly repeated "*Why?*" of infancy; with this difference though that with us the child finds an instructed person to supply the answer to his question while there it is the child itself who gives the answer with the same simplicity. It seems to me as difficult to understand the true point of view of science without having studied those primitive *savants* as to have the lofty sense of poetry without having studied primitive poesy.

A busy civilization like ours is by no means favourable to the glorifying of those speculative wants. Nowhere is curiosity more keen, more disinterested, more attracted by the outward than in the child and the savage. How sincerely and simply interested they are in nature, in animals, without a single second thought or a respect for humanity (2). The busy man, on the contrary, is bored in the company of nature and of animals; those disinterested enjoyments are no part or parcel of his egoism. Unsophisticated man, left to his own thought, often conceives a more complete and far-reaching system of things than he who has only received a conventional and fictitious education. The habits of practical life weaken the instinct of pure curiosity; but there is comfort to the lover of science in the thought that nothing can destroy it, that the monument to which he has added a stone is eternal, that like morality, it has its guarantee in the very instincts of human nature.

As a rule science is only looked at from the standpoint of its practical results and its civilizing effects. There is no great difficulty in finding out that modern society is indebted to it for its principal improvements. This is very true, but nevertheless, putting the thesis in a dangerous way. One might just as well, in order to establish the claims of morality point exclusively to the benefits society derives from it. Science, as well as morality is valuable in itself and independently of all beneficent results.

These results are, moreover, nearly always conceived in a mean and shabby spirit. As a rule people remain blind to anything but practical applications which no doubt have their importance inasmuch as by their rebound they powerfully contribute to mental progress but which in themselves have little or no ideal value. Moral applications, in fact, almost always lead science astray from its true aim. To study history for no other purpose than the lessons of morality or practical wisdom to be deduced from it is simply to revive the ridiculous theory of those poor interpreters of Aristotle who considered the only object of dramatic art to be the cure of the passions it puts into action. The spirit against which I am especially tilting here is that of English science, so lacking in loftiness, in philosophy. I know of no Englishman, Byron perhaps excepted, who has deeply grasped the philosophy of things. To order one's life in accordance with reason, to avoid error, not to embark upon enterprises that cannot be carried out, to provide for one's self a gentle and assured existence, to recognize the simplicity of the laws of the universe, to get hold of a few views of natural theology seems to be to the Englishman who thinks the sovereign aim of science. There is never as much as an idea of lofty and harassing speculation, never a deep glance at that which is. This, no doubt, arises from the fact, that with our neighbours, positive religion, kept under a conservative sequestration, is held to be unassailable, is still considered as capable of giving the key to the enigma of great things (3). But science, in fact, being only of value in as far as it is capable of replacing religion, what becomes of it under such a system? A kind of petty process to knock a little bit of understanding into folk, a kind of help in obtaining a social status, a means of acquiring useful and interesting knowledge. All this is not worth a moment's consideration. As for myself, I only admit of one result of science, namely, the solution of the enigma, the final explanation to mankind of the meaning of things, the explanation of man to himself, the giving to him in the name of the sole legitimate authority which is the whole of human nature itself, of the creed which religion gave him ready made and which he can no longer accept. To live without a system whereby to explain things is not to live the life of man. I certainly understand scepticism, it is a system that may be as good as any other; it has its grandeur and noble qualities. I understand faith, I envy its possessors and regret perhaps not possessing it myself. But what seems to me most monstrous in humanity is indifference and flippancy. As intelligent as you please, he who face to face with the infinite fails to perceive that he is surrounded by problems and mysteries is to me nothing better than an imbecile.

It has become a hackneyed truth by now to say that the world is governed by ideas. Still, it is after all but saying what ought to be and what will be rather than what has been. There is no gainsaying that in history we should make large allowances for force, for whim, and even for what is called accident, that is to say, to that, the moral cause of which is not proportionate to the effect (4). Philosophy, pure and simple, scarcely had any immediate influence on human progress until the eighteenth century, and it would be much nearer the mark to assert that it is the historical period which creates the philosophy than that the philosophy creates the period. But what admits of no doubt is that humanity amidst its oscillations ever tends to a condition of greater perfection, that it has the right and the power to make reason more and more predominant over whim and instinct in the government of things. It is of no use arguing with him who has not recognized by now that history is not merely so much aimless agitation, a movement without a result. One will never succeed in proving the onward course of humanity to him who has not managed to find it out for himself. The first word of the creed of the nineteenth century is contained in the immense results achieved by the science of humanity during the last hundred years. Above the individual stands collective humanity which lives and develops like every other organic being and which like every organic being tends towards perfection, that is, to the plenitude of its being (5). After having groped for many long centuries in the darkness of infancy without consciousness of itself and by the mere motive power of its organism, the grand moment came when, like the individual, it took possession of itself, as it were, when it became aware of its own strength, when it felt itself to be a living unity; a moment for ever to be remembered, a moment we do not see because it is too near to us, but which, it seems to me, will be considered by future generations as a revolution comparable to that which has marked a new era in the history of all the nations. Barely half a century has elapsed since humanity began to understand and reflect upon, itself; and still we profess to be surprised at the consciousness of its unity and mutual adherence being so weak (6). The French Revolution is the first attempt of humanity to take the reins in its own hands and to drive itself. It is the advent of the power of reflection in the government of humanity. It is the moment corresponding to that in which the child hitherto led by spontaneous instinct, by mere whim, by the will of others, takes up his stand as a free, moral being, responsible for his acts. All that went before may be called in the words of Robert Owen "the irrational period of human existence;" and one day this period will only count in the history of humanity and in that of our nation in particular as a curious preface, something like that chapter on the history of the Gauls which generally precedes the history of France. The real history of France begins at 1789, all that goes before is but the slow preparation of '89 and is of no interest except as viewed in that light. In fact, study history as you may you will find nothing analogous to that immense fact presented by the whole of the eighteenth century; of philosophers, men of wit in no way concerned with actual politics, radically changing the whole of previously received ideas and carrying the greatest of all revolutions, conscientiously, and with deliberation on the faith of their systems. The revolution of '89 is a revolution wrought by philosophers. Condorcet, Mirabeau, Robespierre are the first instances of theorists meddling with the direction of affairs and endeavouring to govern humanity in a reasonable and scientific manner. All the members of the Constituent Assembly, of the Legislative Assembly and the Convention were literally and almost without exception disciples of Voltaire and Rousseau. I will show by and bye how this chariot driven by such hands could not be driven so well at first as it was when it rolled along by itself and how it was almost bound to be shattered to pieces in an abyss. For the present it is sufficient to note the matchless audacity, the marvellous and bold attempt to reform the world according to the dictates of common sense, to attack everything savouring of prejudice, the blind established thing, habits to all appearances irrational, in order to replace them by a system calculated like a formula, combined like an artificial machine (7). This, I repeat, is a thing unique and without example in all the preceding centuries, nay, it constitutes in itself an age in the history of humanity. Surely, an attempt such could not be without blemish in every respect. Because those institutions that seem so absurd are not altogether so absurd as they seem, those prejudices have their sensible side which you fail to see. The principle involved in all this admits of no controversy, intelligence alone must reign, intelligence alone; in other words, sense must govern the world. But are you sure that your analysis is complete, that you are not led into denying what you do not understand and that a more advanced philosophy will not succeed in justifying the spontaneous work of humanity? There is nothing easier than to show that the majority of prejudices on which the old society was based; the privileges of the nobility, the law of primogeniture, legitimacy, etc., are irrational and absurd from the point of view of abstract reason, that in a society constituted regularly such superstitions would find no abiding place. There is an analytical and seductive clearness about this such as the eighteenth century loved. But is that a reason to blame absolutely these abuses in the old structure of humanity in which they enter as an integral part? It is certain that the criticism of those first reformers was on many points harsh, that it showed the non-intelligent side of spontaneity, the arrogant pride of the easy discoveries of analytical reason.

As a rule the philosophy of the eighteenth century and the policy of the first Revolution show the errors inseparable from crude reflexion, the non-intelligent side of mere mother-wit, the tendency to consider as absurd that of which one fails to see the immediate reason. That century only understood itself and judged all the others by its own. Dominated by the idea of the inventive power of man, it extended too much the sphere of deliberate invention. In poetry it substituted artificial composition for the innate inspiration that wells up from the recesses of the heart without troubling about literary composition. In politics man was supposed to create freely and deliberately society and the authority that rules it. In morality man found and established the principle of duty as a useful invention. In psychology he seemed to be the creator of the results most necessary to his constitution. In philology the grammarians of the period spent their time in showing the inconsistency, the errors of speech such as the people had made it and in correcting the deviations that had become habit by logical argument without perceiving that the terms which they want to suppress are as a rule more logical, clearer and easier than those they want to substitute for them. That century did not understand nature, spontaneous activity. No doubt man produces, in a certain sense, all that comes out of his nature. He spends all his active energy upon it, he supplies the brute strength which brings about the result, but the directing of all this does not belong to him. He supplies the material, the shape of it comes to him from on high; the real author is that living and truly divine force, secreted as it were by the human faculty, which is neither convention, nor calculation, but which produces its effect out of its own self and by its own tension. Hence arises the tendency towards the artificial, towards the merely mechanical with which we are still so deeply smitten. We fancy that we shall be able to foresee all possible cases, but the work is so complex as to set all our efforts at naught. The holy horror of the arbitrary is carried so far as to become destructive of all initiative. The individual is so swaddled round with rules and regulations, cramping his every limb that a lay-figure could do as much as he if we fashioned it to move to the turning of a handle. The difference between mediocrity and distinction in the individual has in that way become almost insignificant, administration has become a soulless machine which will accomplish the work of a man. France is too apt to believe that one may supply the private impulse of the soul by mechanism and extraneous process. Nay, there have been attempts to apply this detestable spirit to even more delicate things, to education, to morality (8). Have not we seen ministers of public instruction who pretended to turn out great men by means of suitable regulations? Did not they conceive a process of making man moral as one makes fruit ripe by squeezing it between one's fingers? Ye of little faith in nature, why not leave it to the sun?

Such then are the excusable and necessary errors of centuries in which reflection takes the place of spontaneity (9). And although this first degree of consciousness is an immense progress, the condition resulting from it may have seemed in some of its aspects inferior to that which went before, and the enemies of humanity have been enabled to take advantage of it by combating the dogma of progress with some show of plausibility (10). In fact during the blind and irrational state, affairs proceeded spontaneously and by themselves by virtue of established order. The world had institutions made of one piece, the origin of which was never questioned, dogmas that were accepted without criticism. The world was a huge machine organized so long ago and with so little thought that people believed it had been put together by God Himself. Such was no longer the case the moment humanity wanted to govern itself and to underpin the instinctive structure of ages. Instead of old institutions whose origin was lost and which merely seemed the necessary result of the equilibrium of things, it had constitutions made by the hands of men, brand-new, with corrections and sentences struck out and from this very fact shorn of the old prestige. And seeing that it knew the authors of the new work, that mankind considered itself their equal in authority, that the improvised machine had visible defects, and that the whole of the business was henceforth transferred to a field open to discussion there was no reason why it should ever be declared closed again. The result was an era of upheaval and instability during which the dull but honest minded might well regret the old order of things. One might just as well prefer the positive assertions of bygone science which was never at a loss to the prudent hesitations and fluctuations of modern science. No doubt, the uncontrolled reign of absolutism in politics as in philosophy is the one conducive to the greatest amount of rest and the *grands seigneurs* who are fond of rest are most likely to be fond of such a *régime*. Oscillation on the other hand is the necessary condition of true human development, and modern constitutions are perfectly consistent in laying down periodical terms for their further modification.

Hence it is not surprising that after the disappearance of the primitive state and the destruction of the old edifices built up by the blind conscience of ages, there should be some regrets, and that the new structures should be by no means equal to the old. Imperfect thought cannot reproduce at the first attempt the works of human nature, acting with all its innate forces. Combination is as powerless to reconstruct the works of instinct as art is powerless to imitate the blind work of the insect that spins its web or builds its honey-cells. Is that a reason to give up reflective science, to go back to blind instinct? Certainly not. It is a reason for continuing to inquire to the end, with the assurance that perfected thought will reproduce the same works but with a higher degree of clearness and reasoning. We must hope, march onward, and keep on marching always, despising meanwhile the objections of sceptics. Besides, the first step has been taken already, humanity has once for all begun the task of its emancipation, it has attained its majority, wishing to govern itself, and supposing even that advantage might be taken of a moment's sleep to impose fresh chains upon it, it will be mere child's play to break them. The only means to reconstitute the bygone condition of things would be to destroy its consciousness by destroying science and intellectual culture. There are people who know this, but I pledge you my word that they will not succeed in doing it.

Such then is the condition of the human intelligence. It has overthrown gothic edifices, constructed one knows not well how, but which sufficed nevertheless to shelter humanity. Then it tried to rebuild the edifice on better proportions, without, however, succeeding; for the old temples raised by humanity had some wonderful subtleties of architecture and design which were not perceived at first and which the modern engineers with all their geometry cannot contrive to compass. Besides, the world has become more difficult to please and does not like to fatigue itself in sheer waste. The preceding centuries did not complain of the organization of society because there was no organization to speak of. The evil was accepted as emanating from fate. What would arouse an outcry nowadays did not provoke a murmur then. The neo-feudalistic school has taken a singularly unfair advantage of this misunderstanding. What are we to do? Reconstruct the old temple? That would be more difficult still, for even if its original plan were not lost, the materials could never be found again. What is wanted, is to look for the perfect beyond, to push science to its furthermost limits. Science and science alone is capable of restoring to humanity that without which it cannot live; a creed and a law.

The dogma which must be maintained at any cost is that the mission of intellect is the reforming of society according to its own principles; that it is by no means conspiring against Providence to attempt to improve its work by well considered efforts. True optimism can only be conceived under such conditions. Optimism would be a mistake if man were not susceptible of being made more perfect if it were not given to him to improve the established condition of things by science. The formula; "Everything is for the best," would without this be only a bitter mockery (11). Yes, everything is for the best thanks to human intelligence capable of reforming the *necessary* imperfections of the first establishment of things. Let us rather say, "Everything will be for the best when man, having accomplished his legitimate task, shall have restored harmony in the moral world and conquered the physical world." As for bygone conceptions of Providence in which the world is conceived as made once for all, and bound to remain as it is, in which the effort of mankind against fate is considered as so much sacrilege, these conceptions are vanquished and obsolete. What is very certain, at any rate, is that they will not hinder man in his task of reformation, that he will persist *per fas et nefas* in correcting Creation; that he will pursue to the end his holy work, viz. to fight blind causes and the fortuitous establishment of things, *to substitute reason for necessity.* The religions of the East enjoin man to put up with evil; European religion is summed up in the few words; "Fight against the evil." This race is verily the offspring of Iapetus; it is bold against God.

The keen observer will notice that this is the kernel of the problem, that the whole of the struggle at this moment lies between the old and the new ideas of theism and morality. It is sufficient that he should see it. We have reached the sacred line where the doctrines divide; one point of divergence between two rays starting from the centre places the infinite between them. But this much should be remembered; that the theories of progress are irreconcilable with the ancient doctrine of divine justice, that their only meaning lies in attributing divine action to the human intellect, in one word, by admitting as the primordial power in the world the reforming power of *the spirit.*

The secret link of these doctrines is nowhere more apparent than in the last book of M. Guizot, a book of the greatest value and which will retain the privilege of being read by posterity, because it sets forth in a highly original manner a curious intellectual moment. Will people five hundred years hence believe that one of the foremost minds of the nineteenth century could have said that since the emancipation of various classes of society the number of distinguished men has not increased in France, "as if Providence," he adds, "did not permit human laws to affect the intellectual order of things, with regard to the extent and magnificence of his gifts" (12). The Aristarchi of that time will consider this passage an interpolation and will advance peremptory proofs to that effect. They will say that so narrow-minded a conception of the government of the world could have never entered the mind of the author of "The History of Civilization." But how will they manage to excuse an argument like the following; "Up to this time society has always presented three types of social condition; men living on their income, men exploiting their income, men living by their work. Hence we may take this to be the natural condition of humanity, and thus it will ever be." Equally valid would the argument of antiquity have been; "Society has up till now always had three classes of men, an aristocracy, freedmen, slaves. Hence we may take this to be the natural condition of humanity, hence it will always be." It would have been just as reasonable to say in 1780, "Till now the State has always contained three classes of men, the governors, the aristocracy, limiting the power of the latter, the plebeians; hence we may take this to be the natural condition of humanity; hence, you who wish to change this condition of things are nothing better than a parcel of dangerous lunatics, of Utopians."

Assuredly no one is more profoundly convinced than I am of the impossibility of reforming human nature. But narrow-minded and dictatorial people have a strange way of interpreting human nature. To them human nature means that which they see of it existing in their own times and the preservation of which they ardently desire. There are a great many better reasons for maintaining that a privileged aristocracy is essential to every society than for maintaining that a moneyed aristocracy is necessary to it. The truth is that human nature is only made up of instincts and very general principles which by no means sanction or sanctify one social condition in preference to another, but only certain conditions of the social fabric, such as for instance the family, individual property. The truth is that with the eternal principles of his nature at work, man can reform the political and social edifice; he can do this, seeing that undoubtedly he has already done it, seeing that there is no one who does not admit actual society to be better organized in certain respects than that of the past. "It is the result of religion;" people will say. Granted that it is, but what is religion if not the most beautiful and energetic creation of human nature? The appeal to human nature is the final argument in all social and philosophical questions. But we should be careful not to invoke this nature in a petty and narrow-minded manner with regard to the habits and customs, the order of things we are actually witnessing. It is a much deeper ocean, the bottom of which is not so easily reached, and which the weak-sighted cannot even perceive. How numerous are the ridiculous errors in commonplace psychology springing from neglect of this principle! They are nearly all due to the narrow ideas that prevail with regard to the revolutions already undergone by the moral and social system, to ignorance of the profound difference existing between the various literatures, and the feelings of various peoples.

Without altogether pinning his faith to one particular system of moral reform no lofty and penetrating mind will be able to deny that the very question of that reform is of kind different from that of political reform, the legitimacy of the latter being, I trust, beyond controversy. The social fabric like the political has been shaped under the influence of blind instinct. It is the mission of human intelligence to correct it. It is not a bit more illegal to say that society may be improved than to wish the Shah of Persia to improve his government. The first time this terrible problem was assailed; to reform political society by human reasoning there was no doubt an outcry at the boldness of the thought, at the unheard-of attempt. The conservatives of '89 could oppose to the revolutionaries what the conservatives of 1849 opposed to the socialists. You attempt to do a thing which has no precedent, you are attacking the work of ages, you do not take into account history and human nature. The cheap bombast of the middle classes against the hereditary nobility; "You only had the trouble to be born, etc.," may with advantage be retorted against the moneyed classes. It is very patent that the existence of a nobility is not rational, that it is the result of the blind ordering of humanity. But if we are to argue in that way, where are we to stop? There is no great merit in twitting it with its want of rationality; it is simply an indefensible truism. I am even bound to admit that, all things considered, the attempt of the political reformers of '89 seems to me a great deal bolder in its aim and above all more wonderful than that of the social reformers of our days. Hence, I fail to understand how people who admit '89 can reject as a matter of right social reform. As for the means, I understand, I repeat, the most radical diversity. No general difficulties have been advanced against the socialists which may not be equally advanced against the constituents. It is a bold thing to assign limits to the reforming power of human reason and to reject no matter what attempt on the plea that it is without precedent. Every reform was characterized by the same defect originally, and besides, they who prefer that reproach do so nearly always because they have not a sufficiently extensive idea of the various forms of human society and of its history.

In the East thousands of people die of starvation or of wretchedness without ever having thought of revolting against the established powers. In Europe, rather than die of hunger a man thinks it simpler to snatch up a rifle and to attack society, guided as he is by that profound and instinctive view that society has duties with regard to him which it has never fulfilled. We find at every page of our actual literature some remarkable turn, which is perhaps not thirty years old; it is merely a way of looking upon individual suffering as a social evil and making society responsible for the wretchedness and degrading condition of its members. A novel idea, a thoroughly novel idea, assuredly. We have ceased to consider those evils as emanating from fate (13). Well, we had better remember that humanity has never taken up a standpoint to relinquish it immediately afterwards.

Hence, by every way open to us we are beginning to proclaim the right of human reason to reform society by means of rational science, and the theoretical knowledge of existing things. It is, therefore, no exaggeration to say that science contains the future of humanity, that it alone can give us the explanation of its destiny and teach it the way to attain its object. Until now it is not human reason that has governed the world, but whim and passionate impulse. The day will come when reason enlightened by experience will resume its legitimate sway, the only one that can claim the title of "right divine," and will lead the world, no longer at haphazard, but with a clear perception of the goal to be reached. Our period of passionate impulse and error will then appear as so much pure barbarism, or as the capricious and fantastical age which in the child divides the charms of tender age from the rational existence of the mature man. Our mechanical politics, our blind and selfish parties will seem like so many monsters of another age. People will no longer understand how a century could have accorded the title of "able" to a man like Talleyrand, who looked upon the government of mankind as upon a mere game of chess, without having an idea as to the object to be attained, without having as much as an idea of humanity itself. The science which will govern the world will not be politics. Politics, that is, the way to govern humanity like a machine will vanish as a special art as soon as humanity shall cease to be a machine. The master science, the then sovereign, will be philosophy, that is to say, the science which will investigate the aim and conditions of society. "In politics," says Herder, "man is a means, in morality, he is an end. The revolution of the future will be the triumph of morals over politics.

Hence, the *scientific organization of humanity* is the final word of modern science, that is, its bold, but legitimate pretension. I will go further still. The universal task of all that breathes being to *make God perfect*, that is, to further the grand final result which will close the circle of humanity by the unity of the whole, it admits of no question that *human reason* which until now has had no share in this work, the latter having been accomplished blindly and by the mere tendency of everything that is, it admits of no question, I repeat, that human reason will one day take in hand the management of this work and after having organized humanity will *organize God* (14). I do not insist upon this point and am willing that people should treat it as a mere illusion, because to many worthy minds which I wish to please, it would seem questionable form; besides, I do not require it in support of my thesis. I will confine myself to saying that nothing should astonish us, considering that the whole of the progress accomplished up to the present moment is perhaps no more than the first page of the preface of a work without end.

CHAPTER III.

PEOPLE may, if they like, consider the whole of the foregoing as absurd and chimerical, but I beg of them, in the name of Heaven, to grant me this, that science alone can supply mankind with those vital truths, without which life would be unbearable and society impossible. If we could conceive the possibility of arriving at those truths in any other way than by the patient study of things, higher science would have meaning no longer. We should have erudition, the curiosity of the amateur, but not science in the noblest acceptation of the term, and noble natures would assuredly forbear engaging in researches, having neither horizon nor future. Thus those who think that metaphysical speculation, pure reasoning, can, without the *pragmatic* study of what is, supply us with the higher truths must necessarily despise that which to them is nothing more than useless lumber, an unnecessary and cumbersome burden to the intellect. Malebranche has not been too severe upon those *savants*, "who make their brain a storehouse in which they pile up without discernment everything that presents a certain character of learning, and who pride themselves on their likeness to those collections of curiosities and antiquities and which have neither a monetary nor an archæological value, and the price of which simply depends upon fancy, accident or passion." Those who think that matter of fact sense, common sense is a sufficiently efficient teacher to mankind must look upon the *savant* in about the same way that Socrates looked upon the sophists; as useless and subtle disputators. Those who think that feeling and imagination, the spontaneous instincts of human nature can get at essential truths of life by a kind of intuition will be equally consistent in considering the researches of the *savant* as ponderous and of no use, or else as frivolous superfluities not having the merit even of being amusing. In short those who think that human reason cannot attain to the higher truths and that a superior power only has the mission of revealing these to them also contribute to the destruction of science by depriving it of what constitutes its life and its true value.

What, in fact, does there remain, if you deny science its philosophical aim? Trifling details, capable, no doubt, of whetting the curiosity of inquiring minds and of providing a pastime to those who have nothing better to do; very indifferent to those who look upon life as a serious matter and who above all, concern themselves with the moral and religious needs of man. Science is of value only in as far as it can investigate what revelation professes to teach. If you take away that which constitutes its worth you leave it only an insipid *residuum*, fit at best to fling to those who feel the want of a bone to gnaw. I sincerely congratulate the good souls that are content with this, as for myself I will have none of it. The moment a doctrine intercepts my horizon, I declare it to be false, the infinite only is to be my background. If you offer me a system ready-made, what then remains there for me to do? To verify by rational research what revelation teaches me? That would indeed be a useless exercise, a frittering away of time most frivolous, for if I know beforehand that what I have been taught is absolutely true there is no need of my tiring myself in looking for its demonstration. It is tantamount to wanting to observe the stars with the naked eye when one might use a telescope. It is tantamount to appealing to men when one may claim the authority of the Holy Spirit. I only know of one contradiction more flagrant than that; a constitutional pope.

I shall be told that there remains a vast field of inquiry in the natural truths which God has given for disputation by mankind. Vast you call it, when you take away God, man, mankind, the origins of the universe. I myself think it very narrow and at best fit for those who to their need of believing add the need of disputing. You think I ought to be very thankful for allowing me to exercise my mind on a few not clearly defined points by flinging to me the world as a bone of contention and by warning me distinctly that from the first word to the last I shall not understand a syllable. Science is not a dispute of a few otiose minds on a few questions left to them as food for their taste for controversy. Lives there the lofty mind that would devote his life to such humble and debasing labour? I feel reluctant to answer because to remove beforehand the objections that might be addressed to me here, would require long explanations and numerous reservations; profane science in any system of *frankly* admitted revelation can only be a *disputation* (15). That which is essential has been given; the only serious science will be that of commenting on the revealed word; no other will be of any value except in connection with this. Orthodox people have as a rule very little *scientific honesty*. They do not *investigate*, they try *to prove* and this must necessarily be so. The result has been given to them beforehand; this result is true, undoubtedly true. Science has no business with it, science which starts from doubt without knowing whither it is going, and gives itself up bound hand and foot to criticism which leads it wheresoever it lists. I know the theological method very well and may safely affirm that its process is opposed to the true scientific spirit. Heaven forbid I should deny that among the most sincere believers there have been men who have rendered the most eminent services to science; and to go no further back than our contemporaries it is among the sincerest Catholics that I should perhaps find the men most sympathetic to my intellect and heart. But if I were allowed to come to a very close understanding with them, we should soon see how far their scientific ardour partakes of the character of a noble inconsistency. Let me be allowed to cite an instance in point. To my mind Silvestre de Sacy is the type of the orthodox *savant*. Undoubtedly, one could not possibly demand science of a higher standard as far as correctness and criticism of detail go. But if one looks higher, we have the strange fact of one of the most learned men of modern times never having arrived at a single idea of lofty criticism. When for critical or other purposes I am studying the man's works—an eminently respectable man, I am always tempted to ask him, "What is the good of it all? what is the use of knowing Hebrew, Arabic, Samaritan, Syriac, Chaldaic, Ethiopian, Persian, what is the good of being the foremost in Europe in the knowledge of the literatures of the East, if one has not grasped the idea of humanity, if the whole of all this has been conceived without a higher and religious aim." True higher science only commence when human reason conceives its task seriously, when it says to itself; "Everything else fails me; my salvation depends upon myself." It is only then that one resolutely sets to work; it is then that everything reassumes its value in view of a final result. We have done with playing at science, with making it a theme for insipid and pointless paradoxes (16); we are upon the great business of man and mankind; hence, there is begotten a seriousness, an attention, a respect unknown to those who only embarked upon science with but one part of themselves. One must not be too exacting; to work out one's salvation is the only thing needful, one will lend one's self to the rest as to a secondary matter, one will not be comfortable in it; if one's taste leans too much that way one will reproach one's self for it as for a weakness, one will be only semi-profane, one will do like Saint Augustine and Alcuin who accuse themselves of being too fond of Virgil. They are not as guilty as they think they are. Human nature, in reality stronger than all the religious systems hits upon some secret modes of taking its revenge. Has not Islamism, by the most flagrant of contradictions nourished in its own bosom a development of purely rationalistic science? Kepler, Newton, Descartes and the majority of the founders of modern science were believers. Truly, a strange illusion, which proves at least the good faith of those who undertook that work, but more still the fate that impels the human intellect, entering the paths of rationalism, to an absolute breach, which at first it repels, with all positive religion. With some of those great men this was explained by the limited view they had of science and its aim; with others as with Descartes (17), whose pretension was really to deduce from reason the truths essential to mankind, there was a manifest superfetation, the use of two mechanisms to attain the same end. I beg the reader to remember that there is no need for me to take my stand here as a controversialist, to prove that science and revelation contradict one another; it is sufficient for me that there is double employment to prove my actual thesis. In a revealed system, science has only a very secondary value and does not deserve devoting one's life to it, because that which constitutes its worth is given elsewhere in a much more eminent manner. No one can serve two masters, nor worship a double ideal.

As for me, I say it with the candour which, I trust, the reader will not question (he who is not candid at twenty-five is a wretch) that I cannot conceive the higher science, the science understanding its aim and its end, except as outside all supernatural belief. It is the pure love of science that made me break the bonds of all revealed belief, and the day I declared myself to be without any other master than human reason, I felt that I was laying down the conditions of science and philosophy. If on reading these lines some religious soul should fancy that I am insulting him, I should tell him; "No, I am your brother; I would insult nothing that belongs to the soul. It is because I am in earnest that I speak like this, it is because I look upon things religious in the most serious light." If like so many others I looked upon religion as merely a machine, a dyke, a useful prejudice, I should assume that indescribable semi-tone which in reality is only so much indifference and flippancy. But seeing that I believe in truth, the same that I believe Christianity to be a serious and important thing, there is a quasi-air of the controversialist about me and certain squeamish minds will I am certain raise the outcry about a recrudescence of Voltairianism. I am glad to have the opportunity of telling people once for all that if I import into religious discussion a frankness and heaviness of hand which are no longer the fashion, it is because I take it in sober earnest and with the deepest respect. You have no more dangerous enemy, gentlemen than those wary, half-hearted and merely insinuating critics. The least controversial age is after all the most incredulous and frivolous one. If, therefore, I am more candid and pointblank, it is because I am more deferential to, more anxious for *intrinsic truth.* Of course people will say that it argues a want of tact to take things in that way.

I shall often, in my life, have occasion to speak of Christianity. How could I do otherwise? The glory of Christianity lies in the fact of its engrossing still half of our earnest thoughts, of engrossing the attention of all thinkers, of those who struggle as well as of those who believe. I managed for a long while to write and to think as if there were no religions in the world, like so many rationalistic philosophers who have written volumes upon volumes without broaching a word about Christianity. But this abstention seemed to me subsequently so irreverent towards history, so partial, so great a denial of all that is most sublime in human nature, that, at the risk of offending inquisitors and philosophers alike, I have made up my mind to take the human intellect as it is and not to deprive myself of the study of its more beautiful half. In my opinion religions are worth speaking about and there is as much philosophy in the study of them as in a few chapters of dry and insipid moral philosophy.

The day is not far distant when with a little candour on both sides and by the removal of misunderstandings that divide those best fit to understand one another, the world will be bound to admit that the lofty perception of things, higher criticism, deep love, truly divine art, and the sacred ideal of morality are impossible except on the condition of taking one's stand at the very outset in the divine; of declaring that everything which is pure, beautiful and lovable is equally sacred and worthy of worship, of considering everything that is as appertaining to one sole order of things which is nature itself, as the variety, the blossoming, the germination of a selfsame and living substratum.

Science really worthy of the name is therefore impossible except on the condition of most perfect autonomy. Criticism is no respecter of things; it neither stops at mystery or prestige, it breaks every charm, it pulls aside every veil. This power, utterly lacking in reverence casting an unflinching and scrutinizing glance on everything alike, is from its very essence guilty of high treason against the divine and the human. It is the sole authority without control; it is the spiritual man of Saint Paul "who judgeth all things, yet he himself is judged of no man." The cause of criticism is the cause of rationalism, and the cause of rationalism is the cause of the modern spirit itself. To curse rationalism is to curse the whole development of human intelligence from Petrarch and Bocaccio, that is, from the first appearance of the critical spirit. It is crying back to the Middle Ages. No, it is not that, because even the Middle Ages had their bold attempts at rationalism. It is tantamount to proclaiming the uncontrolled sway of superstition and credulity. The real question at issue is to know whether we are to go back five centuries and to blame a development which was evidently called for by the necessity of things. And *à priori* and independently of all examination such a development carries its own legitimacy with it, for though the present century may not be infallibility itself, it is symbolic of the moment, and if there be an appeal against it, it must be an appeal to the future, not to the past. If, in fact, we study the march of modern criticism since Petrarch and Bocaccio we shall find it always following the line of its inflexible progress, overthrowing one after the other all the idols of incomplete science, all the superstitions of the past. First of all it is Aristotle, the god of mediæval philosophy who succumbs beneath the blows of the reformers of the fifteenth and seventeenth centuries with their grotesque procession of Arabs and commentators; then comes the turn of Plato, who, set up for a little while against his rival, his doctrines preached like the gospel, recovers his dignity in dropping once more from the rank of prophet to that of man. After that it is the whole of antiquity which resumes its real significance and its importance—wrongly understood at first—in the history of human intellect; then comes Homer, the ideal of ancient philology, who one day was missing from his three thousand year old pedestal, and went to drown himself in the fathomless ocean of humanity; then comes the whole of primitive history, accepted up till then in its grossest literalness, which finds ingenious interpreters, rationalistic hierophants who lift the veil from before the old mysteries. Then come writings hitherto considered sacred and which tested by an ingenious and subtle exegesis become a most curious literature. Would you know what the work of modern criticism is? Well, it is simply an admirable deciphering of a superstitious hieroglyph, it is the bold march from the letter to the spirit.

The modern spirit, then, means criticism, well-weighed intelligence. The belief in a revelation, in a supernatural order of things is the negation of criticism, it is the remains of the old anthropomorphic conception of the world, conceived at a time when man had not arrived at a clear perception of the laws of nature. When speaking of the supernatural we should say what Schleiermacher said of angels: "We cannot prove their impossibility. Nevertheless, the whole of this conception is such as to be impossible in our days, it belongs exclusively to the idea of the world as conceived by antiquity" (18). The belief in miracles, in fact, is the consequence of an intellectual view in which the world is considered as being governed by fantasy and not by immutable laws. No doubt, this is not the way in which the modern supernaturalist looks at it. Compelled by science which he dare not seriously offend to admit a stable order of things in nature he falls back upon the supposition that the free action of God may change now and then, and thus the miracle is conceived as a *deviation* from the established laws. But this conception, I repeat, was by no means that of primitive man. The miracle in those days was not considered as supernatural. The idea of the *supernatural* only appears when the idea of the laws of nature has been clearly formulated and makes its influence felt even upon those who timidly attempt to reconcile the marvellous and that which is proved by experiment. It is one of the half-hearted compromises between primitive ideas and the data of experiment which are neither poetical nor scientific. To primitive man the miracle was, on the contrary, perfectly natural and confronted him at each step in life, or to speak correctly, neither laws nor nature counted for much with those naïve souls, perceiving everywhere the immediate action of free agents. The idea of the laws of nature only appears very much later on and only becomes accessible to cultivated intellects. It is completely wanting in the savage and even nowadays the simple minded admit the miracle without the smallest difficulty.

It is not from one argument only but from the whole of modern science that the tremendous result is derived. "There is no such thing as the supernatural." It is impossible to refute by direct arguments one who persists in believing in it; he will snap his fingers at every *à priori* argument; you might just as well argue with the savage on the absurdity of his fetishes. You cannot convert a believer in fetichism; the only means to bring him to a superior religion is not to preach it to him directly, for if he accepts it in that condition, he will only accept it as another kind of fetichism, but to civilize him, to raise him to the rung of the human ladder to which that religion corresponds. The orthodox supernaturalist is equally unassailable. No logical or metaphysical argument has the slightest effect on him. But by taking one's stand on a high level of the development of human nature supernaturalism appears only as a conception that has been left behind. The sole cure of this strange malady, which to the disgrace of civilization has not disappeared as yet from humanity, is modern culture. Raise the intellect to the level of science, nourish it according to the rational method, and without a struggle, without arguments those superannuated superstitions will drop away. Since the dawn of existence everything that has happened in the phenomenal world has been but the regular development of the laws relating to such existences, laws that constitute but one sole order of government which is nature. Whosoever speaks of anything as being above or beyond nature in the order of facts is guilty of contradiction, just as one would be in speaking of the superdivine in the order of substances. It is simply a vain attempt to rise above the highest. All facts have for their stage space or mind. Nature is simply human reason, it is the immutable, the exclusion of everything savouring of the whimsical; and our modern task will not be accomplished until we have destroyed the belief in the supernatural, no matter in what shape—the same that we have destroyed the belief in magic, in witchcraft. All these belong to the same order of things. Posterity will look upon those who are fighting supernaturalism in our days as we look upon those who fought against the belief in magic in the sixteenth and seventeenth centuries. No doubt, the latter rendered eminent service to human reason, but their very victories have consigned them to oblivion. It is the fate of all those who fight against prejudices to be forgotten the moment those prejudices disappear. Positive and experimental science only, by imbuing man with a strong sentiment of the reality of life, is capable of destroying supernaturalism. Metaphysical speculation is far from attaining that object. India shows us the curious phenomenon of perhaps the most powerful metaphysical development ever realized by the human mind side by side with the most exuberant mythology. Speculations of the Kantian and Schelling order have co-existed in the Brahmanic brain side by side with fables more extravagant than those sung by Ovid.

In endeavouring to account for the motives that caused me to cease to believe in the Christianity that held my childhood and early youth spellbound, it seems to me that the system of things, such as I understand it to-day only differs from my first conceptions in that I consider all the real facts as belonging to the same order and that I have restored to nature that which I formerly looked upon as superior to nature. One is bound to admit that there was something admirably powerful and lofty in primitive supernaturalism, in the supernaturalism which created the mythologies of India and Greece (19). I willingly forgive that supernaturalism, nay regret it at times, but it is no longer possible; reflection has made too many strides, and imagination has cooled down too much to allow of such superb anti-common sense. As for the half-hearted compromise that seeks to reconcile an enfeebled supernaturalism with an intellectual condition incompatible with a belief in the supernatural, such a compromise only succeeds in outraging the most imperious scientific instincts of modern times without reviving the marvellous ancient poesy, henceforth and for ever impossible. Everything or nothing; absolute supernaturalism or unreserved rationalism.

Simple faith has its charms, but semi-criticism will never be anything but a burden to the spirit. The man who ponderously discusses fables shows himself as much of a simpleton and a gull though far less of a poet than he who accepts them "in a lump." We look upon the hagiologists of the seventeenth century who in writing the "Lives of the Saints" accepted certain miracles while rejecting others as too extravagant as upon so many barbarians —and not unjustly. It is very evident that with such principles they should have rejected everything, and from the artistic point of view we prefer for instance the "Sainte-Elisabeth" of M. de Montalembert where everything is unreservedly accepted. In that case the line between the belief in nothing and the belief in everything is very vague both as regards the author and the reader. One may lean towards the one or the other according to the mood being rationalistic or poetic, while the work itself preserves at least an indisputable value as a work of art. That also was the beautiful and poetic way of Plato, that is the secret of the matchless charm which the half-sceptical, half-believing use of popular myths imparts to his philosophy. But the acceptance of a part can only proceed from a narrow mind. Nothing can be less philosophical than to apply semi-criticism to narratives conceived beyond all criticism.

Hence the task of modern criticism is to lay the axe to every system of belief tainted with supernaturalism. Islamism which, by a strange fate scarcely constituted as a religion in its earlier years has since then marched onward constantly acquiring new degrees of strength and stability, Islamism, I say will perish without striking a blow by the sheer influence of European science, and history will point to our century as the one in which the first causes of that immense event began to appear on the horizon. The Turkish and Egyptian youth coming to our schools in search of European science will take back with them that which is its inseparable corollary, the rational method, the spirit of experiment, the sentiment of the real, the impossibility of belief in religious traditions evidently conceived beyond all sphere of criticism. Rigidly orthodox Musulmans are already growing uneasy at this and pointing out the danger to the emigrating younger generation. Sheikh Rifaa in the interesting narrative of his journey in Europe lays great stress on the deplorable errors that disfigure our books on science, such as for instance, the motion of the earth, etc.; and still deems it not utterly impossible to cleanse them of this poison. It is, however, patent that these heresies will shortly prove stronger than the Koran with minds initiated to modern methods. I fancy that there also will occur a Renaissance analogous to that of Europe in the fifteenth century, and which will be due, not to our *literature*, which has no more meaning to the Oriental than had the literature of the Greeks to the Arabs of the ninth and tenth centuries, but to our *science*, which, like that of the Greeks, having no stamp of nationality, is a pure work of the human intellect (20).

I am aware that man has weakling, humble, feminine instincts, a kind of yielding softness, if I may so term it, possessing wide-spread analogies at which one guesses without wishing to define them and which are as much perhaps the concern of the physiologist as that of the psychologist (21), instincts that suffer from the manly and firm attitude of rationalism which sometimes looks like stiffness (22). In the life of individuals, as in that of humanity there is a kind of mediævalism, a movement in which reflection becomes veiled and obscured and in which instinct takes the upper hand for the time being. There are certain highly, delicately-strung souls to whom it will ever be impossible to submit to this severe system and austere discipline. Seeing that those instincts are part and parcel of human nature one should not blame them too much, and the true moral and intellectual system will assign their part to them; but that part must never be despondent depression or superstition. Great calamities by humiliating man and blunting the edge of his keenest and boldest faculties become in that way a downright danger to rationalism and inspire humanity, as illness inspires the individual with a certain tendency to submission, abasement and humiliation. A moisture-laden and enervating breeze, relaxing our rigidity, loosening that which held firm is wafted over us. One feels almost tempted to strike one's breast in atonement for the boldness felt while in good health, the mainsprings are weakened, generous and vigorous instincts drop, one feels a nameless inclination to become converted and to kneel down. If the calamities of the Middle Ages came back once more, the monasteries would become peopled again, the superstitions of the Middle Ages would return. The old beliefs have no resource left but ignorance and public calamities. Faith will always be in inverse ratio to vigour of mind and intellectual culture (23). It is there in the rear of humanity espying its weak moments to clasp it in its arms and to pretend afterwards that it is humanity which gave itself freely. As for ourselves we will not yield, we will hold out like Ajax against the gods, if they count upon driving us back by striking us they are mistaken. Shame upon the timid ones that are afraid. Shame above all on those cowards who take advantage of our misery, who watch for the moment to conquer until misfortune has already half-conquered us.

The everlasting objection which keeps away from rationalism certain distinguished natures which from their very delicacy of disposition feel the most urgent need of belief, is the brevity of its creed, the contradiction of its systems, the appearance of negation, which imparts a look of scepticism to it. But scantily gifted in the matter of intellect and critical faculty they would like a ready made system uniting a great many suffrages and lending itself to acceptance without intrinsic examination. "How," say they, "can we put faith in those philosophers; no two speak in the same way" (24). All these objections are but the scruples of small minds, incapable of rational discussion and only too glad of the opportunity of stopping at those outward characteristics; scruples, deserving of respect nevertheless, for they are honest and imply faith in the truth. To reply to these good and noble souls that it is a pity it should be so, but that after all, rationalism is not to blame for the inability of man to affirm so few things, that it is better to affirm little with certainty than to affirm that of which one has no legitimate knowledge, that if the best intellectual system were that which affirms most, none would be preferable to primitive credulity which admits everything alike without criticism, to reply all this to these gushing and unsophisticated souls would be tantamount to arguing with an over excited appetite to prove that the craving it feels is a morbid one. One must reply only one thing and that one thing is the truth, namely, that the brevity of the creed of science is only so in appearance, that its contradictions are only contradictions in appearance, that its negative form is only so in appearance. Rational people contradict one another because they do not treat of the same things, or because they treat the same things from a different point of view. It is certain that two men having had exactly the same education, having gone through the same studies in exactly the same manner would look at things exactly in the same way, though they might feel differently. No doubt science does not state its results like dogmatic theology, it does not count its propositions, it does not stop at a given number of its articles of belief. Its acquired truths are not ponderous theorems, ostentatiously "showing themselves off" for the benefit of the coarsest intellects. They are merely delicate perceptions, undefinable and even fugitive glimpses, ways of framing one's idea rather than positive data, ways of looking upon things absolutely undefinable; science is a culture of subtlety and delicacy rather than a positive dogma. But the true form of moral truths is, in reality, such that to apply to them those inflexible moulds of mathematical sciences which are only fit for truths of another order, and acquired by different processes is tantamount to warping their method. Plato has no symbol, no definite propositions, no fixed principles in the scholastic sense we attach to the word; to attempt to extract a dogmatic theory from him is tantamount to warping his idea. Still, Plato represents a *spirit;* Plato is a religion. A *spirit*, that is the thing essential. The spirit is everything, the positive dogma is little or nothing, and the chances are a thousand to one on its being contradictory; nay more, it must be fatally narrow, if it be not contradictory. A *spirit* does not express itself by an analytical theory, in which every point of science is successively elucidated. It is neither by *yes* nor by *no* that it solves the delicate problems it puts to itself. A *spirit* expresses itself in its entirety at one and the selfsame time, it shows itself in twenty pages as it would show itself in a book; in a book as it would show itself in a complete collection of works. There is not a single dialogue of Plato which is not a philosophy in itself, a variation on the selfsame, ever identical theme. The word *Voltairien* expresses as clearly distinct and as easily understood a shade as the word *Cartesien;* nevertheless Descartes has a system and Voltaire has not. Descartes may be reduced to propositions, Voltaire cannot be so reduced. But Voltaire has a spirit, a way of taking things, the result of a whole *ensemble* of intellectual habits. Read through the whole of his works and then say if it is not sufficiently characteristic, whether the man has not deliberately and definitely chosen his vantage point to depict in his own way the grand landscape, whether the man had not a system of life, a manner exclusively his own of looking at things. When, oh when shall we cease to be ponderous schoolmen, insisting upon bits of phrases in geometrical fashion on God, on the soul, on morality. We will suppose those phrases to be as exact as possible, still they must be false, radically false by reason of their absurd attempts to *define*, to assign a limit to the infinite. I would sooner you read me a dialogue of Plato, a meditation of Lamartine, a page of Herder, a scene of "Faust." These if you like, contain a philosophy, that is, a way of looking at life and things. As for individual propositions each one arranges them according to his own tendencies, and this is the least important. But it worries small minds who only like formulas of two or three lines, easy to learn by rote. Then when they find that each philosopher has his own, and that all this is far from coinciding, they get grievously vexed in spirit and wondrously impatient. "It is the tower of Babel," they say, "each one speaks his own tongue. Let us go to those whose propositions stand better on their legs and to a creed settled once for all."

When I wish to initiate young minds to the science of philosophy, I begin by no matter what subject, I speak in a certain sense and in a certain tone; I care little or nothing about their remembering the positive data I give them, I do not even attempt to prove them, but I insinuate a certain spirit, a way, a turn; then when I have inoculated them with this new sense, I leave them to search in their own way, to build their temple according to their own style. For there begins individual originality which we are bound to respect rigorously. The positive results of this order teach nothing, produce no effect, have no value if transmitted in that way and merely accepted by rote. One must have been led to them, one must have discovered them or guessed at them almost before they left the lips of him who propounded them. Positive propositions are every one's business; the spirit only can be transmitted. I frankly admit that I do not possess an *ensemble* of settled and clearly limited propositions constituting a natural religion, nor do I think that science possesses such. But there is an intellectual standpoint capable of being expressed in a book, not in a single phrase, which is in itself a religion; there is a lofty and religious way of taking things, and that way is my way. Those who have breathed—if only once in their lives—the air of the unseen world, and tasted of the ideal nectar, they will understand me (25).

It seems to me that it need not take long to find out that too great a precision in things moral is as unphilosophical as it is unpoetical. Every system is assailable by the very precision of it (26). How far removed, for instance, are those admirable funeral orations of Bossuet, in which he has commented upon death in such magnificent language, how far removed are they from that upon which our actual mode of feeling would insist by reason of the cramped and precise framework to which theology has reduced the ideas concerning the future life. Nowadays, we do not conceive an eloquent discourse over a tomb without a doubt expressed in it, without an attempt to draw the veil on what is beyond, with the mere expression of a hope, left in the clouds as it were, a less eloquent standpoint perhaps, but certainly more poetical and philosophical than a too definite dogmatism, supplying if I may be permitted to say so, the map to the life to come. The savage of the Pacific looks upon his island as the whole of the world. Those who pretend to limit the lines of the infinite are still more foolhardy. That is why of all studies most brutalizing, most destructive of all poetry, theology is the first.

A system means an epic on things. It would be as absurd for a system to contain the last word on reality as it would be for an epic to attempt to exhaust the whole range of the beautiful. An epic is the more perfect in proportion as it corresponds best with the whole of humanity, nevertheless after the most perfect epic, the theme is still new and may lend itself to infinite variations according to the individual character of the poet, of the century or nation to which he belongs. How can one have a feeling for nature, how can one unrestrictedly inhale the perfume of things if one only sees them in the narrow and moulded forms of a system. I felt this divinely one day on entering a small wood. But an unseen hand repelled me as it were, because at that moment I pictured nature under I know not what physical aspect and I only became reconciled to the idea by saying to myself that all this was but a trait seized from the infinite, a vapour on a pure sky, a fluting on a vast curtain. We must dismiss the narrow conception of the schools, looking upon the human mind as a machine perfectly exact and adequate to the absolute. Views, glimpses, gleams of light, perceptions, sensations, colours, physiognomies, aspects, these are the forms under which the intellect perceives things (27). Geometry alone can be reduced to axioms and theorems. Elsewhere the vague is the true view.

The activity of human intelligence is such that to confine it in too narrow a circle is to force it into alienation. The right to think for one's self is imprescriptible, if you bar vast horizons to man, he will resort to subtle arguments in revenge, if you impose a text upon him he will escape from it by using it in the wrong sense. The wrong sense or nonsense is the revenge of the human intellect in periods of authority on the fetters imposed on it ; it is a protest against the text. This text is infallible, very well. But it lends itself to various interpretations, and there begins anew the diversity, the simulacrum of freedom with which one puts up, in default of another. Under the *régime* of Aristotle, as under that of the Bible people were permitted to think as freely as they are nowadays but on the condition of proving that such and such a thought was really in Aristotle or in the Bible, which was after all, not very difficult. The Talmud, the Masora, the Cabbala are curious proofs of the capability of the human intellect when fettered to a text. One begins to count its letters, its words, its syllables, the material sound gets to count far more than the sense, one goes on multiplying the exegetical subtleties, the modes of interpretation, like the starving wretch who, after having devoured his hunk of bread, carefully collects the crumbs thereof. All the commentaries on sacred writings are like one another, from those of Manu to those of the Bible, from those of the Bible to those of the Koran. All are a protest of the human intellect against the enslaving tendency of literal interpretation ; a miserable attempt to fertilize a barren field. When the mind does not find an object commensurate with its activity, it is fain to create one by a thousand tricks.

That which the human intellect is apt to do before a text imposed, it does before a settled dogma. Why were people so terribly bored in the seventeenth century ? Why did Madame de Maintenon die of *ennui* at Versailles ? Alas ! because there was no horizon. What resource is there left to the prisoner chained up opposite a dead wall of which he has counted the bricks over and over again ? That is the very reason why this century of orthodoxy and rule was the century of equivocation. It is the narrow rule that breeds equivocation. Why is law the science of equivocation? Because one is cribbed and confined on every side by formulas? Why was equivocation so universal in the Middle Ages? Because Aristotle was there. Why is theology from beginning to end nothing more than a long-drawn subtlety? Because the authority is ever present; one rubs shoulders with it constantly, its uncomfortable pressure is felt at every moment. It is a perpetual struggle between liberty and the divine text. A spout of water, left free uprises in a straight line, compressed, confined it swerves and deviates. Similarly the intellect left free operates normally, compressed it subtilizes.

I am convinced that if minds cultivated by the study of rational science were to interrogate themselves, they would find that, without formulating any proposition capable of being reduced to one sentence, they have sufficiently clear views on vital matters, and that these views variously expressed for every one come to about the selfsame thing; only they do not happen to be fixed into hard and fast forms and settled once for all. Hence springs the individual shade of all philosophies, and above all of German philosophies. Each system is merely the way in which an eminent mind has looked at the world, a way always deeply stamped with the individuality of the thinker. I do not doubt that each of these systems assumed the shape of a very great truth in the brain of its author, but the very fact of their individuality renders them incapable of being communicated and demonstrated (28). They are pure explanatory hypotheses, like those applied to physics, but which do not prevent the subsequent trial of others. We must not absolutely say that it is thus for we cannot have a conception adequate to primordial causes; all we can say is that things happen as if it were thus (29). It is impossible for two well ordered minds to look upon the same things and to come to a different conclusion. If the one says, "Yes," and the other, "No," it is evident that they do not speak of the same thing or that they do not attach the same sense to the same words (30). That is what Hegel meant when he averred that every thinker is at liberty to create the world in his own fashion.

It is not surprising, then, that the orthodox man should be able to lock up his beliefs in a safer way than the philosopher. Orthodoxy puts, if I may be allowed to say so, the whole of its vital provision in a hard and resisting tube, which is an outward and palpable fact; revelation, a kind of carapace that protects, but at the same time, makes, it heavy and ungraceful. The faith of the philosopher, on the contrary is always nude, clad in nothing but its simple beauty. One may fancy the temptation it affords to brutal outrage. But the day will come when the stiletto of criticism will in its turn, pierce the carapace of the believer and reach the human nature within.

Truth, to the thinker, is only a more or less advanced but always incomplete form, or at any rate a form capable of being perfected. Orthodoxy, on the other hand, petrified, stereotyped in its forms, can never cast off its past. Seeing that it pretends to have been made at one sitting and in one piece it places itself beyond the pale of progress, it becomes rigid, overbearing, unbending, and while philosophy is always abreast with humanity, theology at a certain period lags behind. For it is immutable and humanity marches onward. Not that theology has not been forced to march now and then like the rest. But it denies this, it lies to history, it warps all criticism in order to prove that its actual state is its primitive one, and it is obliged to do this on the penalty of forfeiting the conditions of its existence. The philosopher on the contrary, conceives under no circumstances either absolute retrogression or predetermined immobility. He recommends concession to the successive modifications brought about by time, without ever categorically severing one's self from the past any more than being its slave. He has no wish to deny that past, but endeavours to explain it in a new sense, to show the part of ill-defined truth it contained. There is nothing contradictory in a philosopher exceeding the limits of his own philosophy, in using several systems, by which I mean several not equally perfect expressions of the truth; this simply redounds to his honour.

The problem of philosophy is ever new, it will never attain a definite formula, and the day we shall be satisfied to abide by the assertions of the past by accepting them as so many absolute truths, incapable of reconstruction, that day will sound the death-knell of philosophy. The orthodox man is never more annoying than when, pluming himself upon his immobility, he twits the thinker with his fluctuations and philosophy with its constant modifications (31). It is exactly those very modifications which prove that philosophy is truth; through them it is in harmony with human nature, always in travail and happily condemned to obtain all its conquests in the sweat of its brow. Only that which is not progressive does not vary. There is nothing more motionless than the nonentity who has never lived the intellectual life, or the dullard who has never seen aught but one side of things. The way not to vary is not to think. If orthodoxy is immutable it is simply because it has placed itself beyond the pale of human nature and reason.

And I pray you not to say that this is scepticism. It is criticism, that is, the ultimate and transcendental discussion of what at first was admitted without sufficient inquiry, in order to deduce from it a purer and more advanced truth. It is high time that we should accustom ourselves to call sceptics all those who do not believe as yet in the religion of the modern spirit, and who still lingering around effete systems, deny with blind hatred the acquired dogmas of the living century. We accept the inheritance of the three great modern movements; protestantism, philosophy and the revolution, without having the slightest inclination to become converts to the symbols of the sixteenth century, or Voltairians or to recommence another 1793 or 1848. There is not the slightest need for us to recommence what our fathers accomplished. Their work is summed up in liberalism, we shall know how to continue that work.

In logic, in morality, in politics man aspires to the attainment of something absolute. Those who base human knowledge, and duty and government on human nature appear to deprive themselves of such a foundation; for unhampered inquiry means dissent, variety of views. It seems easier, therefore, to seek for knowledge, morality and politics a basis outside man, a revelation, a right divine for instance. It is unfortunate that there exists nothing of the kind; that such a revelation would first of all have to be proved, that it is not proved, and that if it were proved, it could only be proved by reason, that consequently diversity would uprise again with regard to the appreciation of those proofs. Hence, it is better to remain within the field of human nature, to look for the absolute only in science and to do away with all those timid palliatives which only produce illusions and postpone the difficulty.

Nowadays there are only two systems confronting one another; one portion of humanity, despairing of reason, believing it to be condemned to eternal contradiction of itself, frantically embrace an outside authority and become believers through scepticism (a Jesuitical system; authority, the director, the pope substituted for reason, for God). The other portion from a more profound view of the march of human intellect perceive progress and unity beneath apparent contradictions. But let us remember this, for it is essential; unless one believes from instinct, like the most simple minded, one can only believe from scepticism; to despair of philosophy has become the first basis of theology. I love and admire this grand despairing scepticism the expression of which has endowed modern literature with so many admirable works. But I can only laugh at and be disgusted with that petty irony of human nature which results in superstition and pretends to cure Byron by preaching the Pope to him.

A great deal is said nowadays about the accordance of human reason with faith, of science with revelation, and some pedants who wish to make themselves interesting and to pose as impartial and superior minds have made this a theme full of ambiguity and frivolous nonsense. The principal thing is to understand one another. If revelation be really what it pretends to be, the word of God, it is but too clear that it is master, that it has no need to enter into a pact with science, that the latter can only pack up its traps in presence of this infallible authority, and that its role (that of science) would be reduced to that of *serva et pedissequa*, to comment upon or to explain the revealed word. From that moment also the custodians of this revealed word will be superior to the investigators of human science, or rather they will be the only power before which the other will vanish, like the human before the divine. No doubt, truth not admitting of self-contradiction, one would be bound to conclude that *sound science* could not very well contradict revelation. But seeing that the latter is infallible and more clear, if science appears to contradict it, one will conclude that it is not *sound science* and impose silence on its objections. But if, on the contrary, the fact of the revelation is not real, or if at any rate, there is nothing supernatural about it, then religions are nothing more than human creations and the whole matter is reduced to the finding of the reason for the different fictions of the human intellect. On such an hypothesis man himself has done everything by means of his natural faculties, in one case spontaneously and in the dark; in the other scientifically and reflectingly; but in sum man has done everything, he finds himself everywhere face to face with his own authority and with his own work. The theologians are right when they say that before everything one should discuss the fact; is this doctrine the word of God? And whether we reply yea or nay, the so-called problem of the agreement between faith and reason, supposing as it does two equal powers which it is necessary to reconcile, is utterly devoid of sense; for in the first case, reason vanishes in the presence of faith, like the finite before the infinite and the strictest orthodoxy has right on its side; in the second case there remains nothing but reason, manifesting itself in various ways, but nevertheless always remaining identical with itself (32).

It is you who are the sceptics, and we are the believers. We believe in the work of modern days, in its sanctity, in its future, it is you who curse it. We believe in reason and you insult it; we believe in humanity, in its godlike destinies, in its imperishable future, and you laugh at it; we believe in the dignity of man, in the goodness of his nature, in the rectitude of his heart, in his right to attain the perfect state, and you shake your head at these comforting truths and you complacently lay stress on the evil, and the most saintlike aspirations towards a divine idealism, you denounce them as the works of Satan, and you talk of rebellion, of sin, of punishment, of expiation, of humiliation, of penitence, of the hangman to him who should hear no words from your lips save those of deification and expansion. We believe in everything that is true, we love everything that is beautiful (33); and you, wilfully blind to the infinite charm of things, you pass through this beautiful world of ours without deigning to bestow a smile upon it. Is the world a cemetery and life a funeral procession? Instead of the reality you cherish an abstraction. Which of us denies, you or we? And he who denies is he not the sceptic?

Our rationalism, therefore, is not that *hauteur*, analytical, dry, negative, incapable of understanding the things of the heart and the imagination which was inaugurated by the eighteenth century; it is not the exclusive use of what has been called "the acid of reasoning;" it is not the *positive philosophy* of M. Auguste Comte, nor the irreligious criticism of M. Proudhon. It is the acknowledgment of human nature, hallowed in all its parts, it is the simultaneous and harmonious use of all the faculties, it is the exclusion of all exclusiveness. According to our views M. de Lamartine is a rationalist, still in a more restricted sense, he would no doubt deny this title, seeing that he himself tells us that he attains his results not by combination or reasoning but by instinct and direct intuition. Until now criticism has been conceived as being only a dissolving agent, an analysis destroying life; from a more advanced point of view it will be understood that higher criticism is only possible on the condition of giving the whole of human nature full play, and that reciprocally a higher love and a great admiration are only possible on condition of criticism. The pretended poetical natures who imagined they could get to the true sense of things without science will then turn out to be so many chimera-mongers, and the austere *savants* who shall have neglected the more delicate gifts whether from scientific virtue or from a compulsory contempt of what they did not possess will remind us of the ingenious myth of the daughters of Minyas who were changed into bats for having been unable to do anything but argue in the presence of symbols to which a more generous method of elucidation should have been applied.

History appears to raise an objection against science, criticism, rationalism, civilization—synonymous terms after all—which it is expedient to explain. It seems in fact to show us the most cultured people always a prey to the most barbarous people; Athens to Macedonia, Greece to the Romans, the Romans to the barbarians, the Chinese to the Manchoos. The process of thinking is a wearing one. Our middle class families which in reality have only been conscious of their own strength for one or two generations are already wearing out. The half-century that has elapsed since 1789 has exhausted them to a greater extent than the innumerable generations of primitive darkness. Too much knowledge apparently weakens humanity, a people of philologists, thinkers and critics would probably be too weak to defend its own civilization. The Germany of the beginning of the century gave way disgracefully to France and yet how superior from a mental point of view was the Germany of Goethe and Kant to the France of Napoleon. Barbarism being unconscious of its own powers, is obedient and passive; the individual, unaware of his individual worth is lost in the masses, and obeys the command as he would fate. Passive obedience is only possible on the condition of stupidity. The man who thinks for himself, on the other hand, calculates his interests too well and asks himself with that feeling of positiveness which he applies to all things whether it is really his interest to get himself killed. Besides, he clings more tenaciously to life and the reason is plain enough. His individualism is much stronger than that of the barbarian, civilized man says *I* with unparalleled energy; with the barbarian on the contrary, life is scarcely raised one degree above the level of the dull sensation that constitutes the life of the animal. He does not resist, for the reason that he barely exists. Hence the contempt for human life (for his own as well as that of others) which is the secret of the barbarian's heroism. The cultivated man whose life has a real value, sets too much store by it to stake it casually (34). Brutal strength appears to him such an extravagant idea that he revolts against such absurd means and cannot make up his mind to pit himself against weapons which a savage handles better than he does. In those rude struggles the most benighted conscience is the best; personality, reflection are simply so many causes of inferiority. Hence, the liberty to think for one's self has up till now been by no means favourable to enterprises requiring the abdication of their individuality by masses of men in order to yoke themselves to the vehicle of a grand idea and to drag it majestically through the world. What would Napoleon have done with an army of reasoners?

This is a real contradiction which like so many others, cannot be cyphered away except by acknowledging that humanity is as yet far removed from its normal condition. While one portion of humanity is still leading a brutal life, misunderstanding and evil passions will succeed in exploiting barbarian humanity against civilized humanity and in letting loose the ferocious brute on reasonable men. The critics are right, whether they are the stronger or the weaker does not prevent them from being in the right, and if they fall, it simply proves that the actual condition of humanity is still far distant from the point when justice and reason will be the only real forces as they are the only legitimate ones.

Bear in mind, I pray you, that this is not a mere academic question, a dream discussed in an idle hour. It is the question of humanity itself and the legitimacy of its nature. If humanity is so constituted as to require necessary illusions, if too much refinement leads to dissolution and weakness, if too great a knowledge of the reality of things becomes injurious to it, if it wants superstitions and incomplete views, if the legitimate and necessary development of its own being prove its own degradation, then humanity is badly constituted, it is based on false foundations, it is only travelling towards its own destruction, because those who have conquered thanks to their illusions will be forcibly brought to their own disillusion afterwards by civilization and rationalism. Under such circumstances our symbol is destroyed, for our symbol is the legitimacy of progress. And on this hypothesis, humanity would find itself in a blind alley, its line of route would not be the straight one, proceeding towards the infinite, seeing that pushing forward and forward it would discover in the end that it had gone backward. The law, which in such a case one would have to enjoin on human nature would no longer be to exert all its strength in the attainment of the absolute, civilization would have its *maximum*, arrived at by an equilibrium of opposites, and wisdom would consist in stopping at that. The question, in one word, comes to this; the law of humanity is either an expression such that by increasing all its variable quantities, one increases the total value; or else it must be assimilated to those expressions that attain a *maximum*, beyond which any addition brought to the several factors results in a decrease of the total value.

Happy will they be who by means of a definite experiment will be enabled to oppose a proved answer to these terrible apprehensions. Our affirmations in that respect may perhaps possess some of the merit of faith which believes without having seen, and truth to tell, when one looks at isolated facts, optimism appears a very gratuitous liberality to God. As for myself, if I were to see humanity collapse on its own foundations, mankind slaughter one another in some fateful darkness, if I were to see all this, I should still go on proclaiming the rectitude of human nature, that perfection is its final aim, that misunderstandings will disappear, and that the day must come when reason and perfection shall reign supreme. Then we shall be remembered, and some will say; "Oh, how they must have suffered." We must take care not to assimilate our civilization and our rationalism to the fictitious culture of antiquity and above all to that of degenerate Greece. Our eighteenth century was no doubt an epoch of moral depression, nevertheless it closed with the greatest eruption of devotion, of abnegation of life recorded in history. Those philosophers, those Girondists who so proudly marched to the scaffold were they nothing more than quaking rhetoricians? Was it a superstitious illusion that strengthened those noble souls? There exists, I know, a generation of egotists, a generation that has grown up in the shadow of a prolonged peace, a sceptical generation, born under the star of Mercury, without faith or love, which at the first blush, seems to be governing the world. But even if this were so, we should not despair of humanity, without a doubt, for humanity does not die, but we should have to despair of France. But, after all, are these the men whom in good faith, we ought to oppose as an objection to science and philosophy? Is it too much knowledge that has taken the muscle and marrow out of them? Is it too much thought that has destroyed all feeling of patriotism and honour in them? Is it their too frequent excursions to the realms of the intellect that have made them unfit for great things? They whose minds are closed to every idea, whose only science is that of a fictitious world, whose philosophy consists of frivolity. For Heaven's sake, do not talk to me of these men when we are discussing philosophy and civilization. Even if it be proved that the tone of the society which became more and more powerful under Louis-Philippe was calculated to hamstring all noble effort, it would be no argument against the society which will be brought to the front by reason and human nature in its frankest and truest development. Even if the final impotency of the official world were proved, if it were proved incapable of creating aught original and strong, it would not justify us in despairing of humanity, for humanity disposes of unknown sources whither it goes constantly to renew its youth. Is it too much rationalism that has ruined unhappy Italy, which at the present moment presents to us the deplorable spectacle of a member of humanity stricken with paralysis? Is it too much criticism that has dried up the vessels that gave it life? Was it not stronger and more beautiful in the fifteenth, and in the first half of the sixteenth, centuries, when it was the pioneer of the whole of Europe along the roads of civilization, and spread its sheltering wings over the boldest of rationalism? Is it its religious beliefs that have preserved its vigour? Was not the pagan Italy of Julius II. and Leo X. worth the exclusively Catholic Italy of Pius V. and the Council of Trent? To knock down the Capitol or the Temple of Jupiter Stator would have been tantamount to overthrow Rome. Such things should no longer be possible among modern nations, seeing that the rest and be thankful in the matter of religious creeds is sufficient to enervate a nation (35). A few months ago the people of Rome cast their church bells into the melting pot to make coppers of them. No doubt if the religion of the moderns were like that of the ancients, the spinal marrow of the nation itself, this would have been a great piece of absurdity. One might as well profess to enrich France by converting the Vendôme Column into money. But what are people to do when the gods are departed? Symmachus asking for the restoration of the altar dedicated to Victory was simply playing the rhetorician.*

Seeing that antiquity never understood the great object of literary culture, having always looked upon it as a kind of mental drill with a view to "speaking well," it is not very surprising that the strong and energetic natures of those days should have so severely condemned the puerile system of the rhetoricians and the meretricious and sophistical education they imparted to the young. The ideal of virtue as conceived by serious minded men was the rough and uncultivated character and their ideal of a society consisted in a development exclusively tending to devotion to the country and *well-doing* (Sparta, ancient Rome, etc.). And seeing that literary culture was found to be subversive of such a state, that culture was denounced as offering the greater facility to the

* Symmachus, Prefect of Rome 384 of the Christian Era, Consul in 391, reported to be the last adherent and advocate of Paganism in the West.—Transl.

enemy to vanquish. Hence those commonplaces about the superiority of *well-doing* over *eloquent talking*, of rough hewn virtue over refined civilization, the contempt of the *Græculus*, primed with grammar, etc. In our days all this would be so much nonsense. From our point of view, Sparta and ancient Rome represent, in fact, one of the most imperfect conditions of humanity, seeing that one of the essential elements of our nature, thought, intellectual perfection was utterly neglected there. No doubt the simple and genuine cultivation of the love of country is superior to that artificial culture of the latter days of the empire, and if anything could inspire one with fear for the future of modern civilization it would be the fact that the so-called classical education given to our young generation resembles that of that lamentable epoch. But there is nothing superior to science and to the grand and purely human civilization, and it is only the superficial mind that could compare this grand form of complete life to the artificial centuries in which a man could have no noble sentiment apart from a rhetorical reminiscence, when he sent for a philosopher to hear the reading of a *Consolation* when he had lost his nearest and dearest, and when people on their death-beds pulled from their pockets a speech prepared for the occasion.

Thus, if civilization were to founder once more in presence of barbarism, it could not be argued as an objection against it. It would have right on its side even beyond that. It would once more vanquish its conquerors, and so it will be always until the day when there will be no longer any one to conquer, and when sole mistress it will reign in its own right. What does it matter by whom the work of civilization and the welfare of humanity be accomplished? In the eyes of God and posterity, Russians and French are only so many human beings. We only appeal to the principle of nationality when the nation oppressed is superior intellectually to that which oppresses her. The absolute partisans of nationality can only be narrow-minded people. Humanitarian perfection is the aim, and from that point of view civilization is always triumphant; and it would be strange indeed if an invisible weight dragged down humanity in that sense, if that sense meant only degeneration.

From the point of view of humanity there is no such thing as decline. Decline is a word that ought to be banished once for all from the philosophy of history. Where does the decline of Rome begin? The narrow-minded, ever concerned with the preservation of ancient habits and customs will aver that it is after the Punic wars, that is, just at the very moment when the preliminaries having been laid down, Rome begins her mission and gets rid of the habits of her infancy, which henceforth have become impossible to her. Those who are preoccupied with the idea of the republic will place the fatal line at the battle of Actium; they are the poor folk who would have committed suicide in company with Brutus; they fancy they can see death in what was after all but the crisis of ripe age. Can that decline be placed with greater justice in the fourth century, when the work of Roman assimilation is at its height, or in the fifth, when Rome imposes her civilization on the barbarians that invade her? And when we look at Greece from the Homeric times to the days of Heraclius, where is her decline? Is it at the epoch of Philip, when she is on the eve of making her brilliant first appearance in the work of humanisation through Alexander? Is it during the domination of Rome, when she becomes the cradle of Christianity? So true is it that the word decline has no sense except from the narrow point of view of politics and nationalities, not from the grand and wide point of view of the work of humanisation. When atrophy takes hold of certain races, humanity has always a sufficient reserve of living forces left to make good such deficiencies. And if it be apprehended that humanity having exhausted all its reserve stock be one day in the position of each nation in particular, my answer must be that before then humanity will no doubt have become stronger than all the destructive causes put together. In our actual condition, criticism carried to extremes causes moral and physical weakness, in the normal condition science will become the mother of strength. Seeing that hitherto science has only appeared in the guise of criticism, it is difficult to conceive its ever becoming a powerful active motor. Such will be the case, nevertheless, the moment it has succeeded in creating in the moral world a conviction equal to that produced of yore by religious faith. All the arguments deduced from the past in order to prove the impotence of philosophy are no proofs at all with regard to the future, for the past has only been a necessary introduction to the grand era. Reflection has as yet not shown itself as a creative power. Let us wait a while, let us wait a while. . . .

Many of my readers will no doubt be surprised at my frequent appeals to the future. It is because I am really convinced that the majority of the arguments advanced apologetically in behalf of science and modern civilization are very faulty and lay themselves open to the attacks of the retrogressive schools if they, the arguments, are to be considered by themselves and irrespective of the ulterior condition of things to which they will have contributed. The only means of understanding and justifying the modern spirit is to look upon it as a necessary stage towards the perfect, in other words, towards the future. And this appeal is not the mere act of blind faith which falls back upon the unknown. It is the legitimate result emanating from the whole of the history of the human intellect. "Hope," says George Sand, "is the faith of this century."

Side by side with theological dogmatism which makes science useless and robs it of its dignity we must place another dogmatism, still more narrow and more absolute, that of superficial common sense, which in reality is only so much self-sufficiency and emptiness, and which not perceiving the difficulties of problems, thinks it strange that their solution should be looked for outside the beaten track. It is too evident that the common sense in question is not that resulting from straightforward action of the human faculties on a sufficiently well-known subject. The one I am tilting against is a somewhat rather equivocal quality the exclusive possession of which small minds claim for themselves and which they liberally grant to those who agree with them, the subtle triviality which succeeds in investing everything with an appearance of evidence. Now, it is very evident that common sense, thus understood, cannot make up for science in its search for truth. First of all, let us remember that the superficial minds who are constantly appealing to common sense designate by that name the very special and very limited form of customs and habits under which they happen to have been born. Their common sense is the way of looking at things of their century or of their province. Only he who has systematically compared the various facets of humanity would have the right to make this appeal to universal opinions. Besides is it common sense which will give me philosophical, historical, philological knowledge, all of which are necessary for the criticism of the most important truths? Common sense has every right to speak when it comes to establishing the bases of morality and psychology, seeing that in that case it is only necessary to note that which appertains to human nature, which should only be looked for in its general, consequently in its most commonplace expression, but common sense is only ponderous and clumsy when it pretends to solve by itself problems, to the elucidation of which divination is more essential than sight, in connection with which one has to catch a thousand shades almost imperceptible, to pursue hidden and secret analogies. Common sense is partial; it always looks at its opinion from the inside, it never emerges from it to judge it from the outside. It so happens, however, that nearly every opinion is true in itself, but only relatively true with regard to the point of view whence it is conceived. The really true in moral and historical sciences appeals to subtle and refined intellects only, the same that mathematics appeal to the systematic intellect only. The truths of criticism do not lie on its surface, they almost look like paradoxes, they do not plant themselves, visible from all sides before the ordinary understanding like theorems of geometry, they are fugitive flashes of which the eye just catches a glimpse, which one perceives in an absolutely individual way, and which it becomes almost impossible to communicate to others. The only resource left is to bring the other minds to the same point of view in order to show them things from the same side. What business has this vulgar common sense, with its blustering ways, loud voice and self-satisfied laughter in that world of *finesse* and subtle thought? *I can make neither head nor tail of it* is its last and sovereign sentence, and how very easy indeed is it to say that much. The self-sufficient tone it assumes when face to face with the results of science and thought is one of the most aggravating nuisances the thinker has to contend with. It unhinges him, and if he be not a philosopher at heart, he cannot help getting annoyed with those who abuse their privilege in that way against his delicate and feeble voice.

Hence it is inadmissible to appeal from science to common sense, seeing that science is only enlightened common sense operating knowingly. The really true is no doubt the voice of human nature, but of nature fittingly developed and brought by culture to its utmost capacity.

CHAPTER IV.

SCIENCE has no enemies save those who consider truth as useless and making no difference, and those who granting to truth its priceless value profess to get at it by other roads than those of criticism and rational investigation. The latter are, no doubt to be pitied as having strayed from the right method of the human intellect, but they at any rate recognize the ideal aim of life; they may come to an understanding and to a certain extent sympathize with the man of science. As for those who despise science as they despise lofty poetry, as they despise virtue, because their degraded soul only understands the perishable, we have nothing to say to them. They belong to another world, they do not deserve the name of men, seeing that they are without the faculty which constitutes the noble prerogative of humanity. We are rather proud that they should look upon us as men of another age, as fools and dreamers, we glory in knowing the routine of life less well than they do, we delight in proclaiming our studies to be *useless;* their contempt for them makes them valuable to us. These men are the immoral ones, the atheists who are impervious to every breath coming from on high. The atheist is the indifferent, the superficial and frivolous who has no cult save that of interest and self-gratification. England to all appearance one of the most religious countries of the world is in fact the most atheistic; for it is the least ideal. Unlike some of the Latin orators I do not wish to fall into the *convicium saeculi.* I believe that there exist in the souls of the nineteenth century just as many intellectual needs as in those of any other epoch, and am convinced that at no time were there so many minds open to criticism. The misfortune is that the prevailing frivolity condemns them to form a world apart, and that the aristocracy of the century, which is that of wealth, has as a rule lost the ideal sense of life. I am only speaking conjecturally; for that world is utterly unknown to me and it would be easier to me to quote illustrious exceptions than exactly point out those at whom in this instance my reproach is aimed. Nevertheless it seems to me that a society which *de facto* only encourages a wretched literature in which everything is reduced to measure and adjustment, that a society which finds no middle course between an absence of moral ideas and a religion which it has first "*boned*" to make it more acceptable, that such a society, I repeat, is far from the true and grand sentiments of humanity. The future is with those who taking life seriously, come back to the eternal foundation of truth, that is to human nature, taken in the mass and not in its extreme refinements. For humanity will always be serious, believing, religious and the frivolity which believes in nothing will never hold the foremost place in human affairs.

It seems to me that we ought not to attach too much importance to all that speechifying which has become trite against the *utilitarian* and realistic tendencies of our days, and if aught could prove the lack of sincerity of those lamentations it would be the strange resignation with which those who utter them submit to the inevitable necessities of the century. In fact nearly all seem disposed to wind up with the line:

> "The good old time, the iron age."

Whatever one's opinion may be with regard to the tendencies of the century, it would at any rate be fair to admit that, the sum total of activity having increased there may have been increment on the one side, without a falling off on the other. It cannot be disputed that there is more commercial and industrial activity in our days then there was, for instance, in the tenth century. And must we therefore conclude that the latter was better endowed in respect of intellectual activity?

There is a kind of optical illusion in history which is very dangerous. The actual century is always seen through a cloud of dust raised by the whirl of real life and one can scarcely distinguish amidst this whirlwind the pure and beautiful forms of the ideal. On the other hand, this cloud of petty interests having vanished from before the past, it appears to us grave, severe, disinterested. Looking at it by means of its books and monuments only, in other words, in the manifestation of its thought, we are tempted to believe that people did nothing else but think. The noise of the street, the stir of the mart do not come down to posterity. When the future shall see us freed from that deafening tumult, it will judge us as we judge the past. The race of egoists who have no feeling either for art, science or morality is "of all times." But they die without leaving a trace, they have no place in that grand piece of historical tapestry-work which humanity weaves and leaves to be unrolled behind it. They are the noisy waves that plash beneath the paddle wheels of the steamer in its course, but become silent behind it.

Therefore, let those who dread to see the efforts of the mind stifled by material preoccupations take heart. Intellectual culture, speculative research, in one word, science and philosophy possess the best of all guarantees, I mean, the needs of human nature itself. Man will never live by bread alone; the disinterested pursuit of the true, the beautiful and the good, the realization of science, of art, of morality is as imperative a want to him as the need of satisfying his hunger and his thirst. Besides, the activity, the apparent aim of which is only material improvement has nearly always an intellectual value. What speculative discovery has affected civilization as much as the discovery of steam? A railway does more for human progress than a work of genius, which, from purely extraneous circumstances may be deprived of its influence.

It cannot be denied that Christianity has done a great wrong to humanity in representing the actual life as a matter of no moment and consequently dissuading man from the idea of improving it. For though "it is the spirit that quickeneth," and "the flesh profiteth nothing," the grand reign of the spirit will not commence until the material world shall be completely under man's control. Besides, the actual life is the stage of that perfect life which Christianity relegated to the world beyond. There is nothing exaggerated in the spiritualism of the Gospel nor in the exclusive preponderance it gives to a higher life. But it is here below and not in a fantastic heaven that this life will be realized. It is essential therefore that man should commence by assuming the mastership of the bodily world in order to be at liberty afterwards for the victories of the spirit. That is where the injustice lies of the anathema flung by Christianity against the present life. All the great material and social improvements of this life have been accomplished outside Christianity and have even been prejudicial to it. Hence the annoyance of the actual representatives of Catholicism at all the most rational reforms of the abuses of the past, judicial reforms, penal reforms, etc. They are well aware that all this hangs together, and that one step on that road necessarily entails all the rest. Posterity will, no doubt, not wholly approve our materialistic tendencies. It will, no doubt, judge our work as we judge that of Christianity and find it equally one-sided. But it will at any rate admit that unconsciously we have laid down the condition of future progress, and that our "industrialism" has been, with regard to its results, a holy and meritorious work.

A certain number of social doctrines are often taunted with concerning themselves solely with material interests, with the supposition that there exists only one kind of work for man and one kind of food, with the ideal conception of an easy life for all. This is unfortunately the case, nevertheless, one is bound to observe that if those systems could really bring about the material improvement of a considerable portion of humanity the reproach would not hold water. Because the improvement of the material condition is the necessary condition of moral and intellectual improvement, and that item of progress will like every other have to be accomplished by a special work; humanity cannot do two things at once. It is evident that a man who has not the necessities of life, or in order to procure them is compelled to devote himself to some kind of mechanical labour every minute of the day is forcibly condemned to dependence and utter insignificance. At the present period, the most signal service one could render to the human intellect would be to discover a system which would insure material comfort to every one. The human intellect will only be free when completely emancipated from those material needs that humiliate it and obstruct its development. Such improvements possess no ideal value in themselves, but they are the essential condition of the dignity of humankind and the perfection of the individual. That protracted labour by which the middle classes managed to accumulate wealth during the whole of the Middle Ages is apparently something sufficiently vulgar. We cease to look at it in that way by reflecting that the whole of modern civilization which is the work of the trading classes would have been impossible without it. The secularization of science could only be accomplished by an independent, consequently a well-to-do class. If the urban populations had remained poor or fettered by incessant manual labour like the peasant, science would be up to this day, the monopoly of the priestly class. Everything that contributes to the progress of humanity, however humble and commonplace it may appear, is by the very fact entitled to respect and sacred.

It is odd that the two classes which nowadays have divided French society between them fling reciprocally the charge of materialism. In common fairness we are bound to say that the materialism of the opulent class only is blamable. The striving of the poorer classes after material welfare is just, legitimate and sacred, seeing that the poorer classes cannot attain real holiness, by which I mean moral and intellectual perfection unless they acquire a certain degree of material welfare. When a well-to-do man still seeks after greater wealth, he does a thing which to say the least is profane, seeing that his only aim can be indulgence of self. But when a poor wretch strives to lift himself above want he attempts a virtuous task, for he tries to carry out the condition of his redemption. He does the right thing at the right moment. When Cleanthes spent his nights in drawing water he did as saintly a work as when he spent his days in listening to Zeno. I cannot help getting angry whenever I hear the fortunate ones of the century stigmatize the feeling with which the proletarian looks upon the more distinguished existence of the superior classes as one of vile jealousy and disgraceful lust. You think it disgraceful that they should desire that which you enjoy. You would preach to the people monastic confinement and abstinence from all pleasure, when pleasure is the alpha and omega of your life, when you have poets who sing of nothing else. If the life is a good one, why should not they desire to lead it also? If it is bad, why do you enjoy it?

The tendency towards material improvements is therefore far from prejudicial to the progress of the human intellect provided it be fittingly ordained to its end. That which debases and degrades is the mean spirit brought to bear upon it; the petty combinations, the shabby processes resorted to in the race for wealth. I honestly believe that it would be better to leave the people to their poverty than to educate them in that way. Ignorant and uncultivated, they blindly aspire to the ideal by virtue of the powerful and inarticulate instinct of human nature, they are energetic and true like all great masses whose consciousness is still benighted. Inspire them with those paltry instincts of lucre and you lower them, you destroy their originality without making them more moral or educated. The science of *Bonhomme Richard* has always seemed to me a sufficiently bad science. Just fancy a man who sums up his life in the words; *to make a fortune honestly;* (and even then there is a suspicion that *honestly* is only recommended in order *the better* to make the fortune) the last thing of which one needs to think, a thing of no value save as a means to an ulterior and ideal end. This is immoral, this is a narrow and finite conception of existence; this can only proceed from a soul void of religion and poetry (36). Great Heavens, what is the good, I ask you, what is the good of having realized at the termination of the short life of ours the more or less complete type of outward happiness? What is better is to have thought and loved a good deal, to have cast a resolute glance at all things, to be able to criticize death itself while on one's death-bed. I prefer a Jogui, I prefer an Indian Mouni, nay Simon Stylites himself gnawed at by the worms on his strange pedestal to a prosaic trader capable of pursuing for a score of years the one selfsame idea of wealth.

Ye heroes of the disinterested life, saints, apostles, cenobites, ascetics of all epochs, sublime poets and philosophers who preferred to have no inheritance here below; sages who went through life having the left eye fixed on earth, the right on Heaven, and above all you, godlike Spinoza who remained poor and forgotten for the sake of cultivating your idea and in order the better to worship the infinite, how much better did you understand life than those who treat it as a narrow calculation of interest, as an insignificant struggle of ambition and vanity. Better would it have been, no doubt, not to have put your God in so remote an abstract, not to have placed Him on such a nebulous height where to contemplate Him you were compelled to take up so uncomfortable a position. God is not only in Heaven, He is near every one of us; He is in every flower you crush beneath your feet, in the breath that wafts its odour over you, in that small life which buzzes and murmurs everywhere around you, in your heart above all. How much more proof of the needs and the extremely sensitive instincts of humanity do I find in your sublime madness than in those colourless existences never lighted up by a single ray of the ideal, of those existences which from their first to their last moments have proceeded day by day, exact, ruled like the leaves of a counter ledger.

Certainly we ought not to regret seeing the peoples pass from spontaneous, blind aspiration to clear and well considered perception, but it must be on the condition that the object proposed to this thoughtful consideration be not unworthy to occupy it. The tendency which at certain epochs of civilization impels certain minds to be smitten with admiration of barbaric and original peoples is a logical and in a sense a legitimate one. For the barbarian with his dreams and his fables is better than the positive man who understanding stops at the finite. Perfection would be the aspiration towards the ideal, that is religion, not operating in the world of chimeras and fantastic creations, but in that of the reality. Until we have got to understand that the ideal is near every one of us we shall not be able to prevent certain natures (and these the most beautifully constituted ones) from seeking it beyond the vulgar existence, and finding their delight in asceticism. The sceptic and the frivolous minded may shrug their shoulders as much as they like and as long as they like at the folly of those beautiful natures, it will make no difference to them. Religious and pure natures understand them, the philosopher admires them, like every energetic manifestation of a real need which mistakes its road for want of criticism and rationalism.

Nothing is easier with our positive spirit than to point out the absurdity of all the sacrifices a man makes of his welfare in obedience to the suprasensible. To the realist the man on his knees before the invisible looks uncommonly like an imbecile, and if antique libations were still in usage (37), there are many people who would say with the apostles; *Ut quid perditio hæc ?* Why waste this liquor thus ? You would have done better to drink or to sell it, which might have given you pleasure or profit, than to sacrifice it to the invisible. Saint-Eulalia fascinated by the charm of asceticism escapes from the paternal home, takes the first road she comes to, wanders about at haphazard, gets lost in the bogs and cuts her feet in the brambles. "The girl was mad;" I hear people say. You may call her mad, as long as you like, I would give everything I possess to have seen her at that moment. Our judgments on a life of asceticism all start from the same principle; the ascetic sacrifices himself to the useless; hence he is absurd. Or else, if an apology for the life is attempted, it will be solely for the material services it has been able to render accidentally, without the consideration that those services were utterly foreign to its aim, and that those works that are accounted a title to glory, had no value attached to them, save in so far as they contributed to his asceticism. Assuredly he who would embrace a useless life, not from the need of contemplation but simply for the sake of idleness (and this was what happened during the degeneracy of the institution) would be thoroughly contemptible. As for pure asceticism, it will always remain, like the pyramids, one of those grand monuments of the inmost needs of man, manifesting themselves with energy and grandeur, but with too little conscience and reason. The principle of asceticism is eternal in humanity, the progress of thought will impart to it a more rational direction (38). The ascetic of the future will not be the Trappist monk, one of the most imperfect types of man, he will be the lover of the purely beautiful, sacrificing to this dear ideal all the personal needs of the lower life.

The English have imagined that they were furthering the cause of sacred morality by prohibiting in India the processions stained with the blood of voluntary sacrifices, the *suttee.* A strange mistake indeed. Do you really believe that the fanatic who joyfully lays down his head under the wheels of the car of Juggernauth is not happier and more beautiful than you, insipid merchants. Do not you think that he honours human nature more by attesting—in an irrational, but none the less powerful manner, no doubt—that man has within him instincts superior to all the cravings for the finite and the love of self. Undoubtedly if we looked upon those acts as the mere sacrifice to a chimerical deity, they would be simply absurd. But we look upon them as the fascination which the infinite exercises on man, as impersonal enthusiasm, the cult of the suprasensible. And it is upon those magnificent outbursts of the grand instincts of human nature that you would impose limits, with your paltry morality and your narrow common sense. . . . In those sublime and picturesque exaggerations of human nature there is a foolhardiness, a spontaneousness which the healthy and regular exercise of reason, do what it will, will never equal, and which the poet and the artist will always prefer (39). A morbid and exclusive development is more original and shows in greater relief the energy of nature, like an injected vein which stands out more clearly to the inspection of the anatomist. Go and have a look in the Louvre at the marvellous Spanish collection; it is ecstasy and the superhuman incarnated, saints whose feet scarcely touch earth; virgins with necks craned, haggard eyes, staring into space, martyrs who wrench their hearts from their bodies or lacerate themselves, monks undergoing all kinds of self torture, etc. Well, I love those monks of Ribeira and Zurbaran, without which one would fail to understand the Inquisition. It is the moral force of man exaggerated, off the track, but original and bold in its excess. The apostle is certainly not the pure type of humanity, nevertheless, where shall we find a more powerful manifestation for the psychologist from which to study the inmost energy of human nature and its divine outbursts?

We must make due allowance for everything. Some of the charges preferred against the middle class civilization by the adversaries of the modern spirit are unquestionably true. The Middle Ages which assuredly understood the reality of life less well than we do, understood in some respects the suprasensible life better than we do. The error of the neo-feudal school lies in its non-perception of the fact that the faults of modern society are necessary to its character of transition, that these faults arise from a perfectly legitimate tendency operating under an exclusive and partial form. And this very partial form is also necessary; for it is one of humanity's laws that it must proceed with its phases the one after the other, and in setting aside temporarily all the rest; whence arises the incomplete appearance of all its successive developments.

If aught could inspire doubt in the mind of the thinker with regard to the future of reason, it would no doubt be the absence of grand originality and the small initiative displayed by humanity as it proceeds on its roads of reflection. When we compare the timid works brought forth with so much travail by our reasoning age with the sublime creations engendered by primitive spontaneity, engendered without so much as a feeling of their difficulty; when we ponder the strange facts which must have taken possession of men's consciences in order to create a generation of martyrs and apostles, we might be tempted to regret that man has ceased to be instinctive in order to become rational. But we are comforted with the thought that if his inward potentiality has diminished, his creation has become much more personal, that he is much more the master of his own work, that he is its author by virtue of a more lofty title; with the thought in sum that the actual condition of things is only a painful, difficult condition, full of effort and hard striving through which the human intellect will have had to pass towards a superior state; with the thought that the progress of the rational state will bring about another phase, in which the intellect will be once more the creator, but freely and with more conscience of its work. It is no doubt very sad for the man of intellect to have to pass through those ages of "little faith," to see sacred things railed at by the profane and to be exposed to the insulting laughter of triumphant frivolity. But it matters not; he is the custodian of the sacred deposit, he is the standard bearer of the future, he is man in the grand and extended signification of the word. And he knows it, hence his joy and his sadness; his sadness because he is permeated with the love of the perfect, it grieves him to think that so many consciences should for ever be closed to it; his joy, because he knows that the mainsprings of humanity never wear out; that, though momentarily overcome with sleep, they are there nevertheless, deep down his inmost being and that one day they will wake up to astonish by their proud originality and their invincible energy both their timid apologists and their insolent despisers.

Let us suppose for a moment that a thought as original, as powerful as that of Christianity were to appear to-day. At the first glance it looks as if it would make no headway at all. Selfishness is dominant, the sense of grand devotion and of disinterested apostleship is lost. The century seems to obey two motives only; fear and interest. At such a spectacle a deep sadness comes over the soul. It is all over then? We must bid farewell to great things; noble thoughts will have no existence save in the recollections of the orator; religion will no longer be aught but a check in the hands of the frightened wealthier classes. The sea of ice is for ever spreading and getting more solid. Who shall be able to break through it?

Timid friends who thus despair of humanity, let us go back together eighteen hundred years. Fancy that you are living at that period when a handful of obscure men founded in the East the dogma which since then has governed humanity. Cast a glance at the contemptible world that obeyed Tiberius, and tell me whether the world is really dead. Chant once more the funeral hymn of humanity; for humanity is no more, its heart is still in the cold grip of death. How are these poor enthusiasts to endow it with life once more; how, without a lever, can they uplift a world? Well, they have done it nevertheless; three hundred years later the new dogma was master, and four hundred years after that it became tyrant in its turn.

This is our triumphant reply. The condition of humanity will never be so desperate as to preclude us from saying; "Many a time it was believed to be dead. The gravestone seemed fastened down for ever, and the third day it uprose from the dead."

CHAPTER V.

It is not altogether inadvertently that I designate by the name of *science* that which is generally called *philosophy*. To philosophize is the word by which I would most willingly sum up my life; nevertheless, seeing that the popular use of the word only expresses a still partial form of the inner life, that, besides, it only implies the subjective fact of the solitary thinker we must employ the more objective word; to *know* when assuming the standpoint of humanity. Yes, the day will come when humanity will no longer believe; but when it shall know; the day when it shall know the metaphysical and moral world as it already knows the physical; the day when the government of humanity will no longer be given to accident and intrigue, but to the rational discussion as to what is best, and to the most efficacious means of attaining that best. If such be the aim of science, if its object be to teach man its final aim and its law, to make him grasp the true sense of life, to make up, with art, poetry and virtue the divine ideal which alone lends worth to human existence, if such be its aim, then is it possible that it should have its serious detractors?

"But," it will be asked, "will science accomplish these marvellous destinies?" All I know is this, that if science does not accomplish them, nothing else will, and that humanity will for ever be ignorant of the significance of things; for science is the only legitimate means of knowing, and that if it has been possible for the religions to exercise a salutary influence on the march of humanity it is solely because science was obscurely mixed up with them, that is to say, the regular exercise of the human intellect.

If we were to confine ourselves to what science has done hitherto without considering the future we might well ask ourselves whether it will ever carry out this programme, whether it will succeed in providing humanity with a symbol comparable to that of the religions. Up till now science has done little else but destroy. Applied to nature itself, it has destroyed its mystery and its charm by showing mathematical forces there where the popular imagination saw life, moral expression and liberty. Applied to the history of the human intellect, it has destroyed those poetical superstitions of privileged individuals, in the admiration of which semi-science took so great a delight. Applied to moral things it has destroyed the comforting beliefs which nothing can replace in the heart that was at rest in them. Where is the man who after having given himself up to science *with all his heart* has not cursed the day when thought became his birthright, who has not hankered after some fond, lost illusion? As for me, I admit having often been torn with regret; yes, there were days when I still wished to be at rest with the simple-minded, when I should have felt annoyed with criticism and rationalism, if it were possible to be annoyed with fate. The first feeling of him who deserts the ranks of naïve belief for those of critical investigation, is that of regret and almost anathema against that inflexible power, which from the moment it seizes upon him, compels him to accompany it through every stage of its irresistible march, until the final goal where one stops to weep (40). Unhappy like the Cassandra of Schiller through having seen too much of the reality, he is almost tempted to exclaim, "Restore me to my blindness." But are we to conclude from this that science is only fit to take the colour out of life, and to destroy beautiful dreams?

Let me first of all frankly admit that if such be the case, it is an evil beyond remedy, necessary and for which we should blame no one. If there be aught fatal on earth, it is reason and science. It is useless to grumble at it and to get impatient, and the anger of the orthodox folk against the freethinkers really makes one laugh; because it looks as if the latter had really had it in their power to develop themselves in a different fashion, as if a man were free to believe what he likes. It is impossible to prevent reason from taking up every object of belief, and all these objects lending themselves to criticism, reason is fatally bound to declare that they do not constitute absolute truth. There is not a single link in this chain which one was free to shake off for a single moment; the only culprit in the matter is human nature and its legitimate evolution. And the unquestionable principle is that human nature is without reproach and proceeds towards the perfect by means of forms successively and diversely imperfect.

The fact is that science will only have destroyed the dreams of the past to put in their stead a reality a thousand times superior. If science were to remain what it is we should have to submit to it while cursing it, for it has destroyed and not builded up again; it has awakened man from a sweet sleep, without smoothing the reality to him. What science gives me is not enough, I am still hungry. If I believed in any religion, my faith, I admit, would have the greater wherewithal to satisfy its hunger, but a small modicum of good science is better than a great deal of haphazard science. If we had to admit literally all that legend-mongers and chroniclers tell us about the origin of peoples and religions we should know a great deal more about them than with the system of Niebuhr and Strauss. The ancient history of the East in its ascertained facts might be reduced to a few pages; if we were to put our faith in Hebrew, Arabic, Persian and Greek histories, etc., we should have a library full. Folk with whom the craving to believe is very developed may afford themselves the pleasure of swallowing all that. The critical intellect is the sober man, or if you like, the fastidious man; he makes sure above all and beforehand, of the quality. He prefers abstention to indiscriminate acceptance, he prefers truth to himself, and sacrifices his most lovely dreams to it. Think you then that it would not be much more pleasant to us to sing in the temple with the women or to dream with the children than to be pursuing on those inhospitable mountains a truth which is for ever fleeing. Do not, therefore, twit us with knowing so very few things, for you, you *know* nothing at all. The little we do know is at any rate perfectly acquired and will ever go on increasing. Our guarantee is the most invincible of inductions, derived from the example of the sciences of nature.

"If," as Burke maintains, "our ignorance of the things of nature, is the principal cause of our admiration of them, if this ignorance became to us the source of the feeling for the sublime," we might ask ourselves whether modern science by rending the veil that hid from us the forces and the agents of physical phenomena, and by showing us everywhere a regularity subject to mathematical laws and consequently devoid of mystery, has furthered the contemplation of the universe, been of service to æstheticism, while furthering at the same time the knowledge of truth. The patient investigations of the observer, the figures accumulated by the astronomer, the long enumerations of the naturalist are scarcely calculated to awaken the sense of the beautiful, but the genuinely beautiful, that which is not based upon fictions of human fantasy is hidden in the results of analysis. To dissect a human body is to destroy its beauty, and still by means of this dissection, science arrives at the recognition of a much superior order of beauty, which superficial examination would not have as much as suspected. No doubt the enchanted world in which humanity dwelt previous to its entrance upon life guided by thought, the world conceived as being moral, replete with passion, life, and sentiment, that world possessed an ineffable charm; and it is likely that face to face with that inflexible and severe nature created for us by rationalism, some may be tempted to regret the miracle and to reproach scientific experiments with having banished it from the universe. But such a reproach could only be based upon an incomplete view of the results of science. For the real world revealed to us by science is by far superior to the fantastic world created by the imagination. If the human intellect had been challenged to conceive the most surprising marvels, if the limits which realisation imposes on the ideal had been knocked down in favour of the human intellect, if all this had been done even then it would not have dared to conceive a thousandth part of the splendours revealed by observation. We may go on inflating our conceptions as long as we like, we only bring forth atoms at the expense of the reality of things. Is it not a strange fact that all the ideas which primitive science had formed on the world appear narrow, trivial, and ridiculous to us after that which has been proved to be true. The earth resembling a disc, a column, a cone, the sun as big as the Peloponnesus, or else conceived to be a meteor, lighting itself every day, the stars trundling about at a few leagues distance on a solid vault, concentric spheres, *a universe closed*, stifling, walls, a narrow hemisphere against which the instinct of the infinite is shattered (41), these were some of the most brilliant hypotheses at which the human intellect had arrived. Beyond these, truly, was the world of angels with its everlasting splendour; but even there what narrow limits, what finite conceptions. Has not the temple of our God become enlarged since science has revealed to us the infinity of worlds? And still people were free then to create marvels; there was plenty of stuff to cut from and to spare, if I may so express it; observation placed no check on fantasy; but it was the experimental method which many delight in representing as narrow and without idealism which had the honour of revealing to us not that metaphysical infinite the idea of which is the foundation itself of man's reason, but that real infinite which he never attains in his boldest flights of fancy. Hence, we may fearlessly contend that if the marvellous in fiction has up till now seemed necessary to poetry, the marvellous in nature, when it shall be unveiled in all its splendour will constitute a poetry a thousand times more sublime, a poetry which will be reality itself, which will be science and philosophy at the same time. If the experimental knowledge of the physical universe has exceeded by far the dreams of imagination, are we not justified in believing that the human intellect by investigating more and more closely the metaphysical and moral sphere and applying its severest method to it without any consideration for chimeras and desirable dreams—if there be such, are we not justified in believing that by doing this it will simply shatter a narrow and paltry world to open another world of infinite marvels? Who knows but what our metaphysics and theology are not to those to be one day revealed by rational science as the Cosmos of Anaximenes or Indicopleustes to the Cosmos of Herschel and Humboldt.

The above consideration, it appears to me, is as eminently calculated to reassure us on the future and eventual results of science, as to justify all boldness and to condemn all timid restriction. However destructive a criticism may seem, we should give it free scope, provided it be really scientific; salvation never comes by going backward. It is, first of all too evident that the very consciousness of having retreated before the salutary method, and the permanent sentiment of a fictitious objection would cast over the whole of ulterior existence a scepticism more distressing than denial itself. We must either never discuss at all, or discuss to the end. Besides, it is certain that the true moral system of things is infinitely superior to the wretched hypotheses overtoppled by severe reasoning, that one day science will unfold a reality a thousand times more beautiful, and that criticism in that way will have been a first step in the direction of beliefs more comforting than those which it seems to destroy. Yes, if I were to see all the truths constituting what is called natural religion, a personal God, Providence, prayer, anthropomorphism, personal immortality, etc.; if I were to see all these truths, without which there is no happy life go to wreck beneath the legitimate effort of critical examination, I should clap my hands for joy over their ruin, thoroughly convinced that the real system of things, of which I may be still ignorant but in the direction of which this very denial is a step, infinitely surpasses the poor imaginations without which we cannot conceive the beauty of the universe. The gods only go to make room for others. That infinite beauty which we perceive only in vague outlines and which we endeavour to reproduce by paltry images exists, truly exists. It is more beautiful, more comforting a thousand times by far than that which my imagination may have conjured up. When the old anthropomorphic conception of the world disappeared before positive science, one might have been for a moment tempted to exclaim: "Good-bye to poetry, good-bye to the beautiful," and behold, the beautiful has revived more beautiful than ever. In the same way the moral world is far from having received a mortal blow by the destruction of old chimeras, the most realistic method being that which will lead us to its most dazzling marvels, and until we have discovered ineffable splendour, intoxicating truths, delightful and comforting beliefs we may rest assured that we are not within the truth, that we are merely passing through one of those fatal periods of transition when humanity ceases to believe in chimerical beauties ere it arrives at the discovery of the marvels of the reality. We should never get frightened at the onward march of science, seeing that we may be sure that it will only lead to the discovery of incomparably beautiful things. Let us leave vulgar natures to exclaim with Micah when they had taken away his idols; "Ye have taken away my gods." Let us leave them to say with Serapion the converted anthropomorphist of Mount Athos; "Alas, they have taken away my god, and I no longer know what I worship." As for us, when the temple topples down, instead of weeping on its ruins, let us think of the temples, which more magnificent and vast, will uprise in the future until the day when for ever shattering their narrow walls, thought will only have one temple, the roof of which will be the sky.

Hence, science must pursue its road without minding with whom it comes in collision. Let the others get out of the way. If it appears to raise objections against received dogmas, it is not for science but the received dogmas to be on the defensive and to reply to the objections. Science should behave as if the world were free from preconceived opinions, and not heed the difficulties it starts. Let the theologians come to an arrangement with one another to come to an agreement with science. We may as well take it for granted that that which is infinitely exceeds in beauty any and everything that which may be conceived; that the Utopian who sets his fancy to work in the creation of the best possible world only brings forth child's work in comparison with the reality; that when positive science only reveals triviality and finiteness, it is because it has not reached its final result. Fourier, scattering broadcast belts, crowns and *auroræ boreales* over the various worlds is nearer the truth than the physicist who thinks his small universe equal to that of God, and still, one day Fourier will be surpassed by the realists who will know the truth of things through unquestionable scientific observation.

The reader will allow me to cite an example.

The old manner of looking at immortality is in my opinion a "survival" of the conceptions of the primitive world and seems to me as narrow and as unacceptable as the anthropomorphic God. Man, in fact, is not to me, a composite of two substances, he is a unit, an individual resultant, a great persistent phenomenon, a thought prolonged. On the other hand, as soon as we deny immortality in an absolute manner, the world becomes colourless and sad. Now, it is an unquestionable fact that the world is beautiful beyond expression. Hence, we are bound to admit that everything which has been sacrificed in the furtherance of progress will be recovered at the end of the infinite, by a kind of immortality which moral science is sure to discover one of these days (42), and which will be to the fantastic immortality of the past as the palace of Versailles to a house of cards put together by a child. One may say as much of all the dogmas of our natural religion and morality, so colourless, so narrow, so lacking in the poetical that I should be afraid of offending God by believing in them. One may compare the old dogmas to those hypotheses of the physical sciences which afford sufficiently exact formulas for the representation of facts, though their expression be very faulty and contains a considerable part of fiction. One cannot say; "things are thus;" but one may say that, "things proceed as if it were thus." In calculating by the light of these hypotheses, one will get at exact results, because the error lies only in the expression and the illustration not in the *schema* and the category itself.

For the sake of the ultimate welfare of humanity there are centuries condemned to scepticism and immorality. To get from the beautiful, poetical world of the naïve peoples we have had to pass through the atomic and mechanical world. In the same way, in order to create for itself a new world of beliefs humanity must destroy the old, which can only be done by passing through an age of incredulity and speculative immorality. I say speculative, for no one has the right to throw the blame of his personal immorality on the century; elevated souls are under the happy necessity of being virtuous and the eighteenth century has proved that the most hideous doctrines may go hand in hand with the purest conduct and the most honourable character. This is an inconsistency, if you like. But there is no condition of humanity which can do without, and the first step of him who would think is to grow bolder in the face of contradictions, trusting to the future to reconcile them. A man who is consistent in his system of life is assuredly a narrow-minded one. For I defy him, in the actual condition of the human intellect to make all the elements of human nature agree. If he wants a system without a joint, he will be reduced to deny and to exclude.

Paltry and absolute criticism always springs from the fact of looking at every development of philosophical history by itself and not from the point of view of humanity. All the conditions through which humanity passes are faulty and assailable. Every century proceeds towards the future, carrying its objection in its side like a bullet in a wound. The destruction of ancient beliefs and the formation of new ones is not always accomplished in the most desirable order. Science often destroys a belief when it is still necessary. Let us suppose that the day will come when humanity shall have no longer any need to believe in immortality, can we imagine the anguish the premature destruction of this comforting faith will have caused to the unfortunate ones sacrificed to fate during our age of sorrow. In the definite constitution of humanity science will be happiness; but in the imperfect state through which we are passing it may be dangerous to know too soon.

My inmost conviction is that the religion of the future will be pure *humanism*, that is, the cult of everything that appertains to man, the whole of life sanctified and raised to a moral value. *To tend one's beautiful humanity* (43) will then be the Law and the Prophets, and that without any particular form, without any limit which will remind one of sect, or of exclusive brotherhood. The general characteristic of religious works is their specialness, that is, they require to be understood a special sense which every one has not; separate beliefs, separate feelings, separate style, separate figures. Religious works are for adepts; for they, the works, assume the existence of the profane. Assuredly St. Paul was an admirable genius; still, does the beauty of his letters consist in the grand instincts of human nature taken in the most general form as, for instance, in the dialogues of Plato? No. Seneca and Tacitus, perusing these curious compositions would not have considered them beautiful, at any rate not to the same extent that we do, initiated as we are in the data of Christian æsthetics. Several religious sects of the East, the Druses, the Mendaïtes, the Ansarians have sacred writings affording them a very substantial pabulum, but which to us are ridiculous or utterly insignificant. The sectarian's mind is closed to half of the world. Every *sect* presents itself to us circumscribed by limits; and no matter what limit is most antipathetic to our expansion of mind. We have seen so many of them that we cannot resign ourselves to the belief that the one any more than the other has got hold of the absolute truth. While willingly admitting that grand originality has up till now been sectarian or at any rate dogmatic, we cannot fail to see with equal certainty the absolute impossibility of confining the human intellect in the future in any of those vices. With a consciousness of humanity as developed as ours, we should soon bring about the reconciliation, we should judge ourselves as we judge the past, we should criticize ourselves whilst alive. Sectarian dogmatism is irreconcilable with criticism, for how can one help testing on one's self the laws observed in the development of other doctrines, and how can one reconcile absolute belief with such a reflective view? We may therefore say without hesitation that no religious sect will henceforth spring up in Europe unless new and ingenuous races strangers to thought, stifle once more all civilization; and even then one may affirm that this form of religion will be far less energetic than in the past and will not result in anything very characteristic. People are not converted from *finesse* to stupidity. One always remembers having been a critic and is often taken with laughter—even at one's adversaries in default of others. And apostles never laugh; laughter means scepticism already, for after having laughed at others, if one be consistent, one will also laugh at one's self.

For a religious sect to become henceforth possible we should want a deep moat of oblivion like that dug by barbaric invasion in which all the recollections of the modern world must be shot. Keep but one library, one school, one more or less significant monument, and you preserve criticism, or at least, the remembrance of a critical age. And, I repeat, there is but one means of being cured of criticism as of scepticism, namely, to forget radically its previous development and to recommence on another footing. That is why all the religious sects which for the last half century have endeavoured to establish themselves in Europe have struck against the spirit of criticism which took them on their ridiculous and irrational side to a degree such that the sectaries in their turn choose the wiser part of laughing at themselves. The century is so little religious as to have been unable to give birth even to a heresy (44). To attempt a religious innovation is to perform an act of belief, and it is because the world knows well enough that nothing is to be done in that order of things that it becomes bad taste to change anything in the *statu quo* of religion. France is of all countries the most orthodox for it is the least religious country in the world. If France had to a greater extent the sentiment of religion, it would have become Protestant like Germany. But not understanding anything of theology, and nevertheless feeling the need of a belief she thinks it easier to take the ready made system readiest to hand without caring in the least to make it more perfect; because to attempt to make it more perfect would be to take it earnestly, it would be assuming the part of theologian; and it is the correct thing with us to profess not to concern one's self with that kind of thing. Nothing is nearer to indifference than orthodoxy. The heresiarch has then, nothing to hope for nowadays, either from the severe orthodox who would anathematize him, or from the freethinkers who would smile at the attempt to reform that which cannot be reformed.

There is a very delicate line of demarcation beyond which the philosophical school becomes a sect; woe to him who crosses it. In a moment the language becomes altered; one no longer speaks to the world at large, there is an affectation of mystical form, a kind of incredulity and superstition creeps into doctrines—one knows not whence—that seemed altogether rational, reverie becomes mixed with science in an indistinguishable tissue. The school of Alexandria presents the most curious instance of this transformation. Saint-Simonism has renewed it in our days. I am convinced that if that school had kept to the lines of Saint-Simon, who, though very superficial on account of his defective education had really the scientific spirit, and under the direction of Bazard, who was certainly a philosopher in the best acceptation of the word, it would have become the original philosophy of the France of the nineteenth century. But the moment the less earnest spirits take the upper hand, the dross of superstition appears, the school turns to religion, only arouses laughter and breathes its last at Menilmontant amidst extravagances which wind up the history of all sects. Assuredly a tremendous lesson for the future.

Large-minded and unfettered science without any bond but that of reason, without a defined creed, without temples, without priests, living at ease in what is called the profane world, that is the form of the beliefs which henceforth will carry humanity with them. The temples of this doctrine will be not schools like those of to-day, childish, cramped, scholastic, but resorts of leisure as in ancient times (*scholæ*) where men foregather to partake together of the food provided for supra-sensible minds. The priests are the philosophers, the *savants*, the poets, the artists, that is, the men who have accepted the ideal as part of their inheritance and have relinquished the earthly portion (45). In this way the poetic priesthood of the first pioneers of civilization will come back to us. Many admirable intellects have often expressed regret at philosophy not having its temples and pulpits. Let us have them by all means, provided that nothing shall be taught in them except what is taught at the Sorbonne and the College of France, that, in one word, they shall be schools, shorn of their pedagogic varnish. Schools are the real competitors with temples. If you raise altar against altar, you will be told; "We prefer the old ones, not because we have greater faith in them, but because our fathers worshipped that way." If we were entrusted with the religious education of the people, we should have to begin by its so-called profane education, by teaching them history, science, and languages. For real religion is only the culmination of intellectual culture and it will not be accessible to the masses, until education shall be accessible to all. Our glory lies in always appealing to light; we pride ourselves in not being understood save on the condition of superior culture, on our strength being in direct proportion to our civilization. The eighteenth century must in this respect, always remain our model; the eighteenth century which has changed the world and inspired energetic convictions without attempting to become a sect or a religion and by remaining purely scientific and philosophical. Social and religious reform will assuredly come, seeing that every one is wishing for it, but it will not come from any one sect; it will come from the grand science, common to all, and operating in the unrestricted midst of human intelligence.

Hence the question of the future of religion must be resolved in divers ways according to the meaning attached to the word. If by religion we understand an *ensemble* of doctrines traditionally bequeathed, assuming a mythical form, exclusive and sectarian, then we are bound to say unhesitatingly, that the religions will have been the distinguishing mark of an epoch of humanity, but that they are in no way an integral part of human nature, and that one day they will disappear (46). If on the other hand we take the word to mean a belief accompanied by enthusiasm, crowning conviction with devotion and faith with sacrifice, then there is no doubt that humanity will never cease to be religious. But it is equally certain that henceforth no doctrine will have a chance of making headway without being solidly interwoven with humanity at large, and eliminating all speciality of form, without appealing to every one without distinction of adepts and profane. It causes me genuine grief to see distinguished intellects desert the grand audience of humanity in order to enact the easy part, so flattering to their estimate of self, of grand priests and prophets in conclaves which up till now are nothing more than clubs. What a difference between the philosopher whose name was formerly Pierre Leroux and the patriarch of a small church, surrounded by a knot of affiliated converts, concerning whom one hesitatingly puts the question; "Are they stupid enough to be believers?" In Heaven's name, if you do happen to have got hold of the truth, do address yourself to the whole of humanity. The man of secret societies is always narrow-minded, suspicious, one sided. The familiarity with such a small world destroys the familiarity with the great one; one ends by becoming suspicious of human nature and by founding the hope of success on factitious means, and obscure manœuvres. Great things are accomplished in open daylight. I do not mean to offend those whom present necessity compels to lock themselves in conclaves; very often, I am bound to say, they cannot help themselves. When the majority of the public is egotistical and immoral, we must pardon those who constitute themselves into secret committees, however prejudicial a blow such a life may strike at their intellectual development. Who shall blame the first Christians for having made unto themselves a world apart amidst the corrupt society of their times? Nevertheless, such a necessity is always a misfortune. One of the results of my historical studies, as far as I am personally concerned, has been to make me understand the apostle, the prophet, the founder of a religion; I am thoroughly aware of the sublimity and the errors inseparable from such an intellectual position. It seems to me that I have succeeded now and then in reproducing within myself by means of reflection the psychological facts that must have naturally perturbed those lofty souls. Well, I do not hesitate to say that the time for this kind of parts is gone by. The *universal*, that is, the human must henceforth be the outward criterion of a doctrine that solicits the faith of human kind. All that is sectarian must be placed on the same level with those products of a puny, weak literature which cannot live outside the atmosphere of the drawing-rooms in which they blossomed. We must be on our guard against those people who can only be understood by a committee. Common sense has relegated to its proper place that curious æsthetic school of *irony*, brought into fashion by Schlegel, where the artist proudly draping himself in his *virtuosity* and *geniality* made it a point to present nothing but insignificant and tasteless truisms, then shrugged his shoulders at the denseness of the public which did not care for these platitudes. All that partakes of the nature of monopoly in the world of thought, all that requires, in order to be understood a kind of special revelation, a sense apart, not possessed by humanity, must be fatally driven towards the like excess.

Hence, science is a religion, science alone will henceforth make the creeds, science alone can solve for men the eternal problems, the solution of which his nature imperatively demands.

CHAPTER VI.

WHY then is science whose fate is so closely bound up with that of the human intellect so badly understood as a rule? Why does it seem only a pastime or a supplementary thing? Why is the learned man in France I do not say an object of chaff to the frivolous minded—that would be an honourable title as far as he himself is concerned—but a useless piece of furniture in the opinion of many refined minds, something analogous to those literary abbés who were part of the furniture of a nobleman's seat much in the same way as the library itself. In fact, *literature* proper is a great deal better understood. There is no one who, from a more or less elevated standpoint, does not admit the necessity of people to write plays, novels, and periodical articles. Truly, there are few who have a conception of the earnest side of literature and poetry; the literary man is, in the opinion of the majority, only fit to amuse them, and the *savant* not having the same privilege is for that very reason, voted useless and a bore. People are apt to think that this is the case because he investigates, edits and comments on the work of others. Besides it is so very easy to ridicule his patient researches. One must be dull indeed not to be able to coin a feeble joke on the subject of a man who spends his life in deciphering old marbles, in guessing at unknown alphabets, at interpreting and commenting on texts, which to the ignorant, are only ridiculous and absurd. Those jokes have that false appearance of common sense, so powerful in France and which too often governs public opinion. A journalist, a manufacturer is considered a "serious" person. But the *savant* is of no account, unless he be a professor. Science should not put its head out of the college or special school, the public at large has no business with it. Let the professor occupy himself with it, that is right enough, it is his trade. But every one else who devotes his life to it meddles with what does not concern him, something like the man who learns a trade without ever intending to practise it. Hence the discredit attached to every branch of study not contributing directly to classical or pedagogical education, the necessity of which is accepted in good faith, without much knowledge of the reason why. The best judges acknowledge that all the branches of philological studies, the East and India above all may afford the history of the human intellect its most precious data. Then why is this California so little exploited? Alas, let us give the reason in its most prosaic harshness. It is because there is no outlet.

Whence arises this ignoble mistake? Let us first of all admit that a feeling of enthusiasm for science is much more rare and difficult in an age like ours which has witnessed such unquestionable progress in every branch of human knowledge than at an epoch when all the sciences were merely being created. Conquest and discovery imply a waking up and entail an exertion of strength which must remain foreign to those who need only march along in the already beaten track. Where is the philologist of our days who brings to his researches the intoxication of the first classical students, Petrarch, Boccacio, Poggio, Ambrose Traversari, the men who were so powerfully moved by the desire to know, who carried the worship of the new studies by which they enriched the human intellect to a degree of lofty mysticism, who suffered persecution, who starved in the pursuit of their ideal object. Where is the Orientalist who raves on his subject like Guillaume Postel? Where is the astronomer capable of the ecstacies of Kepler, the student of physical science capable of the prophetic excitement of the two Bacons? It was the heroic age of science when one philologist haply counted among his *Anecdota* Homer, another Livy, a third Plato. It is very easy to stigmatize those noble follies with the rather equivocal term of pedantry, it is easier still to show that those passionate lovers of science had neither the good taste nor the severe method of our century. But might we not also envy them their powerful love and their disinterestedness?

It does not come within the scope of my plan to inquire in how far the system of public education in France is responsible for the decay of the scientific spirit. Nevertheless, it seems to me that the small importance attached among us to higher education, the total lack of an institution corresponding to that of the German *universities* is one of its principal causes (47). I am the least likely man to libel the teaching given at the faculties. Germany has nothing to compare to the Sorbonne or to the College of France. In fact, I am not aware of the existence outside Paris of an institution whither *savants* and thinkers come pretty well without a programme to entertain regularly a public solely attracted by the charm or the importance of their lectures. They are two admirable institutions, essentially French, but they are not like the German universities. They are far beyond them, but they do not answer the same purpose. Apart from a few courses of a character altogether special, the lack of a constant and compulsory audience does not allow of demonstration of a very scientific nature. Face to face with a public the majority of which wishes above all to be *interestingly amused*, the lecturer is bound to enunciate ingenious views, to afford ingenious glimpses, rather than rely upon scientific discussion. I am aware that those views are the principal aim which we should set ourselves in scientific research; but however excellent the manner in which they are propounded, it can scarcely be denied that the courses which necessarily attract a great number of listeners and which exercise the greatest influence on the culture of the intellect, are the least likely to contribute to the spread of the scientific spirit. A great number of theories can, therefore, only find room in the curriculum of the lyceums, where science does not occupy its dignified position (48). How can public opinion be favourable to science, when the majority only know it from their old college recollections, which besides they are in a hurry to drop and which cannot make them conceive it in its true light? Hence, serious books and studies seem to have no significance save with a view to education, while on the contrary education should be only one of the least applications of science. This ridiculous prejudice is one of the greatest troubles in the path of him who devotes his life to pure science.

Thus by a strange reversal, science with us is only made for the schools, while the school should only be made for science. No doubt if the school in modern days were like that of antiquity, a gathering of men solely impelled by the desire to know and united by a common method of philosophizing, one might allow science to lock itself into it. But the school with us having as a rule a practical or *pedagogic* purpose, the reduction of science to narrow proportions, the supposition that philology is only of use in so far as it aids classical education is the greatest humiliation imaginable and most opposed to good sense. The department of science and serious research becomes in this way that of public education as if such things had no value except in so far as they are aids to education. Hence the idea that as soon as one's education is finished, one need not concern one's self any longer with them, and that they only concern the professors. In fact, it would be difficult indeed, I believe, to find among us a philologist who does not in some way belong to the scholastic profession and a philological book, not written for the use of schools or with some other scholastic purpose. It is a strange vicious circle, for if these things are of no use save for the purpose of teaching, if they are to be studied only by those who are to teach them, what is the good of teaching them?

Heaven forbid that we should try to lower those noble and useful functions that help to prepare the seriously-inclined intellect for every career; but it appears to me that we should thoroughly distinguish between science and instruction and give to the former, apart from the latter a religious and philosophical aim. The *savant* and the professor differ as much from one another as the manufacturer and the retailer. The confusion that has arisen in peoples' minds with regard to them has contributed to throw a kind of slight on the most important branches of science, on the very ones which on account of their importance were considered worthy of being selected as the bases of classical studies. Fashion is not half so severe with regard to studies of lesser importance which do not happen to remind us in so awkward a way of our college days.

Hence, we must accustom ourselves to consider the application of certain branches of science and especially of philology to classical studies as something accessory and secondary from the standpoint of science. It is only in connection with positive philosophy that everything has its price and its value. The frivolous mind which does not understand science, pedantry which understands it badly and causes its depreciation both spring from an absence of the philosophic spirit. We must accustom ourselves to look for the value of knowledge in knowledge itself and not in the use one may make of it for the education of childhood or youth.

No doubt, by the natural force of things, the men most eminent in each branch of science will be called upon to teach them, and reciprocally the professors will always have a gift apart. It is even worthy of note that all the most illustrious names in modern science are those of professors. One would look in vain among the free contingent of amateurs for Heynes, Bopps, Sacys, Burnoufs. Nevertheless, one cannot willingly blind one's self to the grave danger of science becoming too exclusively a matter of schools. It would contract habits of pedantry, which by investing it with a particular colour would drag it away from the grand midst of humanity. No one is more profoundly convinced than I that science cannot live without what we call the *technique*, no one has less sympathy than I have with the kind of drawing-room science, emasculated in form, trying to be interesting, a science of semi-scientific, semi-fashionable reviews. True science is that which belongs neither to the school, nor to the drawing-room, but which responds exactly to the want of man; that which shows no trace of institution or fictitious custom, in one word that which reminds us most of the schools of ancient Greece, which in this as in everything has given us the model of the pure and sincere. Look at Aristotle. The scientific apparel takes up more space with him that with no matter what modern *savant*, Kant perhaps excepted. It is evident that the human intellect delighted at the discovery of those orderly pigeon holes of thought which dialectics bring to light, attached at first too much importance to them and naïvely believed that every idea might advantageously be moulded in those forms. Still, is Aristotle, though so eminently *technical*, exactly *scholastic?* No. Compare his "Rhetoric" to the modern rhetorics, which after all and in reality are but so many weak reproductions of it, and you will on the one hand, have an original work, though strange in form, a true, albeit somewhat idle analysis of one of the facets of the human mind, on the other, books utterly insignificant, and absolutely useless outside the college. Compare the "Analytics" with the "scholastic logics" of the old school and you will be confronted with the same contrast. Therefore, in forbidding science to assume the scholastic air, we make no concession to the superficial intellect which should never be considered. We merely wish to bring science back to its grand and beautiful form which the French intellect understands so well. There is "good taste" in science as well as in literature which our countrymen have caught every now and then in a very superior and delicate way. German science is in this respect not bound to observe so many precautions. It may assume the scholastic air and surround itself with a scholastic perfume which with us would be considered scandalous. Are we to congratulate it on that account? Grave intellects readily make excuses for pedantry. They know that this form of intellectual labour is often necessary, always pardonable. No one objects to it in the classics of the Carlovingian restoration, nor in those of the Renaissance; the human intellect is sure to amuse itself for some time with its discoveries and the new results it introduces into science, it is bound to make a pleasant pastime, nay a toy of it, before making it an object of philosophic meditation. The same tone will be sure to crop up once more and will be equally pardoned in the exclusive and pre-occupied *savant* who works his mine with passion, especially if a powerful spirit does not come to the rescue to animate his patient researches, and if the simplicity of his outward life reduces him to the everlasting and unchanging rôle of *savant*. The higher philosophy, intercourse with society or the practical pursuit of business can alone preserve science from pedantry. But for many, many long years we shall still have to pardon the *savant* for being neither a philosopher, nor a man of the world, nor a statesman, even when they bear the title of *Court Councillors* as in Germany.

Our sensitiveness in this respect is perhaps one of the reasons why philology, though represented in

France by so many illustrious names is always held back by a kind of shamefacedness, and dares not proclaim itself openly. We are so terribly afraid of ridicule that everything which can possibly lend itself to it arouses our suspicion; and the most worthy things by slightly changing their name and shade lay themselves open to it.. The term pedantry, which, if not clearly defined, may be so mischievously applied, and which with the unthinking is almost synonymous with everything relating to serious and scientific inquiry has in this way become a scarecrow to subtle and refined intellects who have often preferred to remain superficial rather than lay themselves open to the attack the most painful of all to us. This scruple has been carried so far that we have seen critics of the highest order of intellect deliberately leave their expressions incomplete rather than employ the scholastic word, when the scholastic word was the right one to employ. Scholastic jargon when it merely hides the absence of thought or when merely used to "show off" by the narrow intellect is tasteless and ridiculous. But to deliberately banish the exact and technical style, which alone is capable of expressing certain delicate or deep shades of thought, to do this is to fall into a purism equally unreasonable. Kant and Hegel or even minds as absolutely emancipated from the scholastic traditions as Herder, Schiller and Goethe would, if judged like that, not escape our terrible accusation of pedantry.

We may congratulate our neighbours upon their freedom from all such fetters, which, we are bound to say, would be less hurtful to them than to us. With them the school and science are in touch; with us, every system of higher education, which in its manner, has still the scent of the school about it, is voted "bad form" and unbearable. People think they are showing great subtlety and tact by placing themselves above everything that reminds them of school instruction. Every one indulges in that small conceit, and fancies he is proving in that way that he has got over his schooling long ago. Does it sound credible to us that in the ceremonies, resembling our distributions of prizes, when with us a display of eloquence is THE THING, the Germans confine themselves to the reading aloud of grammatical dissertations of the severest kind and bristling with Greek and Latin words (49)? Should we be able to understand solemn and public sittings, occupied with the following lectures. "*On the Nature of the Conjunction.*"—"*On the German period.*"—"*On the Greek Mathematicians.*"—"*On the topography of the Battle of Marathon.*"—"*On the plain of Crissa.*"—"*On the centuriæ of Servius Tullius.*"—"*On the Vineyards of Attica.*"—"*Classification of the Prepositions.*"—"*Elucidation of the difficult words in Homer.*"—"*Commentary on the portrait of Thersites in Homer;*" etc.; etc. (50)? All this implies among our neighbours a marvellous taste for the serious, and also perhaps a certain amount of courage in being bored without wincing when etiquette requires it. Madame de Stael says that the Viennese of her time amused themselves methodically and as a matter of conscience. Perhaps the German public may also be more patient than ours when it becomes a question of being bored ceremoniously and by official invitation. It will soon become with us a meritorious act to witness a sitting of the Academy of Inscriptions, though the blame may not be laid at the door of the Academies. Our public is too difficult to please, it wants to be interested and even amused where instruction should suffice; and in fact until one has a conception of the lofty and philosophical aim of science, as long as people will look upon it as a kind of curiosity, like any other curiosity, it will be voted a bore and taxed with the *ennui* it produces. Play for play there is no reason why the least attractive should be chosen.

Montaigne who in many respects is the eminently typical intellect of the Frenchman, represents that typical intellect above all by his horror of everything that reminds him of pedantry. It is a treat to see him do the "free and easy," the man of the world who understands nothing of science and "knows everything without having ever learned anything."* "All these," he says; "are but so many dreams of a man who has only tasted the upper crust of science in his infancy, and has only retained a general and vague impression of it; a little of everything and nothing at all, in the French fashion. For in sum, I know that there is a science of medicine, of jurisprudence, four parts in mathematics, and roughly at what they aim. And I may perhaps have an inkling of the pretensions of sciences in general with regard to their use in life, but as for having gone deeper into them, as for having bitten my nails to the quick in the study of Aristotle, the king of the modern method, or having obstinately pursued any science at all, I have never done so, there is no art of which I can do more than lay down the first principles. There is not a middle class child, but what may call itself more learned than I, who could not make it go through its first lesson. And if I were compelled to do so, I should be obliged to draw absurdly from said lesson some matter of general import, on which I would examine his natural aptitude; a lesson as utterly unknown to him, as his is to me."

Nevertheless, he takes good care to show that he understands as much about it as any one else, and to reveal such traits of learning as may do credit to his understanding, provided it be taken for granted that he sets no store by them and that he is above such pedantry. He prides himself upon having no memory (*retention*) and "to be a capital hand at forgetting" *excellent en oubliance*. "I have got no store cupboard;" (*Je n'ai pas de gardoire*) he says; "for it is by that that the learned shine." Upon the whole, it is a nice quiet way of snapping his fingers at the virtues of the *savant* in order to raise himself in

* Molière.—TRANSL.

peoples' estimation by those of the man of sense and the man of wit, a way eminently characteristic of the French intellect and which Madame de Stael designates so very cleverly by the name of "the pedantry of thoughtlessness" (51).

CHAPTER VII.

Just as in the very bosom of religion there are a great many men handling sacred matters without the least idea of their lofty meaning, and only looking at them in the light of vulgar manipulation, so in the field of science there are labourers—very worthy people in the main—often utterly lacking in the sentiment of their work and its value in the furtherance of the ideal. Let us hasten to add that it would be unfair to require of the *savant* the ever immediate consciousness of the aim of his work; it would be bad taste on our part to wish him to speak of it deliberately in and out of season; it would be compelling him to head all his works with the self-same prolegomena. Take the most beautiful scientific works, peruse the works of Letronne, of Burnouf, of Lassen, of Grimm and of the princes of modern criticism generally and perhaps you would seek in vain for a page *directly* and *abstractedly* philosophical. The fact of their authors being thoroughly imbued with the philosophic spirit is made manifest not so much by an isolated flight of rhetoric as by the general spirit and method. Nay, this prudent abstention may often be taken as an act of scientific virtue and the heroes of science are they who, though more capable than no matter whom of indulging in lofty speculation, have the strength of mind to confine themselves to the strict statement of facts and abstaining voluntarily from anticipated generalizations.

Works undertaken without this noble spirit may even powerfully aid in the cultivation of the human intellect irrespective of the more or less paltry intentions of their authors. Is it at all necessary for the marble quarrier to have an idea of the future monument of which the blocks which he quarries will form a part? Among the laborious workers of science who have raised the edifice of science many looked no further than the stone they shaped, or at any rate no further than the limited region where they placed it. Like the ants, they each bring their individual tribute, remove some obstacle, cross one another incessantly, apparently without the slightest attempt at order, and only get in one another's way. Nevertheless, it happens that by the united labour of so many men and without any preconceived plan a science is organized, and organized in noble proportions. An invisible genius has been the architect presiding over the whole and has made all those isolated efforts subservient to a perfect unity.

By studying the origin of each science one will find that the first steps were nearly always taken with no very distinct consciousness, and that among others, philological studies owe a large debt of gratitude to very mediocre intellects, which at the outset laid down their material conditions. Hervas, Paulin de Saint-Barthélemi, Pigafetta who must be considered the founders of linguistic science were certainly not geniuses. What an immense fact in the history of the human intellect is the initiation of the Latin world in the understanding of Greek literature. The two men who contributed most powerfully to it, Barlaam and Léontius Pilatus were according to Petrarch and Boccacio who knew them so very intimately two nonentities who were as surly as they were whimsical. The majority of the Greek emigrants who played so important a part in the development of European intelligence were men of no parts whatsoever, downright labourers who took advantage *per alcuni denari* of their knowledge of

Greek. For every Bessarion there were a hundred Philelphuses. The lexicographers are as a rule not very great philosophers and still the best book on generalities has not had so great an influence on the higher sciences as the dictionary—philosophically an exceedingly poor performance—by which Wilson made the study of Sanskrit possible in Europe. There are works requiring an amount of plodding drudgery to which men impelled by too powerful a craving for philosophical studies would with difficulty submit. Would lofty and energetic minds have succeeded in accomplishing the immense works issued from the scientific workshops of the congregation of St. Maur? * Every scientific work, conducted according to a sound method, keeps its unquestionable value, irrespective of the wider or narrower views of its author. The only useless works are those in which the smatterer or the quack pretends to imitate the bearing of genuine science, and those in which the author, prompted by an interested thought, or by the preconceived dreams of his imagination is bent at all costs upon finding his chimeras everywhere.

Though it is not necessary for the workman to have a perfect knowledge of the work he executes, one could wish those who devote themselves to special labours to have an idea of the whole which alone imparts value to their researches. If the many laborious workers to whom science owes its progress had possessed the philosophical spirit of what they did, if they had perceived in learning something more than the satisfaction of their vanity or curiosity, how many precious moments would have been saved, how many fruitless excursions spared, how many lives given up to insignificant works would have been devoted to more useful researches. When we come to consider that the intellectual labour of whole

* The congregation of St. Maur, a reformed branch of the order of St. Benedict, was founded in France in the seventeenth century and has rightly been called the nursery ground of *savants*.—Transl.

centuries, of whole countries, of Spain for instance, has been wasted, for lack of a substantial object in view, that millions of volumes have crumbled to dust without producing the least result, we cannot help regretting deeply this immense loss of human force, brought about by the absence of guidance and for want of a distinct perception of the goal to be reached. The intensely sad impression one feels on entering a library is due mainly to the thought that nine-tenths of the books crowded together there have missed their mark, and whether through the fault of the author himself, or through the fault of circumstances have never had nor will ever have, any direct action on the onward march of humanity.

It seems to me that science will only then recover its dignity when it takes up its definite stand on the grand and wide point of view of its veritable aim. In former days there was room for that trivial and more or less innocuous character of the *savant* of the Restoration, a semi-courtier-like part, the actor of which had a way of allowing himself to be taken for a man of *solid* attainments who tossed his head at ambitious innovations; it was a way of securing the patronage of some Mæcenas, a duke or peer, who as a mark of high favour admitted him in the capacity of a bit of furniture of his drawing-room, or of an antique *curio* of his collection; and in the whole transaction there was nothing deserving the name of serious, there was nothing but the more or less imbecile laughter of vanity, the laughter so terribly aggravating when it presumes to meddle with serious matters. That is the kind of thing which is doomed to disappear for ever more, that is as good as buried together with the playthings of a society in which shams still played so important a part. To entice science from the grandiose midst of humanity in order to make it an idle plaything of a court or a drawing-room is degrading science, for the day is not far distant when everything that is not serious and true will be ridiculous. Let us, then, be true for Heaven's sake, true like Thales when from his own initiative and impelled from within he took to speculating upon nature, true like Socrates, true like Jesus, genuine like St. Paul, genuine like all those great men which the ideal claimed as its own, took possession of and dragged after it. Let us leave old-world twaddle to plead lukewarmly in apology for science that like everything else, it is a necessity, an ornament, that it confers lustre on a country, etc., etc. All this is so much sheer imbecility. Where is the noble and philosophic soul, eager for perfection, possessing the sentiment of its intrinsic worth that would consent to sacrifice itself to such vanities, that would gladly take up its place in the inanimate tapestry of human nature, to enact in the living world the part of a mummy in a museum. As for myself, I confess frankly that if I could see a form of life more beautiful than that of science, I should run to embrace it. How can one be reconciled to what one knows to be the second-best? How can one consent to put one's self among the waste, to accept a mere part of show when life is so short, when nothing can make up for the moments not devoted to the delights of the ideal? Oh, truth and earnestness of life, oh holy poesy of things, what is there to console us for having no feeling for thee? And to talk for a moment of the serious hour to which everyone must look forward to appreciate things in their true light, who will be able to breathe his last tranquilly, when, in casting a glance backward, he only finds in his life frivolity or gratified curiosity? The end only is worthy of being looked at, the whole rest nothing but vanity. To live does not mean to glide smoothly along a pleasant surface, it does not mean making the world your plaything in the pursuit of your own pleasure; it means to partake of many beautiful things, it means to be the fellow-traveller of the stars; it means knowledge, hope, love, admiration; it means well-doing. He has lived most and most worthily who by his heart and intellect, by his acts has worshipped most.

Hence, to look at science in the light of a mere satisfaction of vanity or curiosity is as great an error as to consider poetry merely the listless exercise of frivolous minds, or literature the amusement of which one tires least, and to which one comes back most willingly. The collector and the amateur may render signal services to science, but they are neither *savants* nor philosophers. They are as far removed from it as the manufacturer. For they amuse themselves, they pursue their own pleasure just as the manufacturer seeks his profit. I am aware of there being various degrees of curiosity. There is a wide difference between the paltry instinct of the collector which scarcely differs from the attachment of the child to his toys, and the more elevated form when it becomes a love of knowledge, that is; a legitimate instinct of nature and may be productive of a very noble existence. Boyle and Charles Nodier are only mere inquirers, and still they closely trench upon the philosopher. It is very rare, in fact, that with the highest exercise of the intellect there is not mixed up a certain amount of pleasure, which though having no value as far as the ideal is concerned, is none the less useful. It would be difficult to say how many discoveries have been simply due to curiosity. How many compilations, precious in view of ultimate researches would have never been made but for that innocent love of labour with which many mildly active natures cheat their craving. It would be cruel indeed to refuse to those humble workers the trivial and not very lofty pleasure, but sweet withal, which M. Daunou has so well defined as *paperasser*.* We are all more or less glad at having felt the same satisfaction, if for no other reason than because it has helped us to devour the barren pages

* *Paperasser* means both a love of scribbling and of ferreting among documents. Daunou was the author of a remarkable series of Historical Essays; died about 1840.—Transl.

of science. Without this the first studies necessary to master the material baggage of a language would be unbearable, and thanks to this they become the most attractive imaginable.

We may go further still and positively affirm that without this attraction the most learned men of modern times who were neither borne up by lofty philosophical views, nor by directly religious motives, would not have undertaken those immense labours which have made the investigations of higher criticism possible to us. He who, with our material wants intensified to a greater pitch, would accomplish a similar act of abnegation to-day, would be a hero. But it is important to maintain for all that, that this curiosity has no immediate moral whatsoever, and that its possession does not constitute the *savant*. There are manufacturers who exploit science for their benefit, the others exploit it for their amusement. No doubt the latter is the better part, but there is not an enormous difference between the two. Pleasure being essentially personal and interested, there is nothing whatsoever sacred or moral about it. All literature, all poesy, all science whose only aim is to amuse or merely to awaken interest is for this very reason frivolous and vain, or to speak correctly, has no right at all to call itself literature, poesy or science. The mountebanks in the streets and elsewhere do as much, and what is more, succeed better. How is it that one considers the perusal of Corneille, Goethe, Byron as a serious occupation, and that the perusal of such and such a modern novel or drama is looked upon as merely a pastime? For the same reason that the *Quarterly* and *Edinburgh Reviews* are serious periodicals and that the *Magasin Pittoresque* is a frivolous book.

Therefore, to take up science as merely interesting and curious is to humiliate it. In that case Christian asceticism would be perfectly right in its opposition. The sole legitimate means of making one's self the apologist of science is to look upon it as the essential element to human perfection. "The Imitation of Christ," after having begun like the "master of those who know" with the words; "Every man is naturally desirous to know," was thoroughly right in adding; "But what is the use of science without love? Better by far the humble peasant who serves God, than the proud philosopher who watches the course of the stars and neglects himself. What is the use of the knowledge of things, the ignorance of which will not cause us to be condemned? All is vanity, save the love of God and to serve Him." There is no gainsaying this, if science be conceived as a simple series of formulas, if perfect love be possible without knowledge. If we place perfection on one side, on the other vanity, how can one fail to follow perfection? But it is this very division which is illegitimate, because perfection is impossible without science. The true way of worshipping God, is to know and to love that which is.

CHAPTER VIII.

Of all branches of human knowledge *philology* is the one of which it becomes most difficult to grasp the aim and the unity. Astronomy, zoology, botany have all a determined object in view. But what is that of philology? The grammarian, the linguist, the lexicographer, the *literary man* in the special meaning of the word have, all of them, a right to the title of philologists, and we observe, in fact, between those various studies a connection sufficiently strong to enable us to call them by a common name. In one respect the words philology, philosophy, poesy and a good many others are very much alike; in that their very vagueness is expressive. When after the manner of the logicians we look for a phrase equivalent to those comprehensive words, and which at the same time gives their definition, we become greatly embarrassed because they have neither in their object nor in their method any unique characteristic. Socrates, Diogenes, Pascal, Voltaire are termed philosophers; Homer, Aristophanes, Lucretius, Martial, Chaulieu and Lamartine are termed poets, but it is not easy to find the family link which unites minds so utterly different under the same name. Such appellations have not been coined according to preconceived and clearly defined notions; they owe their origin to a process less fettered in its application and upon the whole more exact than that of artificial logic. These words designate regions of the human intellect between which we must be careful not to trace lines of demarcation too hard and fast. Where does eloquence end, where does poetry commence (52)? Is Plato a poet, is he a philosopher? Puerile questions, these, no doubt, seeing that whatsoever name we give him he will be none the less admirable, and that genius does not work in the exclusive categories which language coins afterwards, and on the strength of his work. The whole difference lies in a peculiar harmony, in a more or less sonorous ring with regard to which an experienced faculty never hesitates.

Antiquity, wiser in this respect and less distantly removed from the origin of these words, was less embarrassed in their application. The very complex sense of its word *grammar* did not cause it the least hesitation. Since we have drawn up a map of science as it were we persist in assigning to philology, to philosophy each a place apart; and yet they are less special sciences than different ways of treating things intellectual.

At a time when the first question asked of the *savant* refers to the nature of his studies, and the results attained, philology can but find small favour. One understands the physicist, the student of chemistry, the astronomer; the philosopher is less well understood, and the philologist still less. The majority, wrongly interpreting the etymology of his name imagine that he only works on words (and what, say they, could be more frivolous?) and never dream of distinguishing like Zeno the *philologist* from the *logophile* (53). The vague cloud that hangs over the object of his studies, that *sporadic* character as the Germans call it, that almost undefined latitude which embraces so many divers researches under the same name is apt to make one believe that he is only an amateur, who flits about amidst the variety of his works, and undertakes exploring expeditions into the past just as certain burrowing animals construct subterranean passages for the mere pleasure of constructing them. His place in the philosophical organization is as yet not sufficiently defined, his monographs keep on accumulating without any one perceiving their purport.

Philology, in fact, seems at the first glance only to offer an *ensemble* of studies without any scientific unity. Everything that contributes to restore or to illustrate the past has the right to a place in it. Understood in its etymological meaning it should only include grammar, exegesis and criticism of texts. Works of pure learning, of archæology, of æsthetic criticism should be excluded from it. But such exclusion would, however, not be natural at all. For there is the closest connection between these labours, they form, as a rule, part of the studies of the same individual, very often of the same work. To eliminate some of these from the ensemble of philological labours, would be to make an artificial and arbitrary scission in a natural group. Let us take, for instance, the school of Alexandria; a few philosophical and theurgical speculations apart, are not all the labours of that school, even those which do not come directly under the head of philology, stamped with the same spirit, which one may call philological, a spirit it carries even into poetry and philosophy? Would a history of philology be complete if it made no mention of Apollonius of Rhodes, of Apollodorus, of Ælian, of Diogenes Laertius, of Athenæus and other polygraphists, whose works are, however, far from being philological, even in the most restricted meaning? If, on the other hand, we grant to philology its widest possible extension, where are we to stop? If we do not look out, we shall be forcibly brought to include nearly the whole of the literature of thought into it. Historians, critics, polygraphists, the writers of literary history will have to find a place in it (54). Such is the drawback, grave no doubt, but necessary withal and compensated for by great advantages, of separating in that way a group of ideas, belonging to the ensemble of the human intellect and to which it clings with every one of its fibres. Let us add that the bearing of words changes with the revolution of things and that in the appreciation of their meaning we should only be guided by their central notions, without trying to imprison those notions in formulas which are never their perfect equivalents. When it becomes a question of ancient literature, criticism and erudition enter by right into the framework of philology; on the other hand the historian of modern philology will not deem it incumbent upon him to speak of our grand collections of civil and literary history, nor of those brilliant works of æsthetic criticism that have attained the level of the noblest philosophical creations (55).

Therefore, the field of the philologist can no more be defined than that of the philosopher, for both in fact are occupied not with a distinct object, but with all things from a special standpoint. The true philologist must be at once a linguist, a historian, an archæologist, an artist and a philosopher. Everything assumes to him a meaning and a value, in view of the object he sets himself, and which renders serious the most frivolous things distantly or closely connected with it. Those who like Heyne and Wolf have confined the rôle of the philologist to the reproduction in its purely scientific domain, and as in a living library, of every trait of the ancient world (56), do not appear to have understood the full extent of their range. The aim of philology does not lie within itself; it has its value as a necessary condition of the history of the human intellect and the study of the past. No doubt a good many philologists whose learned studies have thrown open antiquity to us, saw no further than the text they interpreted and around which they grouped the myriad spangles of their brilliant learning. In this, as in every other science, the natural curiosity of the human intellect coming to the aid of the philosophical spirit and supporting the patience of investigators has no doubt had its use.

A great many people feel inclined to laugh when

they find grave intellects take an enormous deal of trouble in explaining grammatical peculiarities, collect glossaries, compare the *variorum* editions of some ancient author whose only claim to notice often lies in his oddity or mediocrity. All this arises through not having understood in a sufficiently wide sense the history of the human intellect and the study of the past. Human intelligence after having traversed a certain space likes to retrace its steps to behold once more the road along which it has travelled, to chew the cud of its own thoughts. The first creators did not look behind them, they marched onward, without any other guide than the eternal principles of human nature. But, on the other hand, humanity reaches a certain stage, when books have sufficiently accumulated to be collected and compared, and then the mind will only proceed with a full knowledge of facts; it wants to confront its work with that of bygone ages; that day witnesses the birth of the literature of thought, and parallel with it, the birth of philology. The apparition, therefore, is not, as has been said, a sign of the death of various literatures, it merely bears witness to their having accomplished already *one life*. Greek literature was manifestly not dead in the age of the Pisistratidæ when the philological spirit shows itself already so characteristic. In the Latin and French literatures, the philological spirit was the forerunner of the grand productive epochs. China, India, Arabia, Syria, Greece, Rome, the modern nations have all known the moment when the labour of the intellect from being spontaneous becomes scientific and no longer proceeds without consulting the archives deposited in museums and libraries. The development of the Hebrew people itself which previous to Christ appears to offer less trace than any other of intellectual activity, shows in its decline perceptible vestiges of that spirit of comparison, of collecting, of "patching up" if I may so call it, which terminates the original life of all literatures.

These considerations would, it appears to me, be sufficient as an apology for the philological sciences. Nevertheless, they are, in my opinion, very secondary indeed, when we come to consider the new position which the development of contemporary philosophy ought to grant to these studies. It wants but one step more for science to be proclaimed the true philosophy of humanity, and the science of a being who is in a perpetual condition of "to be" can only be the history of that being. The history of the human intellect not merely inquisitive but theoretical, that is the philosophy of the nineteenth century.

And it so happens that this study is only possible through the direct study of monuments, and that those monuments "cannot be got at" without the special researches of the philologist. Certain forms of the past are each in itself sufficient to occupy a laborious existence. An ancient language, often utterly unknown, an isolated palæography, an archæology and a history painfully deciphered, each of these is assuredly more than enough to engross all the efforts of the most patient investigator, unless humbler workers have given protracted labour to the digging from the quarry and to the presentation in an aggregate for his appreciation of the materials with which he is to reconstruct the edifice of the past (57). In the opinion of posterity, such and such a ponderous, mediocre, but patient intellect, who has contributed one important stone to this gigantic work will perhaps occupy a loftier standpoint than such and such a speculative mind of secondary order, who called himself a philosopher and who did nothing but talk about the problem, without providing a single new datum to its solution. The revolution which since 1820 has completely changed the aspect of historical studies, or which, to speak more correctly has really founded the science of history among us, is obviously as important a fact as the apparition of a new system. Well; would the works of a Guizot, a Thierry, a Michelet, works so full of originality, would they have been possible without the Benedictine collections and other preparatory labours? Mabillon, Muratori, Baluze, Du Cange were not great philosophers, and still they have done more for true philosophy than a good many empty and systematic minds who wanted to build the fabric of things on air, and of whom not a single syllable will remain among our final acquisitions. I am not alluding here to works in which the most solid learning is wedded to a subtle or lofty criticism, such as for instance the last volumes of "L'histoire litteraire de la France" "L'Essai sur le Buddhisme" of M. Eugène Burnouf, "L'Archéologie Indienne" of M. Lassen, "La Grammaire Comparée" of M. Bopp, or "Les Religions de l'Antiquité" of M. Guigniaut. As for myself I have no hesitation in saying that I have got more things philosophical out of each one of these works than from the whole of the collected works of Descartes and of his school. But I am speaking of those works of a severer character which the profane consider as unreadable, such as for instance; Catalogues of Manuscripts, grand compilations, "Libraries" like that of Fabricus, etc.; etc. Well I maintain that such books, almost insignificant in themselves, have a priceless value if looked upon as materials for the history of the human intellect. Ten thousand volumes of philosophy like the "Lessons" of La Romiguière or the "Logique" of Port-Royal might burn before my eyes, and I should prefer to save "La Bibliothèque Orientale" of Assemani or the "Bibliotheca Arabico-Hispana" of Casiri. For philosophy always benefits by taking up things *ab integro*, and after all the philosopher can always say; "*Omnia mecum porto;*" whereas the noblest genius in the world could not restore to me the documents in those collections bearing upon Syriac and Arabic literature, two very secondary facets, no doubt, but still, two facets of the human intellect.

It is very easy to ridicule those attempts at restoring obscure literatures, often of but mediocre value. The error arises from the fact of not understanding the whole extent, the infinite variety of the science of the human intellect. A learned disciple of M. Burnouf, M. Foucaux has for the last few years tried to introduce into France the study of Thibetic. I should not be surprised in the least but that his praiseworthy enterprise has already caused many an epigram to be flung at him; well I declare, I, that M. Foucaux is doing more meritorious work with regard to the philosophy of the future than three fourths of those who claim the position of philosophers and thinkers. When Mr. Hodgson discovered in the monasteries of Nepaul the primitive monuments of Indian Buddhism, he did greater service to thought than a generation of scholastic metaphysicians could have done. He provided one of the most essential elements to the explanation of the Gospel and Christianity by revealing to criticism one of the most curious religious apparitions and the only fact possessing a close analogy to the grandest phenomenon in the history of humanity. He who should bring us back from the East a few works in Zend or Pehlvi, who should make known to Europe the epic poems and the whole of the civilization of the Rajpoots, who should make his way into the libraries of the Djains of Guzerat, who should give us an exact knowledge of the books of the Gnostic sect which is still being preserved under the name of Meudeans or Nasoreans, he who should do all this would be certain to have contributed an imperishable stone to the grand fabric of the science of humanity. Where is the abstract thinker who can make sure of a similar satisfaction?

Hence, it is in philosophy that we must look for the real value of philology. Each branch of human knowledge has its special results which it brings as a tribute to the general science of things and to universal criticism, one of the first needs of thinking man. Herein lies the dignity of all special research and of the last details of learning which have no meaning for the frivolous and light-minded. Looked at in this light, there is no such a thing as trivial or useless research. There is no study, however unimportant its object may seem, but what brings its ray of light to the science of the whole, to the true philosophy of the realities. The general results which alone, we are bound to admit, have a value in themselves, and are the *final aim* of science, become only possible by the *means* of knowledge, and of learned knowledge of details. We go further still and say that the general results not based on the knowledge of the latest details are necessarily hollow and artificial; whereas special researches, even if destitute of the philosophical spirit may be of the greatest value if they are exact and conducted according to a severe method. The spirit of science is that intellectual communion which links the scholar to the thinker, gives to each his deserved glory and establishes the point of convergence at which their several labours meet at last.

The union of philology and philosophy, of learning and thought should, therefore, be the characteristic of the intellectual labour of our age. It is either philology or learning which will provide the thinker with that forest of things (*silva rerum ac sententiarum*, as Cicero has it) without which philosophy will never be aught else than a Penelope web which will constantly have to be recommenced. We must once for all abandon the attempt of the old school, to construct the theory of things by the play of formulas void of all spirit, just as if by manipulating the shuttle of the weaver without putting thread into it, we should pretend to manufacture linen, or by setting a mill into motion without providing corn we should expect to get flour. The thinker implies the scholar; and if it were only for the sake of the severe training of the mind, I should set very little store by the philosopher who had not for once in his life worked at the elucidation of some special point of science. No doubt the rôles may be divided and such division is often desirable. But there should be at any rate a close commerce between these various functions, the works of the scholar should no longer remain buried among the mass of learned collections, where for all the good they do, they might just as well not exist at all; the philosopher on the other hand should no longer insist upon evolving from his own consciousness the vital truths of which the sciences outside possess such enormous wealth for him who explores them intelligently and critically.

Whence come so many novel views on the progress of the different literatures and the human intellect, on spontaneous poetry, on the primitive ages, unless they have sprung from the patient study of barren detail? Would Vico, Wolf, Niebuhr and Strauss have been able to endow thought with so many new views, without the possession of most minute erudition? What if not erudition, has opened to us the worlds of the East, the knowledge of which has made possible to us the comparative science of the developments of the human intellect? Why is one of the noblest geniuses of modern times, Herder, so frequently inexact, chimerical, inaccurate in his treatise on the "Poetry of the Hebrews," in which he has put forth all his soul, unless it be because he failed to support that admirable æsthetic sense with which he was gifted by scientific criticism? From this point of view the study even of the follies of the intellect has its value as regards history and psychology. Many important problems of historical criticism will not be solved until an "intelligent" scholar shall have devoted his life to the dissection of the Talmud and the Cabbala. If Montesquieu in classifying the laws of the Ripuarians, Visigoths and the primitive Burgundians was justified in comparing himself to Saturn devouring stones, what will have to be the strength of the mind capable of digesting such a farrago? And still, there might be extracted from it a number of data exceeding valuable to the history of comparative religion.

Ever since the fifteenth century the sciences having for their object the human intellect and its works have made no discovery to be compared to that which has revealed to us in India an intellectual world of marvellous wealth, variety and depth, in one word, another Europe. If we review our most settled ideas in comparative literature, in linguistic knowledge, in ethnography, in criticism we shall find the whole of them stamped and modified by this grand and capital discovery. As far as I am concerned, I find few elements of my thoughts whose roots are not deeply planted in that sacred ground, and I aver that no philosophical creation has furnished so many living parts to modern science as this patient restoration of a world the existence of which was not so much as suspected. Here then we meet with a series of essential results introduced into the current of the human intellect by philologists, and scholars, men by whom the partisans of the *a priori* would no doubt set little store. What then will it be when this mine, scarcely touched, shall have been worked in every direction? What will it be when every nook and corner of the human intellect shall have been explored and compared in a like way? And philology alone is capable of accomplishing this task. Anquetil-Duperron was undoubtedly a patient and zealous student. Why did all his works have to be "propped up" as it were and radically reformed? Because he was not a philologist.

One might fancy that the very fact of summoning erudition to renewed intellectual activity implies its being exhausted and that we are assimilating our century to those epochs when literature, no longer capable of producing anything original, becomes critical and retrospective. There would be no doubt of this, if our erudition were nothing more than a dead and meaningless letter, if like some narrow intellects we looked for nothing else in the knowledge and admiration of the works of the past than the pedantic right of despising the works of the present. But in addition to our creations being more spirited than those of the ancients, and apart from the fact of every modern nation possessing sufficient sap for two or three engrafted literatures, our manner of conceiving philology is very much more philosophical and fruitful than that of antiquity. Philology is not with us, as in the school of Alexandria, the mere curiosity of the erudite man; it is an organized science having a lofty and serious aim; it is the *science of the productions of the human intellect.* I am not afraid of exaggerating in saying that philology inseparably bound up with criticism is one of the most essential elements of the modern spirit, that without philology the modern world would not be what it is, and that philology constituted the vast difference between the Middle Ages and modern times. If we surpass the Middle Ages in clearness, in precision, in criticism, it is due solely to philological education.

The Middle Ages worked as much as we do, the Middle Ages produced intellects, as active, as penetrating as ours; the Middle Ages had their philosophers, their savants, their poets; but it had no philologists (58); hence that lack of criticism which reduced them to the condition of intellectual infancy. Impelled towards antiquity by that urgent need which impels all the neo-Latin nations towards their intellectual origin, it was unable to get at the truth for want of the necessary instrument (59). There were as many Latin authors and as few Greek authors in the West at the time of Vincent of Beauvais as at the time of Petrarch. And yet Vincent of Beauvais knows nothing of antiquity, he possesses only a few insignificant and detached scraps of it, lacking coherent sense, constituting no spirit. Petrarch, on the other hand, who as yet has not read Homer, but who has a manuscript of him in the original language, who worships him without understanding him (60), has instinctively guessed the spirit of antiquity; he himself has its spirit to a degree as eminent as any *savant* of the subsequent centuries; he understands with his soul that which he cannot grasp literally, his enthusiasm is aroused by an ideal, the worth of which he can only surmise. It is because the philological spirit makes its first appearance in him. That is why he should be considered the founder of the modern spirit in criticism and literature. He is the landmark between inexact fragmentary and mere material knowledge, and comparative, delicate and in one word, critical knowledge. Is the faulty understanding of the philosophy of the ancients by the Middle Ages, for instance, to be attributed to insufficient study? Who would dare to affirm such a thing of the century that produced the vast commentaries of an Albertus Magnus, of a Thomas Aquinas? Is it for lack of sufficient documents? Not at all. It had the complete materials of the Peripatetic philosophy, that is, the philosophical encyclopædia of antiquity; it had furthermore numerous documents on Platonism, and had the works of Cicero, Seneca, Macrobius, of Chalcidius, and the commentaries on Aristotle afforded it almost as much information on the philosophy of the ancients as we ourselves possess. What then did they lack, those laborious workers who devoted so many vigils to that deep study? They lacked that which the Renaissance possessed; philology. If instead of wasting their lives on barbarous translations and second-hand works, the scholastic commentators had mastered Greek and read in the original text, Aristotle, Plato, Alexander Aphrodisiensis, the fifteenth century would have been spared the spectacle of the war between two Aristotles, the one left solitary and forgotten in his original pages, the other artificially created by successive and scarcely perceptible deviations from the primitive text. The original texts of a literature are its true and complete presentment. Translations and second-hand renderings are enfeebled copies and always leave gaps which the imagination takes upon itself to fill up. In proportion to the copies belonging to a period removed from the original and getting reproduced in still more imperfect copies, the gaps increase, conjectures become multiplied and the true colour of things vanishes. The classical translations of the sixteenth century were to antiquity, as the Aristotle and the Galen of the Faculties, for which pupils and professors were referred to the traditional class books, to the real Aristotle, to the real Galen, as Greek culture to the insignificant fragments collected after other compilers by Martinus Capella or Isidore of Seville. It is neither original production nor the inquiring interest into the past, nor the perseverance to work that the Middle Ages lack. The *literati* of the Renaissance were superior neither in penetration nor in application to an Alcuin, an Allain of Lille, an Alexander of Hales, a Roger Bacon. But they were more critical, they enjoyed the advantages of the progress made in their times and of the knowledge acquired, they benefited by the favourable circumstances which successive events had brought about. It has been the fate of philology as well as of every other science to be inseparably linked to the march of things, and to be unable to accelerate by a single day by the effort of its own will the progress to be accomplished.

This ἀκρισία then may be taken as the general character of the knowledge of antiquity in the Middle Ages, or to be more exact, of the whole intellectual condition of that period. Politics as well as literature had its share of it. Those fictions of *kings*, of *patricians*, of *emperors*, of *Cæsars*, of *Augustuses*, transferred to rank barbarism; those legends of Brut, of Francus, the opinion that all authority must necessarily date back to the Roman Empire, as all high nobility to Troy, the way of looking upon Roman law as absolute law, Greek knowledge as absolute knowledge, whence did all this come, if not from the rough "guesswork knowledge" of antiquity, from the semi-fantastic light by which that ancient world was viewed, that ancient world to which they aspired to link themselves? The modern spirit, that is, rationalism, criticism, liberalism, was founded on the same day that philology was founded. *The founders of the modern spirit are the philologists.*

Philology constitutes also one of the claims to superiority of the moderns over the ancients. Antiquity can show no noble type of philological philosopher in the style of Humboldt, Lessing, Fauriel. If some Alexandrian like Porphyrus and Longinus happen to add philology to philosophy, the two realms with them scarcely touch one another; philology does not beget philosophy; philosophy is not philological. What are Dionysius of Halicarnassus, Aristarchus, Aphtonius, Macrobius compared to those subtle and excellent intellects which from a certain point of view constitute the philosophers of the nineteenth century (61)? What are questions like the following: "Why did Homer begin the catalogue of the ships with those of the Boeotians?" "How could the head of Medusa be at one and the selfsame time in the nether regions and on the shield of a God?" "How many rowers had Ulysses?" What are questions like these and other *problems* that supplied the subjects for the wrangling of the schools of Alexandria and Pergamos, when compared to the ingenious, comprehensive and delicate way of examining every aspect of things, of culling the essence of every subject, of strolling through a corner of "the universal" like a many-sided observer, which nowadays we call criticism? Such inferiority is, after all, easy to explain. The ancients lacked the means of comparison; wherever they had sufficiently authentic documents at hand, as in the Homeric question, they left little for us to do, except in the higher criticism to which the comparison of the different literatures is indispensable. That is why their grammar is above all defective, because they only know their own language, and special grammatical systems derive their life from general grammar, and general grammar presupposes a comparison of idioms. The ancients were equal to the modern philologists most enamoured of their subjects in the minuteness of their details and the patience of their comparisons. As for the criticism of texts, their position was far different to ours. They had no inventory of manuscripts of acknowledged authority and settled once for all to guide them, as we have. Hence they were compelled to think less than we do, of comparing and counting them. Aulus Gellius, for instance, in the critical discussions in which he frequently indulges argues nearly always *a priori* and appeals very seldom to the authority of ancient copies. Cicero said that Aristarchus rejected as interpolations such verses of Homer as did not please him (62). The imperfection of lexicography, the infant state of linguistic knowledge caused a great deal of uncertainty with regard to the exegesis of archaic texts. At the philological epoch the ancient tongue had already become a learned idiom, requiring special study, almost like the literal language of the far East, and it is not surprising that the moderns should often censure the interpretations of the ancient philologists, for they were scarcely more competent than we are as regards the scientific theory of their own language, and we have unquestionably hermeneutical means at our disposal which they had not (63). The ancients, in fact, knew no language but their own and only the classical and settled form of that language.

But the inferiority of antiquity was above all perceptible in erudition. The want of elementary books, of manuals containing common and ordinary notions (64), of biographical, historical and geographical dictionaries, etc., threw back everyone upon his own researches, and multiplied errors even with the most skilful writers (65). Where should we be, if in order to learn history or geography, we were reduced to the scattered facts to be picked up in books that do not treat of such science *ex professo?* The paucity of books, the absence of indexes and concordances which so greatly facilitate our researches often compelled them to quote from memory, that is, in a very inaccurate manner. And last of all, the ancients had not the experience of a tolerably great number of literary revolutions to fall back upon; they could not compare sufficiently many literatures to soar very high in æsthetic criticism. Let us remember that our superiority in that way only dates from a few years back. In this respect the ancients were exactly on the level of our seventeenth century. When one reads the *opuscula* of Dionysius of Halicarnassus on Plato, on Thucydides, on the style of Demosthenes, one might well fancy to be reading the "Memoirs" of M. and of Madame Dacier and other worthy *savants* which fill up the first volumes of the Transactions of the *Académie des Inscriptions et Belles-Lettres.* What artificiality, what puerility do not we meet with even in the "Treatise of the Sublime" itself, that is, in the best critical work of antiquity, a work that we may compare with the productions of the French school of the eighteenth century (66)? Perhaps the very ages that know best how to produce the beautiful may know least how to give its theory. Nothing could be more insipid than what Racine and Corneille have left us in the way of criticism. One is tempted to say that they did not understand the beauties of their own works.

To appreciate the value of philology we should not ask ourselves what is the worth of this or that obscure monograph, this or that note which the scholar crams in at the bottom of a page of his favourite author; we have as much right to inquire about the use in natural history of this or that monograph on a certain variety lost among fifty thousand species of insects. We must consider the revolution philology has wrought; we must examine what the human intellect was before the advent of philological culture, what it has become since it has felt the influence of this culture; what changes the critical understanding of antiquity has wrought in the manner of inquiring practised by modern students. And, it seems to me that a carefully written history of the human intellect from the fifteenth century onwards would show that the most important revolutions of thought have been brought about by the men whom we should call *littérateurs* or philologists. At any rate there can be no doubt that such men have had a more direct influence than those properly called philosophers. When posterity shall regulate the ranks in the Pantheon of humanity according to the influence brought to bear on the progress of things, the names of Petrarch, of Voltaire, of Rousseau, of Lamartine will no doubt take precedence of those of Descartes and Kant. The first reformers, Luther, Melanchthon, Eobanus Hessus, Calvin, all the abettors of the Reformation, Erasmus, the Etiennes were philologists, the Reformation was born when philology was in full swing. The eighteenth century, though superficial in erudition gets at its results much more by criticism, history and positive science than by metaphysical abstraction (67). Universal criticism is the only characteristic one can assign to the delicate, fleeting, undefinable thought of the nineteenth century. By what name shall we call so many chosen intellects which without abstractedly dogmatizing have shown thought a new way of exercising itself in the world of facts? Let us take M. Cousin himself; is he a philosopher? No; he is a critic who devotes himself to philosophy, as another devotes himself to history and another again to what we call literature. *Criticism*, then, is the form, in which in every field, the human intellect tends to exercise its faculties; and if criticism and philology are not identical, they are at least inseparable. To criticize is to assume the position of a spectator and a judge amidst the variety of things; and philology is the interpreter of things, the means of entering into communication with them and of understanding their language. The day that philology should perish, criticism would perish with it, barbarism would be born again, credulity would be once more the mistress of the world.

That immense mission which philology has undertaken in the development of the modern spirit is far from accomplished; it is perhaps only at its beginning. Has rationalism which is the general result of the whole of philological culture penetrated among the masses of mankind? Strange beliefs, which cause the critical sense to revolt, are not they gulped down like water even by distinguished intellects? Is there a widespread sense of psychological laws, or at any rate does it sufficiently influence the turn of thought or ordinary speech? The sane view of things, which is not the result of an argument, but of an entire system of critical culture, of a complete intellectual training, is it the view accepted by the greatest number? The mission of philology is to accomplish this task in concert with the physical sciences. To dissipate the mist which envelops the world of thought as well as that of nature to the ignorant, to substitute for the fantastic imaginations of the primitive dream, the clear perceptions of the scientific age, that is the common goal to which those two orders of research so powerfully converge. *Nature* is the word in which they are summed up. I repeat, all this is not the outcome of an isolated demonstration; all this is the result of a clear and unbiassed glance cast at the world, of the intellectual habits begotten of modern methods. Two roads, which make but one, lead to the direct and pragmatic knowledge of things; the physical sciences to that of the physical world, the science of mental facts to that of the intellectual world. And for this science I can find no other name than that of philology. All supernaturalism will receive its death-blow from philology. The supernatural only holds its ground in France because France is not philological.

When I question myself with regard to the most important and the most definitely acquired articles of my scientific creed, I place in the front rank my ideas on the constitution and the mode of government of the universe, on the essence of life, its development and phenomenal nature, on the substantial foundation of all things and its everlasting delimitation in transient forms, on the apparition of humanity, the primitive facts of its history, the laws of its progress, its aim and its end, on the meaning and the value of things æsthetic and moral, on the right of every living being to light and the attainment of the perfect, on the eternal beauty of human nature expanding at every point of space and on the duration of immortal poems (religion, art, temples, myths, virtues, science, philosophy, etc.), in short on the divine part which is in everything, which constitutes the right to existence and which suitably brought to light, constitutes the beautiful. Is it by reading this or that philosopher that I have formulated to myself things in this way? Is it by the *a priori* hypothesis? No; it is by the universal experimentalizing of life,—it is by pushing forward my thoughts in all directions, in scouring every tract, in analyzing and digging deeply into all things, in watching the successive development of the waves of that eternal ocean, in casting a friendly and inquisitive glance here and there. I am convinced of owing everything to experiment; but it is impossible for me to say by which road I arrived at it, out of what elements I have composed this whole (which, no doubt, may possess very little value, but which after all is my life). A balancing of all things, an inmost tissue, a vast equation of which the variable quantity constantly oscillates owing to the accession of new data, such are the images by which I endeavour to represent to myself the fact, without being satisfied. I feel that I have benefited as much in the formation of my general conception of things by the study of Hebrew or Sanskrit as by the reading of Plato; by the perusal of the poem of Job or the Gospel, of Revelations or of a Moallaca, of the

Baghavat-Gita or the Koran, as by Leibnitz, Hegel, Goethe or Lamartine. Nevertheless, it is neither Manu nor Kulluku-Bhatta, Antar nor Beidhawi; it is not the knowledge of the *sheva* and the *virama*, of the *Kal* and the *Niphal*, of the *Parasmaipadam* and the *Attmanépadam* that has given me my philosophy. But it is the general and critical view, it is the universal induction; and I feel that, if I had ten human lives to live in parallel, so as to be able to explore all the worlds, I being there in the centre, sniffing the perfume of all things, judging and comparing, combining and inducting, I should get at the system of things (68). Well, that which no individual can accomplish, humanity will accomplish; for it is immortal, and everyone works for it. Humanity will succeed in *fathoming the true physiognomy of things*, that is; the truth in all order of things. And who, after that, will dare to say that those who have contributed to that immense work, who shall have polished one of the facets of that diamond, who shall have removed a particle of the dross that dims its native brilliancy, are only pedants, idlers, ponderous intellects who waste their time, and who being unfit to carve their way in the world of the living took refuge in that of mummies and graveyards.

To philosophize is to know things; it is according to the beautiful phrase of Cuvier, *to instruct the world in theory*. I believe with Kant that every purely speculative demonstration has no more value than a mathematical demonstration and can teach us nothing with regard to existing reality. Philology (69) is the *exact science* of things intellectual. It is to the sciences of humanity as physics and chemistry to the philosophical science of matter.

This has not been sufficiently understood by an intellect otherwise eminent by its originality and its honourable independence; M. Auguste Comte. It is strange that a man, above all preoccupied with the method of the physical sciences and aspiring to transfer this method to the other branches of human knowledge, should have conceived in the narrowest fashion the science of the human intellect and of humanity and should have applied to it the coarsest method.

M. Comte has failed to understand the infinite variety of this shifting, capricious, multiple, undefinable material, which is human nature. Psychology is to him an aimless science; the distinction between psychological and physiological facts, the contemplation of the mind by itself an illusion. *Sociology* is the summary of all the sciences of humanity, and sociology to him is not the earnest and patient ascertainment of all the facts of human nature; sociology is not (I am quoting M. Comte's own words) that *incoherent compilation* of facts which we call history and over which presides the most radical irrationalism. It merely borrows examples from this indigestible compilation after which it sets to work on its own account without paying heed to *literary knowledge*, considered as very useless. Hence, M. Comte's method with regard to the sciences of humanity is the purely *a priori* one (70). M. Comte, instead of following the infinitely flexible lines of human societies, their offshoots, their apparent whimsicalities, instead of calculating the definite resultant of this immense oscillation aspires from the very first to a simplicity which the laws of humanity present even to a less degree than the laws of the physical world. M. Comte proceeds exactly like the hypothetic naturalists who forcibly reduce to the straight line the numerous ramifications of the animal world. When he has tried to prove that the human intellect proceeds from theology to metaphysics and from metaphysics to positive science his task as far as the tracing of the history of humanity goes is virtually at an end. Morality, poetry, religion, mythology, all these occupy no place whatsoever in his system, all these are pure fantasy without the least value. If human nature were such as it is conceived by M. Comte, every noble soul would hasten to commit suicide; it would not be worth while wasting one's time to turn the handle of such an insignificant piece of mechanism. True, M. Comte believes with us that one day science will endow humanity with a creed; but the science in his "mind's eye" is that of Galileo, of Newton, of Descartes, remaining as it is. On that day, the Gospel, poetry would be superfluous. M. Comte thinks that man lives exclusively upon science; nay, upon little scraps of phrases, like the theorems of geometry, barren formulas. Unfortunately for M. Comte he has a system and he does not take up a sufficiently commanding standpoint on the field of the human intellect, open to every breeze that blows. To pretend to write the history of the human mind, one must have very extensive literary attainments. The laws in this instance being of a very delicate nature and not presenting themselves broadside, the faculty most essential is that of the literary critic, of the delicate turn (it is generally the *turn* which expresses most), the subtlety of perception, in short, the very reverse of the geometrical spirit. What would M. Comte think of a physicist who should be content to observe in the aggregate the physiognomy of natural facts, of the student of chemistry who should neglect the theory of equilibrium? And does not he commit a similar error when he proclaims as useless all those patient explorations in the past, when he declares it to be a waste of time to study those centres of civilization which have no direct connection with ours, when he says that it is only necessary to study Europe in order to determine the law of the human mind and then to apply this law *a priori* to the other developments? In this M. Comte is much more influenced than he thinks by the old historical theory of the *Four Empires*, the germ of which may be found in the non-canonical book of Daniel (71), and which since the days of Bossuet has been the foundation of Catholic teaching. He imagines that humanity has really traversed the three conditions of fetichism, of polytheism, of monotheism, that the first men were cannibals, like the savages, etc. As it happens, this is not to be admitted for a moment. The fathers of the Semitic race had from their origin a secret tendency to monotheism; the Vedas, those matchless songs, really afford the ideas of the original aspirations of the Indo-Germanic race. Among these races, morality dates from the very first beginnings. In one word, M. Comte fails utterly to understand the sciences of humanity, because he is not a philologist.

M. Proudhon, though receptive to every idea, on account of the extreme pliability of his mind and capable of understanding in turns the most diverse aspects of things, does not seem to me to have conceived science in a sufficiently broad manner. No one better than he has understood that science alone is henceforth *possible*, but his science is neither poetic nor religious; it is too exclusively abstract and logical. M. Proudhon is as yet not sufficiently emancipated from the scholasticism of the seminary; he argues a great deal, he does not appear to have understood sufficiently that in the sciences of humanity, logical argumentation means nothing and that delicate mental perception means everything. Argumentation is only possible in a science like geometry, where the principles are plain and absolutely true, without any restriction. But this is not the case in the moral sciences where the principles are only of a "more or less" character, imperfect expressions, founded more or less, but never "in full," on the truth. In this instance the light thrown on to the idea is the only possible demonstration. The form, the style are three-fourths of the idea; and this is not an abuse of the idea as some puritans pretend. Those who inveigh against the style and beauty of form in philosophical and moral science strangely misconceive the true nature of the results of those sciences and the delicate nature of their principles. In geometry, in algebra, a man may fearlessly abandon himself to the play of formulas, without worrying himself, in the course of the argument, as to the realities which they represent. In the moral sciences, on the contrary, it is never allowed to trust one's self in that way to formulas, to combine them indefinitely as did the old theology, being certain that the result emanating from it, must be strictly true. It will only be logically true and may not even be as true as the principles; for the consequence may happen to bear solely on some error or misunderstanding that was in the principles, but sufficiently hidden to make the principle acceptable. Hence, it may happen that while arguing very logically, a man may arrive, in the moral sciences, at consequences absolutely false, though he started from principles sufficiently true. The books written to defend property *by argument* are as bad as those that attack it by the same method. The truth is, that argument in that order of things should not be listened to, that the results of reasoning in that instance are only legitimate on the condition of being controlled at each step by immediate experience. And whenever we are led by logic to extreme consequences, we should not be frightened at them, for the facts "delicately perceived" are in that instance, the sole *criterion* of the truth.

CHAPTER IX.

What then is the meaning of this superficial and idle contempt? Why is the philologist, manipulating as he does, things human in order to extract from them the science of humanity, less understood than the student of chemistry and the physicist, manipulating nature, in order to get at the theory of nature? No doubt the existence of the curious man of erudition who has spent his life in amusing himself learnedly and in treating serious things frivolously has been a profitless existence indeed. Men and women of the world are not altogether wrong in looking upon such a rôle as a mere clever trick of memory, suited to those who have only been endowed with second rate qualities. But theirs is a short and narrow view, in that they fail to perceive that the knowledge of many arts and sciences is the condition of the high æsthetic, moral, religious, poetical intelligence. A philosopher who thinks that he can evolve everything from his own bosom, that is, from the study of the soul and from purely abstract consideration must necessarily despise erudition and look upon it as prejudicial to the progress of reason. From this point of view, the fretful temper of Descartes, of Malebranche and of the Cartesians in general with regard to erudition is legitimate and accounted for by reason. Leibnitz was the first to realize in a magnificent harmony that elevated conception of a *critical philosopher* to which Bayle could not attain through insufficient concentration of mind. The nineteenth century is destined to be called upon to realize it and to introduce the positive method in every branch of knowledge. M. Cousin's glory will lie in his having proclaimed criticism as a new method in philosophy, a method which may lead to results as dogmatic as abstract speculation. His eclecticism has only lost its strength when outward necessities, which he could not resist, have compelled it to embrace exclusively particular doctrines which have made it as narrow as they themselves are, and to screen itself behind certain names, which should be honoured otherwise than by fanaticism. Such was not the grand eclecticism of 1828 and 1829, and of the preface to Tennemann. The new philosophical generation will understand the necessity of transporting itself to the living centre of things, of no longer making philosophy a collection of speculations without unity, of restoring to it at last its ancient and broad acceptation, its eternal mission of giving vital truths to man.

Philosophy, in fact, is not a science apart, it is one side of all sciences. In each science we should distinguish the technical and special part, which has no value except in so far as it contributes to discovery and exposition, from the general results which the science in question provides on its own account towards the solution of the problem of things. Philosophy constitutes the common head, the central part of the grand fabric of human knowledge, the focus where all the rays touch one another in an identical light. There is not a line which traced to the very end does not lead to that focus. Psychology which one has become accustomed to consider as the whole of philosophy is after all one of many sciences, nay, it may not even provide the most philosophical results. Logic understood as the analysis of reason is only a part of psychology; considered as a repertory of processes to lead the mind to the discovery of the truth, it is simply useless, seeing that it is impossible to give recipes for the discovery of truth. Refined culture and the multiple training of the intellect are from this point of view the only legitimate logical methods. Morality and the theory of divine justice are not sciences apart, they become heavy and ridiculous, when one pretends to treat them according to a definite and scientific programme, they should only be the divine resonance resulting from all things or at most the æsthetic education of the pure instincts of the soul, the analysis of which belongs to psychology. By what right then can we constitute a whole, having the right to assume the name of philosophy, seeing that this whole, in the only limits one can assign to it, has already a particular name, that is psychology (72)? Antiquity grasped this lofty and broad acceptation of philosophy in a marvellous manner. Philosophy was to it the sage, the investigator, Jupiter on Mount Ida, the spectator taking up his stand in the world. "Among those who rush to the public festivals of Greece, some are attracted by the wish to contest and to dispute the palm; others come thither to transact their commercial business; a few again come neither for glory, nor for profit, but merely to *see;* and these are the noblest, for the spectacle is provided for them, and they are there for no one's sake but their own. So on entering life, some aspire to mingle in the strife, others are ambitious to make a fortune; but there are some noble souls who despise vulgar cares and while the common herd of combatants rend one another to pieces in the arena, look upon themselves as spectators in the vast amphitheatre of the universe. They are the philosophers (73)."—Never has philosophy been more perfectly defined.

At the origin of rational research, the word "philosophy" might without causing inconvenience represent the whole of human knowledge. But when each of the series of studies became sufficiently extensive to absorb whole existences and to present a side of universal life, each branch became an independent science and left the common trunk impoverished by those successive curtailments. The ripe fruits, having thrived upon the common sap became detached from the stalk and left the tree bereft. In that way philosophy only preserved the least defined notions, those which were unable to group themselves in distinct unities and which had no other reason for being united under one name than the impossibility of arranging each of them under another name. The time has come to revert to the acceptation of antiquity, assuredly not to confine once more all the sciences with their infinite details within philosophy, but to make it the common centre of the conquests of the human intellect, the arsenal of vital stores. Who will dare to say that natural history, comparative anatomy and physiology, astronomy, history and above all the history of the human intellect do not afford to the thinker results as philosophical as the analysis of the memory, of the imagination, of the association of ideas? Who will dare to pretend that Geoffroy Saint-Hilaire, Cuvier, the Humboldts, Goethe, Herder, had not as much right to the title of philosopher as Dugald Stewart or Condillac? The philosopher means a mind sacredly inquisitive of all things; it is the *gnostic* in the primitive and elevated sense of that word; the philosopher is the thinker, no matter what the object be on which he exercises his thought.

No doubt the days are gone long ago when every thinker summed up his philosophy in a Περὶ φύσεως. When we come to reflect that the human intellect in its legitimate impatience and naïve presumption deemed itself able at the very outset to trace the system of the universe in a few pages, then the patient investigations of modern science, the innumerable ramifications of the problems, the limit of research retreating before advancing discovery, in short the infinity of things—will warrant the belief that any "summary" of the world must be infinite as the world itself. An Aristotle would nowadays be an impossibility. Not only has the alliance of psychological and moral studies with the physical and mathematical sciences become a rare phenomenon; but a subdivision, even restricted as to its object, of a branch of human knowledge is often too vast a field for the labours of a laborious life and a deeply penetrating intellect. I do not say this in a critical spirit; this onward march of science is legitimate. The strictness of scrupulous analysis should succeed to primitive syncretism, to vague and approximative study. The superficial study of the whole must make room for the deep and successive investigation of parts; but we must guard ourselves against the belief that the circle of the human intellect closes there, and that the knowledge of particulars is its final term. If the end of science were the counting of the spots on the wings of the butterfly or the enumeration of the diverse species in the flora of a country, in a language often barbarous, it would be better, I think, to come back to the Platonic definition and to declare that there is no such thing as a science of that which passes away. It is no doubt right that experimental studies should be wide enough to include the analysis of all the individualities of the universe, but on the condition of being one day gathered into a perfect synthesis, which will be much superior to the primitive syncretism, because it will be based on a distinct knowledge of parts. When dissection shall have been carried to its utmost limits (and we may believe that in some sciences that limit has been reached), then and not till then will begin the movement of comparison and reconstruction. We shall have had the humiliating and laborious task; nevertheless, when posterity shall have gone far beyond us by profiting by our work, the science of the eighteenth and nineteenth centuries will perhaps incur as harsh a reproach for having been too minute and pragmatic as we reproach the ancients for having been too summary and hypothetical. It will only prove the difficulty of knowing how to appreciate the necessity and legitimacy of the successive revolutions of the human intellect.

One consequence of this fragmentary and partial method of modern science has been to banish from philosophy the cosmology which at its origin almost wholly constituted the former. He whom we look upon usually as the founder of rational philosophy, Thales, would nowadays be no longer called a philosopher. We feel ourselves bound to divide into two or three parts scientific lives like those of Descartes and Leibnitz or even like that of Newton (albeit that with him the part of pure philosophy is already very much weaker), and still those lives have been perfectly "one"; and the word by which their unity was expressed was that of philosophy. The time has, no doubt, gone by to appeal against this necessary elimination; philosophy, after having held in its bosom all the nascent sciences has been compelled to see them separate from it when they attained a sufficient degree of development. Will the day come when they shall return to it, not with the mass of their details, but with their general results; the day when philosophy shall be less a science apart than a focus of all the sciences, a kind of luminous centre in which all branches of human knowledge will meet at their summit, diverging in proportion to their descent into details? The regular law by which their progress takes its start from syncretism and gets to synthesis by way of analysis which alone is the legitimate method, which alone has a philosophic value, may warrant such hope. The appearance of a work like the "Cosmos" of M. von Humboldt, in which a single *savant* renewing in the nineteenth century the attempt of a Timæus or a Lucretius, reviews the Cosmos in its entirety proves that it is still possible to grasp once more the cosmic unity swamped by the infinite multitude of detail. If the final aim of philosophy be the truth on the general system of things, how then could it remain indifferent to the science of the universe? Has not cosmology the same claim to holiness as the psychological sciences? Does not it start problems the solution of which is as imperatively demanded by our nature as that of questions related to ourselves and to the primary cause? Is not the world the first object that excites the curiosity of the human intellect, is it not the first to whet the craving to know which is the marked trait of our rational nature, and which makes of us beings capable of philosophizing? Take the mythologies, which give us the true measure of the spiritual needs of man; they all open with a cosmogony; the cosmological myths occupy a space in them at least as great as those relating to morals or theosophy. And even in our days though the particular sciences are far from having reached their final form, how many precious data have they not afforded to the mind that aspires to know philosophically? He who has not learnt from geology the history of our globe and of the beings who have successively populated it; from physiology the laws of life; from zoology and botany, the laws of form of all breathing things, and the general plan of animate nature (74); from astronomy, the structure of the universe; from ethnography and from history, the science of humanity in its evolution; he who has not learned this can he pretend to know the law of things, nay, to know man whom he only studies in the abstract and in individual manifestations?

I will endeavour to explain by an instance the manner in which one might use the particular sciences in the solution of a philosophical question. I select the problem which from the very first years that I began to philosophize has occupied my mind most; the problem of the *origins of mankind*.

There can be no doubt as to the existence of mankind having had a beginning. It is equally certain that the appearance of mankind on earth was accomplished in accordance with the permanent laws of nature (75), and that the first facts of his psychological and physiological life, though so strangely different from those that characterize his actual condition were the development pure and simple of the laws that are still in force to-day, operating in a medium profoundly different. Hence we are confronted by an important problem, if ever there was one, and from the solution of which would spring data of capital importance on the whole of the meaning of human life. And in my opinion this problem should be divided into six subordinate questions all of which should be solved by different sciences.

1st. *Ethnographic Question.*—If and up to what point the races actually existent are deducible from one another. Were there several centres of creation? which are they, etc.?—The investigator should therefore have at command the *ensemble* of the whole of modern ethnography, in its certain as well as hypothetical parts, also the anatomical and linguistic knowledge without which the study of ethnography is impossible.

2nd. *Chronological Question.*—At what epoch did mankind or each race make its appearance on earth?—This question should be resolved by the collating of two means; on the one hand, the geological data; on the other, the data supplied by the antique chronologies and above all by the monuments. Hence the author must be learned in geology, and well versed in the antiquities of China, Egypt, of India, of the Hebrews, etc.

3rd. *Geographical Question.*—At which point of the globe did mankind or the diverse races take their starting point?—Here the knowledge of geography in its most philosophic part would be necessary and above all the deepest scientific knowledge of antique literatures and the traditions of the various peoples. Languages in this instance supplying the principal element the author should be an able linguist, or if not, he should at any rate have at his disposal the results acquired by comparative philology.

4th. *Physiological Question.*—Possibility and mode of apparition of organic life and of human life. The laws that have produced that apparition, which is still continued in the hidden corners of nature. To deal with this side of the question, a thorough knowledge of comparative physiology is necessary. The author should be able to form an opinion on the most delicate point of that science.

5th. *Psychological Question.*—The condition of mankind and of the human intellect at the first stages of its existence. Primitive languages. Origin of thought and of language. Must have a deep insight into the secrets of spontaneous psychology, an habitual practice of the higher branches of psychology and philosophical sciences. Must be thoroughly versed in the experimental study of the child and the first exercise of its reason, in the experimental study of the savage, consequently must be extensively acquainted with the literature of the great travellers, and as much as possible have travelled himself among the primitive peoples which are fast disappearing from the face of the earth, at any rate in their original condition of spontaneous impulse; must have a knowledge of all primitive literatures, of the comparative genius of the various peoples, of comparative literature, a refined and scientific taste, tact, and spontaneous initiative; a childlike and at the same time serious nature, susceptible of great enthusiasm with regard to the spontaneous and capable of reproducing it within himself, within the very seat of deeply reflected thought.

6th. *Historical Question.*—The history of mankind before the definite apparition of reflected thought.

I am convinced that there is a science of the origins of mankind, and that it will be constructed one day not by abstract speculation but by scientific research. What human life in the actual condition of science would suffice to explore all the sides of this single problem? And still, how can it be resolved without the scientific study of the positive data?

And if it be not resolved how can we say that we know man and mankind? He who would contribute to the solution of this problem, even by a very imperfect essay, would do more for philosophy than by half a century of metaphysical meditations.

CHAPTER X.

Psychology, such as it has been understood up till now, appears to me to have been conceived in rather too narrow a spirit and not to have yielded its most important results (76). First of all, it has generally confined itself to the study of the human intellect in its complete development and such as it appears nowadays. That which is done by physiology and anatomy for organized matter, psychology has done for the phenomena of the soul, of course with the differences of method required by objects so different in their nature. But, just as by the side of the science of the organs and their operations, there is another dealing with the history of their formation and development, so, by the side of psychology which describes and classifies the functions of the soul there should be an *embryogony of the human mind* which would study the apparition and the first exercise of those faculties, the action of which, so regular now, makes us almost forget that at the start, they were only rudimentary. Such a science would no doubt be more difficult and hypothetic than that which simply confines itself to the observation of the present state of the conscience. Nevertheless, there are sure means that may lead us from the actual to the primitive state, and if direct experiments with regard to the latter are impossible, the method of induction bearing upon the present may bring us to the precedent condition of which the former is but the expansion. If, in fact, the primitive condition has disappeared for ever, the phenomena characteristic of it have still their analogies among us. Each individual travels in his turn along the line which the whole of mankind has followed, and the series of the development of human reason is exactly parallel to the progress of individual reason, with the exception of old age, which will never be known by humanity, destined as it is to be for ever blossoming afresh into eternal youth. Hence, the phenomena of infancy present to us the phenomena of primitive man (77). On the other hand the onward march of humanity is not simultaneous in all its parts; while in the one case it rises to sublime heights, in the other it still wallows in the mire which was its cradle, and such is the infinite variety that animates it, that at a given moment one might find in the different countries inhabited by man all the various ages we find disposed "in echelon" in its history. Races and climates produce simultaneously in humanity the same differences which time has shown successively in the series of their developments. The phenomena which, for instance, marked the dawn of human consciousness, may be traced back again in the everlasting infancy of those non-perfectible races, which have remained as witnesses to what happened at the outset of man's existence. Not that we should absolutely maintain that the savage is the primitive man; the infancy of the various human races must have differed very much according to the sky under which they were born. No doubt the wretched creatures who first stammered forth inarticulate sounds on the forbidding soil of Africa and Oceania were very unlike those simple and graceful beings that became the progenitors of the religious and theocratic race of the Semites and the vigorous ancestors of the philosophical and rational race of the Indo-Germanic people. But those differences no more militate against general inductions than the varieties of character among individuals impede the psychologists' progress. Hence the infant and the savage must be the great objects of study of him who would scientifically construct the primeval ages of humanity. How is it that people have failed to understand that there lies a *science* of the highest interest in the psychological observation of those races, which the civilized man superciliously neglects, and that those anecdotes reported by travellers which apparently are only fit to amuse children contain in fact the most profound secrets of human nature?

Science has a still more direct means of communication with those distant ages; the products themselves of the human mind at its different epochs; the monuments in which man has expressed himself, and which he has left behind to trace his footsteps. Unfortunately, they only date from a period too near our own, and the cradle of humanity remains still wrapt in mystery. How, in fact, could man have bequeathed the testimony of an age when he was scarcely in possession of his own powers, and when having no past, he could bestow no thought upon the future? But there is one monument in which are inscribed all the diverse phases of this marvellous Genesis, which in its thousand aspects represents each of the conditions, sketched in turns by humanity, a monument not of one epoch only, but each part of which, if we can only assign a date to it, contains the materials of all the previous centuries capable of being revealed by analysis; an admirable poem which was born and developed with man, which accompanied him at each step and received the imprint of each of his different modes of thinking, of feeling. This monument, this poem is language. The deep study of its mechanism and history will always prove the most efficacious means of mastering primeval psychology. In fact, the problem of its origin is identical with that of the origin of the human mind, and thanks to it, we stand face to face with the primeval epoch like the artist who is to restore an antique statue in accordance with the mould in which its limbs were cast. No doubt, primeval languages have as far as science is concerned, disappeared—disappeared together with the condition of humanity they represented, and no one will henceforth be tempted to tire himself in their pursuit with such ancient linguistic knowledge as he may command. But it is not a mere hypothesis that, among the idioms, the knowledge of which has become possible to us, there are some that more than others have preserved the trace of the various processes that presided at the birth and the development of language and which have undergone a less complicated wear and tear of decomposition and reconstruction ; it is a result proven by the most elementary notions of comparative philology. It is well to remind people of this ; seeing that arbitrary whim could not possibly have played the smallest part in the invention and the formation of language. There is not a single one of our most time-worn dialects which is not connected by a more or less direct genealogy with one of the first attempts which themselves were the spontaneous creation of all the human faculties, "the living product of the whole inward man" (Fr. Schlegel). But who would be capable of finding the trace of the primeval world amidst that immense network of artificial complication with which certain languages have become enwrapped beneath the numerous layers of peoples and idioms which have absolutely been piled upon one another in certain countries? Reduced to such data, the problem would be insoluble. Fortunately there are other languages that have been less worked up by successive revolutions, that are less variable in their forms, and spoken by peoples doomed to remain stationary, with whom the motion of ideas has not necessitated constant modification in the instrument of ideas ; they still remain as witnesses, not by any means, of the primeval *language*, nor even of *a* primeval language, but of the *primitive process* by means of which man succeeded in imparting to his thought an outward and social expression.

Hence, we should have to create a *primeval psy-*

chology showing the tables of facts of the human intellect at its first awakening, the influences by which it was governed at first, the laws that governed its first manifestations. Our vulgar mode of perception scarcely allows us to conceive the difference between that condition and ours, the wondrous activity secretly stored by those fresh and stirring organizations, those powerful but still obscured consciences, giving full and unfettered play to all the native energy of life's spring. Who can, in our reflective state, with our metaphysical refinements, and our senses that have become coarse, form a correct idea of the antique harmony, then existing between the thought and the sensation, between man and nature? Looking back to that horizon where heaven and earth become confounded with one another, man was god, and the god was man. *Alienated from himself*, to use Maine de Biran's expression, man, as Leibnitz says, became the concentric mirror in which was depicted that nature from which he could scarcely distinguish himself. It was not a coarse materialism, only understanding, only feeling the physical; it was not an abstract spiritualism substituting entities for life; it was a high harmony, perceiving the one in the other, expressing by one another the two worlds lying open before man. The *sensitiveness* (the sympathy with nature, *Naturgefühl* as Fr. Schlegel says) was the more delicate, seeing that the rational faculties were the less developed. The savage possesses an amount of perspicuity, of curiosity that astonishes us; his senses detect a thousand imperceptible shades that escape the senses or rather the attention of the civilized man. Unfamiliar as we are with nature we only see uniformity there where nomadic or agricultural peoples have perceived numerous instances of individual originality. We must assume primeval man to have possessed an infinitely delicate tact which enabled him to grasp, with a *finesse* of which we can no longer form an idea, the qualities "to be felt" which were to be the basis of the nomenclature of things. The faculty of interpretation which is simply an extremely great sagacity in perceiving a connection between things was more developed in them. They saw ever so many things at once. Nature spoke to them more intelligibly than she does to us, or rather they found in themselves a secret echo which replied to all those voices from without and reproduced them in articulations, in words. Hence those abrupt passages the trace of which is utterly lost to our slow and laboured systems. Who could once more seize upon those fleeting impressions? Who could once more find the truant paths along which the imagination of primeval man travelled, the association of ideas that guided him in this work of spontaneous production, in which now man, then nature herself "spliced" the broken strand of analogies, and wove their reciprocal action in an indissoluble unity? What shall we say of that marvellous intellectual synthesis, necessary to the creation of a metaphysical system like the Sanskrit language, a sweet and sensuous poem like the Hebrew? What shall we say of that infinite liberty to create, of that boundless fancy, of that wealth, of that exuberance, of that complexity which is beyond our grasp? We should not be capable of speaking Sanskrit, our most eminent musicians would fail to execute the octuple and nonuple quavers of the Song of the Illinois. Ye sacred ages, primeval ages of humanity, who can understand you?

In view of those strange productions of earlier ages, of those facts that seem outside the normal order of the universe, we are inclined to suppose specific laws that have now been abrogated. But there is no temporary government in nature; the same laws that govern the world to-day are those that have presided at its constitution. The formation of the different planetary systems and their preservation, the apparition of organized beings and of life, that of man and of conscience, the first feats of humanity were only the development of an aggregate of physiological and psychological laws settled once for all, without the superior agent, who moulds his action according to these laws, having interposed a specially intentional will in the mechanism of things. No doubt everything springs from the primary cause, but the primary cause does not act upon partial motives, by special manifestations of will, as Malebranche would say. What it has done is and remains the best, the means once established are and remain the most efficacious. But how, it will be said, can we explain facts so diverse by the same system? Why do those strange facts that marked the origin of man no longer repeat themselves, if it be true that the laws which produced them still exist? It is because the circumstances are no longer the same; the incidental causes that determined those laws at their grand phenomenal moments do no longer exist. As a general rule we only formulate the laws of nature with a view to the actual condition, and the actual condition is only a particular case. It is like a partial equation drawn from a more general equation by a special hypothesis. The general equation virtually contains all the others, and its truth lies in the special truth of all the others.

It is the same with all the laws of nature. Applied amidst different surroundings, they produce altogether different effects, if the same circumstances present themselves, the same effects will reappear. Hence there are no two series of laws co-operating with one another in order to fill up their voids and to supply their individual insufficiency; there is no interim in nature; creation and preservation are wrought by the same means, operating under different conditions. Geology, after having appealed for a long while to causes different from those operating to-day, to explain cataclysms and the successive phases of the globe, is coming back from every direction to proclaim that the actual laws were sufficient to produce these revolutions. Those conditions of life which appear to us fantastical because they were different from ours, what strange combinations they must have wrought! And when man appeared on this earth which was still in process of creation, without being suckled by a woman, nor tended by a mother, without the lessons of a father, without ancestors or fatherland, can we form an idea of the astonishing facts that must have happened at the first awakening of his intellect, at the sight of that fruitful nature, whence he was beginning to be divided. Those first apparitions of human activity must have been marked by an energy, a spontaneousness of which nothing nowadays can convey an idea to us. Necessity, in fact, is the real incidental cause of the exercise of all power. Man and nature went on creating as long as there was a void in the scheme of things, they forgot to create as soon as the need of it no longer compelled them. It is not that from that moment they were one power short; but those productive faculties, which at the origin were exercised on an immense scale, henceforth deprived of nutrition, were reduced to an obscure rôle, and as it were, relegated to a back corner in nature. Thus for instance, spontaneous organization, which in the beginning brought forth everything that lives, is still preserved on an imperceptible scale on the lowest rungs of the animal ladder; thus the spontaneous faculties of the human mind still live in the facts of the instinct, but they are lessened and almost stifled by reflected thought; thus the creative spirit of the language is still met with in that which presides at its revolutions; for the force that sustains life is in reality that which brings forth, and to develop is in one sense to create. If man were to lose language, he would once more create it. But he finds it ready made, hence his productive force, in default of an object, withers away like every faculty not exercised. The infant still possesses it before it is able to speak, but loses it as soon as science from without renders the creation from within useless.

How, indeed, is it possible to reduce the science of

man to a system by studying man in his age of reflection only, as Scotch psychology has done, when his originality is as it were effaced by artificial culture, when artificial motors have replaced the powerful instincts under the influence of which he formerly worked at his own development with so much energy?

The second void with which I meet in psychology and which in the same way can be filled up only by the philological study of the works of the human mind is in its application to the mere individual, without ever rising to the consideration of humanity at large. If the immense historical development of the latter end of the eighteenth and the nineteenth centuries has been productive of any result at all, it is that which proves that there is a life of humanity just as there is a life of the individual; that history is not a purposeless series of isolated facts, but a spontaneous tendency towards an ideal goal, that the perfect is the centre of gravity of humanity as of everything that lives (78). Hegel's claim to immortality consists in having been the first to express with perfect clearness this vital force which neither Vico nor Montesquieu had noticed, of which even Herder had but the vaguest notion. Through this he has insured for himself the title of the definite founder of the philosophy of history. Henceforth, history will be no longer what it was to Bossuet, the unfolding of a particular plan conceived and realized by a power superior to man, which leads man, who can only bestir himself according to its designs; it will no longer be what it was to Montesquieu, an interlinked chain of facts and causes; what it was to Vico, a lifeless and almost reasonless movement. It will be the history of a being, developing himself by his inward power, creating himself and attaining by diverse degrees to the full possession of himself. There is, no doubt, a movement, as Vico meant, there are, no doubt causes, in Montesquieu's sense; there is, no doubt, a previously imposed plan, agreeing with Bossuet's theory. But what they failed to perceive was the active and living force impelling that movement, animating those causes, and which without any co-operation from without, and solely by its tendency towards the perfect, accomplishes the providential plan. Perfect autonomy, inward creation, in short, life; such is the law of humanity.

Assuredly Bossuet's plan is simple enough; simple like a pyramid; *commandment* on one side; *obedience* on the other; God and man, the King and the subject, the Church and the believer. It is simple but harsh, and after all it is doomed. We shall, henceforth, have great difficulty to imagine in what manner those who do not believe in progress conceive the world. If there be a notion we have outgrown, it is that of nations succeeding one another, traversing the same periods in order to die in their turn, then to revive under other names, and thus without cessation recommencing the same dream. In that case what a nightmare humanity would be; what a tissue of absurdities the various revolutions! What a colourless, vapid thing life! Amidst such a poor system, would it be really worth while to aspire ardently to the beautiful and the true, to sacrifice one's happiness and peace of mind to them? I can conceive such a paltry conception of actual existence in the severely orthodox, who transports the whole of his present existence to the one beyond; I fail to conceive it in the philosopher. The idea of humanity is the grand line of demarcation between the ancient and the new philosophies. Consider well why the ancient systems no longer satisfy you; you will find that it is because that idea is utterly absent from it. In that, I repeat, lies the whole of the new philosophy (79).

The moment we admit that humanity is conceived as a conscience that shapes and developes itself, we admit the necessity of a *psychology of humanity*, just as there is a psychology of the individual. The irregular and fortuitous appearance of its progress must not hide from us the laws that govern it. Botany shows us that all the trees would be as regular as the coniferæ with regard to their form, the arrangement of their leaves and branches, were it not for abortion and suppression which, destroying the symmetry, impart to them such fantastic shapes. A river would flow straight to the sea but for the hills which compel it to turn so frequently aside. In the same way humanity, apparently given up to chance, yields to laws which other laws may cause to deviate, but which are nevertheless the reason of its movement. History is the necessary form of the science of everything contained in the "*will-be.*" The science of languages means the history of languages; the science of literatures and religions means the history of literatures and religions. The science of the human intellect means the history of the human intellect. To attempt to seize a given moment only of those successive existences in order to dissect and to fixedly examine it is simply falsifying their nature. For they are not complete at a given moment, they are merely tending towards completion. Such is the human intellect. By what right do you select the man of the nineteenth century to illustrate the theory of it? I am aware that there are common elements which the examination of all countries and all peoples will yield to analysis. But these, on account of their very stability are not the most essential to science. The variable and characteristic element is much more important. The only reason why physiology often appears so much tautology and emptiness is because it confines itself too exclusively to those generalizations of small value which make it look like the lesson of philosophy of the "Bourgeois Gentilhomme." Philology commits the same mistake when instead of taking languages in their individual varieties, it confines itself to the general analysis of the forms common to all, to what we call *general grammar*.

How unsuitable, in fact, is our dry and abstract

mode of treating psychology for the purpose of putting into relief the differential shades of the sentiments of humanity. It would lead us to conclude that every race, every century understood God, the soul, morality in an identical manner (80). We never seem to suspect that each nation with its temples, its divinities, its poesy, its heroic traditions, its fantastic beliefs, its laws and its institutions represents a unity, a way of its own of taking life, a separate tone in humanity, a distinctive faculty of the great soul. The true psychology of humanity would consist in analyzing the one after the other those various lives in their complexity, and as every nation has generally tied its suprasensitive life into a spiritual sheaf, which is its literature, true psychology would consist above all in the history of literatures. The second volume of M. von Humboldt's *Cosmos* (the history of a sentiment traced in all its varieties and shades among every race and through the lapse of all the centuries) may be taken as an example of this historical psychology. Ordinary psychology is too much like that literature which by dint of representing humanity in its general characteristics and rejecting local and individual colour will perish through lack of vitality of its own and originality.

I am under the impression that the comparative study of the different literatures has afforded me a much wider idea of human nature than that generally conceived. No doubt there is a good deal that is universal, there are a great many common elements in human nature. No doubt we may see that there is but one psychology the same that there is but one literature, seeing that all literatures live on the same common fund of sentiments and ideas. But this universality is not where we believe it to be, and to apply a rigid and unbending theory to mankind of different epochs is simply to falsify the complexion of facts. That which is universal, is the great divisions and the great needs of nature; they are, if

I may so express it, the natural pigeon-holes, filled up successively by those diverse and variable forms; religion, poesy, morality, etc. Looking at the past of humanity only, religion for instance would seem to be essential to human nature; and still religion in the ancient forms is fated to disappear. That which will remain is the place it occupied, the want to which it corresponded, and which will be satisfied one day by something analogous to it. Has *morality* itself been a form of all ages—provided we attach to the word the complete and quasi-evangelical sense we give to it? A not over delicate analysis, taking no account of the different physiognomy of facts, might affirm such a thing. True psychology which takes care not to designate by the same name facts of a different complexion, though they may be analogous cannot make up its mind to this. Is the word morality applicable to the form which the idea assumed in the ancient Arab, Hebrew and Chinese civilizations, which it still assumes among savage peoples, etc.? I am not making one of those commonplace objections here, which have been so often repeated since the days of Montaigne and Bayle, and which attempted to prove by means of a few divergencies or a few ambiguous terms that in certain peoples the moral sense was entirely absent. I admit that the moral sense or its equivalents appertain to the essence of humanity, but I maintain that to apply the same denomination to facts so diverse is to speak incorrectly. There exists in humanity a faculty or a need, in one word a capacity which in our days is supplied by a code of morals, and which has always been supplied, which will always be supplied by something analogous. In the same way I conceive that in the future the word morality will not be the proper word and that it will be replaced by another. As far as my personal use goes, I prefer to substitute the word æstheticism for it. Face to face with a given action, I ask myself whether it is ugly or beautiful rather than whether it be good or bad, and

I believe that mine is a good *criterion;* because with the simple morality that constitutes the honest man, it is still possible to lead a sufficiently paltry existence. Be this as it may, the immutable should only be looked for in the divisions of human nature themselves, in its compartments, if I may so express it, and not in the forms that are adjusted to it, and which may be replaced by substitutes. This is something analogous to the fact of chemical substitution in which analogous bodies may in their turn fill up the same frames.

China offers me the best example to elucidate what I have just said. It would be altogether incorrect to say that the Chinese are a nation without morality, without religion, without a mythology, without God; in that case they would be a monster among mankind, and yet it is very certain that the Chinese have neither morality, religion, a mythology, nor a God in the sense we understand them. Theology and the supernatural occupy no place in the minds of these people; and Confucius only acted in accordance with the spirit of his nation when he dissuaded his disciples from the study of things divine (81). So vague are the ideas of the Chinese with regard to the Godhead, that since the days of Francis Xavier the missionaries have had the greatest difficulty to find a Chinese term, signifying God. The Catholics, after a good deal of groping about, succeeded at last in agreeing upon a word; but when, about thirty years ago, the Protestants began to translate the Bible into Chinese, there was a repetition of the difficulties. The variety of terms employed by the different Protestant missionaries to designate God became such, that they had to have recourse to a council, which it appears to me, decided nothing at all, seeing that Mr. Medhurst who has recently published a dissertation on the subject which was printed at Shanghai still discusses the sense in which the classic authors employ each of the terms that have been proposed as equivalents for the word God. We might point out similar analogies with regard to morality and religion, and prove that morality with the Chinese is only the observance of an established ceremonial and religion the respect due to ancestors. M. Saint-Marc-Girardin when he compared Voltaire's "Orphelin de la Chine" with the original has pertinently pointed how the pathetic and passionate elements disappear from the Chinese system to become systematic duty, how the family affections disappear by the family becoming an institution (82). A careful study of the different zones of the affections of the human species would reveal everywhere, not an identity of elements, but analogous composition, the same plan, the same disposition of parts in diverse proportions. A given element, prominent in a given race, is only rudimentary in another. "Mythologism" so dominant in India scarcely shows in China, but is nevertheless perceptible on an infinitely reduced scale. Philosophy, the dominant element in the Indo-Germanic races appears to be altogether foreign to the Semites, and still, on looking very closely, one perceives in the latter, not the thing itself, but its rudimentary germ.

At the outset of our scientific career we are apt to imagine the laws of the psychological and physiological world as absolutely non-deviating formulas, but before long the scientific spirit modifies this conception. We meet with individualism everywhere; the genus and the species almost melt into one in the analysis of the naturalist; each fact shows itself as *sui generis;* the most simple phenomenon appears incapable of being reduced; the order of real things is only a vast oscillation of tendencies producing by their infinitely varied combinations constantly varying apparitions. Reason is the only one law that governs the world; it is as impossible to reduce to formulas the law of things as to reduce to a settled number of *schemes* the turns of speech of the orator, as to enumerate the precepts on which the moral man bases his conduct towards the good. "Endeavour to be beautiful (83), and then do at every moment that with which your heart will inspire you;" that is morality in a nutshell. All the other rules in their absolute form are faulty and mendacious. The general rules are only trumpery makeshifts to hide the absence of the grand moral sense, which in itself is sufficient to show man at all times what is most beautiful. It is an attempt to substitute previously prepared instructions for inmost spontaneousness. The variety of cases constantly baffles all previsions. There is nothing, absolutely nothing that can replace the soul; no amount of teaching can make up with man for the inspiration of his nature.

Psychology, as understood up till now, is to true historical psychology what the comparative philology of Bopp and W. von Humboldt is to that "skimpy" part of dialectics, formerly entitled *comparative grammar*. In the latter one treated language like a petrified thing, settled once for all, stereotyped in its forms, as something finished, and which was supposed to have been always, and to remain always as it was. In the former, on the other hand, they take the living organism, the specific variety, the movement, the process of evolution, in short, the history. Its history is the true form of the science of languages (84). It is no doubt useful to take an idiom at a given moment of its existence, if it be an idiom one is learning to speak. But to stop there is as little profitable to science as if we limited the study of organized bodies to the examination of what they are at a definite moment without inquiring into the laws of their development. No doubt if languages were like inanimate bodies condemned to immobility, grammar should be purely theoretical. But they are alive, like man and mankind who speak them, they are constantly being decomposed and reconstructed; it is a real inward growth, a constant circulation from within to without, and from without to within, a continual "becoming." As such they are like everything that lives subject to the laws of changing and successive existence, to their progress and phases, in consequence of that secret impulsion, which allows neither man nor the productions of his mind to remain stationary.

In the same way psychology has insisted too much on considering man from the point of view of his "being" and has not inquired sufficiently into his evolution. Everything that lives has a history; the psychological man as well as the human body, aggregate humanity as well as the individual lives and renews its life. They constitute a moving picture in which the masses of colour, blending with one another by imperceptible gradations should by a constant play tone down and absorb one another, expand and at the same time limit one another. It is a reciprocal action and reaction, a commerce of common parts, a growth on a common trunk. In this eternal evolution one would in vain look for the stable element to which to apply the anatomical process. The word *soul*, so admirably fit to designate the supra-sensitive life of man will always be fallacious and untrue if applied in the sense of a permanent basis which would be the ever identical subject of phenomena. It is this false connection of a fixed *substratum* which has given to psychology its hard and fast forms. The soul is taken for a fixed, permanent being, and analyzed like a natural body, while after all it is only the ever variable resultant of the multiple and complex facts of existence. The soul means individual evolution, just as God means universal evolution. It is certain that if there were an invariable "being" which we might call the soul, just as there are creations we call Iceland spar, quartz, mica, there would be a science called *psychology*, which would be analogous to mineralogy. So true is this that in taking up this standpoint we should cease studying the science of *the soul*, for there are various kinds, and take to studying the science of *souls*. This is how Aristotle understood it, who was far less guilty than people generally think him, for to him the soul is only the persistent phenomenon of life. This above all, is the manner in which it was understood by ancient philosophy which made itself grotesque to the extent of founding a science called *pneumatology* or the science of spiritual beings ("God, man, the angel, and maybe animals," they said) just as, in natural history, we might found a science treating of the horse, the unicorn, the whale and the butterfly. The Scotch psychology avoided those scholastic absurdities, but still, it clung too much to the point of view of the "being" and not enough to the point of view of the evolution, it still conceived philosophy as the study of man in an abstract and absolute manner, and not as the study of the eternal "becoming." The science of man will only then be placed in its true light when students are persuaded that *conscience* evolves itself, vague, feeble, non-centralized at first, in the individual as well as in aggregate humanity, that it only attains its plenitude after having gone through diverse phases. It will then be seen that the science of the individual soul is the history of the individual soul, and the science of the human intellect, the history of the human intellect.

The great progress of modern thought has been the substitution of the category of *evolution* for the category of the "*being*"; of the conception of the relative for the conception of the absolute, of movement for immobility. Formerly everything was considered as "being" (an accomplished fact); people spoke of law, of religion, of politics, of poetry in an absolute fashion (85). At present everything is considered as in the process of formation (86). Not that formerly evolution and development were not, as they are to-day, general laws; but people had no perception of them. The earth revolved before Copernicus, albeit that it was thought to be stationary. Substantial hypotheses always precede phenomenal hypotheses. The Egyptian statue, motionless with its hands "stuck" to its knees is the natural antecedent of the Greek that lives and moves.

But how is it possible to establish the history of the human intellect without the most extensive learning and without the study of the monuments bequeathed to us by every epoch? From that point of view nothing is useless, the most insignificant works are often the most important in so far as they energetically depict one aspect of things. The Talmud is a very curious monument of moral depression and extravagance; but I maintain that no one who has not studied that unique work can form an idea of how far the human intellect may go in its aberration from the paths of common sense. The works of the Latin poets of the decline are insipid enough in all conscience, nevertheless, unless one reads them, it is impossible to conceive the characteristics of a decadence, to get an idea of the exact colour of an epoch in which the intellectual sap is exhausted. Of all literatures, the Syriac is, I imagine, the most colourless. The writings of that nation are pervaded by a suave mediocrity for which I can find no name. Herein lies its very interest, no study affords a better idea of the *mediocre condition* of the human intellect. And, natural, unsophisticated mediocrity being a facet of human life just like any other, it has a claim to our attention. No doubt, such studies possess very little value from the æsthetic point of view, they are very precious from the scientific one. There is certainly very little to be learned from, and to admire in, the Latin poems of the Middle Ages and the scientific literature in general of those days; still, can we pretend to know the human mind if ignorant of the dreams that haunted their sleep of ten centuries' duration.

Among the special works in connection with the Semitic languages I know of none more urgent in the state of actual science than a complete and definitely authentic publication of the books of the small gnostic sect which subsists still at Bassora under the name of Mendaites or Christians of St. John. Those books do not contain a single line of sense; they are simply so much raving composed in a barbarous and indecipherable style. It is that which constitutes their very value. For it is easier to study diverse natures in their crises than in their normal condition. The regularity of life only shows one surface and conceals in its depth the inmost mainsprings; in a state of ebullition, on the other hand, everything rises in its turn to the surface. Sleep, madness, delirium, somnambulism, hallucination afford the study of individual psychology a much more profitable field of observation than the regular condition. For the phenomena which in the latter state are effaced, as it were, by their insignificance show themselves in extraordinary crises in a more conspicuous manner by reason of their exaggeration. The physicist does not study galvanism in the feeble quantity presented by nature, but multiplies it by experiment in order to study it with more facility, being perfectly sure, after all, that the laws thus observed in their exaggerated condition are identical with those of the natural condition. In the same way the psychology of humanity should take its lessons above all from the study of the aberrations of mankind, of its dreams, of its hallucinations, of all those strange absurdities that may be met with at every page of the history of the human intellect.

The philosophical spirit can extract philosophy from no matter what. If I were condemned to make a special study of heraldry, it seems to me that I should cheerfully make up my mind to it and gather a honey that would have its sweetness like the veriest bee among a well stocked flowerbed. If I were incarcerated at Vincennes with the *Anecdota* of Pez or Martène and the Collection of d'Achery, I should consider myself the happiest of mortals. I commenced, and I hope to have the courage to finish a work on the history of Hellenism among the Eastern peoples (Assyrians, Arabs, Persians, Armenians, Georgians, etc.). I can pledge my word that there is no more wearisome business, no more monotonous spectacle, no more colourless and less original page in literary history. Nevertheless, I hope to extract from this insignificant study some curious traits for the history of the human intellect; we shall meet in it with two profoundly different spirits, in presence of, and incapable of penetrating, one another, a superficial education without lasting effects which, by its contrast, will make us understand the immense fact of the Hellenic education of the Western peoples; singular misunderstandings, strange instances of utter nonsense will disclose voids, the knowledge of which will be useful in drawing up in a more exact way the map of the Semitic spirit and the Indo-Germanic spirit.

The history of the *Origins of Christianity* written by a critic who would go to the direct sources would undoubtedly be a work of importance. Well, that marvellous history, which if carried out in a scientific and final manner would revolutionize thought, with what should it be constructed? With utterly insignificant books, such as the book of Enoch, the Testament of the Twelve Patriarchs, the Testament of Solomon, and in general the Apocryphal books of Jewish and Christian origin, the Chaldaic paraphrases, the Mishna, the Deutero-Canonical books, etc. On that day, Fabricius and Thilo who have prepared a creditable edition of these texts, Bruce who brought back from Abyssinia, the book of Enoch, Laurence, Murray and A. G. Hoffmann who have elaborated the text, will have done more for the work than Voltaire with the whole of the eighteenth century by his side.

Thus, from the vast point of view of the science of the human intellect works deemed insignificant at the first glance may prove to be the most important. A given literature in Asia which has absolutely no intrinsic value may afford data for the history of the human intellect more curious than no matter which modern literature. The scientific study of the savage races would be attended with still more decisive results, if it were undertaken by truly philosophical minds. Just as the worst popular jargon is more apt to initiate us in the science of languages than an artificial language polished by the hand of man like French, so may a man be thoroughly versed in various literatures such as the French, the German, the English, the Italian, without having as much as perceived the great problem. Orientalists often make themselves ridiculous by attributing an absolute value to the literatures they cultivate. It would be too painful to have devoted the whole of one's life to deciphering a difficult text and then to have to admit that the text was not admirable. On the other hand, superficial minds smile and joke when they see serious people amuse themselves in translating and commenting books, possessing no form or style, which in our opinions would be only absurd and ridiculous. They are both in the wrong. We ought not to say; "This is absurd; this is magnificent." We should say; "This belongs to the human intellect, consequently it has its value." It is very patent at once that from the point of view of positive science there is nothing to be gained by the study of the East. A few hours given to the perusal of a modern work on medicine, mathematics or astronomy will be more useful as regards the knowledge of those sciences than long years of learned research devoted to the physicians, mathematicians and astronomers of the East (87). Even history is scarcely a sufficient motive to invest those studies with any value. For, first of all, the ancient history of the East is absolutely fabulous; secondly, the moment it becomes more or less trustworthy, the political history of the East becomes almost insignificant. The platitude of the Arabian and Persian historians who have transmitted to us the history of Islamism is absolutely without a parallel. And, in fairness be it said, the history itself is much more to blame than the historians. What could the latter make out of a world of ice, as it were, consisting of whims and freaks of absurd and bloodthirsty despots, revolts of governors, changes of dynasties, successions of viziers, a world from which humanity is apparently completely absent, in which the voice of nature seems dumb, in which there is not a single true or original movement of the people? Certainly, those who are under the impression that a man studies Turkish literature with the same object that he would study German literature; to find something to admire, are right when they smile at those who devote their vigils to it, in looking at them as benighted intellects, incapable of doing aught else. As a rule, the modern literatures of the East are weak, and would not in themselves, deserve the attention of the serious student (88). But they become exceedingly valuable when bearing in mind that they afford important elements for the study of ancient literature and the comparative study of idioms. Nothing is useless when we know how to reduce it to its aim, but we must not lose sight of the fact that mediocrity has only its value in the whole of which it is a part.

But has the study of the ancient literatures of the East a value of its own, and irrespective of the history of the human intellect? I confess that there is real and incontestable beauty in those ancient productions of the East. Job and Isaiah, the Ramayana and the Mahabarata, the pre-Islamite poems are beautiful for the same reason that Homer is beautiful. But, if we analyze the feeling produced in us by those ancient works, what, as far as we are concerned, is their claim to the award of beauty? We admire a poem of M. de Lamartine, a tragedy of Schiller, a canto of Goethe, because we meet with our ideal in it. Do we equally meet with our ideal in the poetical dissertations of Job, in the sweet psalms of the Jews, in the picture of Arabian life of Antara, in the hymns of the Vedas, in the admirable episodes of Nal and Damayanti, of Yadnadatta, of Savitri, of the descent of the Ganga?

Is it our ideal we meet with in the symbolical figure of Um or of Brahma, in an Egyptian pyramid, or in the Caverns of Ellora? Certainly not. We can only admire on the condition of transporting ourselves to the times to which these monuments belong, of placing ourselves in the centre of human intellectuality, of regarding all this as the eternal growth of hidden forces. That is why limited and more or less rigid intellects who judge these ancient productions by clinging obstinately to the modern point of view cannot make up their minds to admire them, or else admire exactly that which is not worthy of admiration, or what is altogether absent from it (89). Just submit the myths of the Maruthas or the visions of Ezekiel to a man who is not versed in strange literatures, he will simply vote them hideous and repulsive. From his point of view, Voltaire was right in ridiculing Ezekiel (90), just as Perrault and some critics of the school of Alexandria were right in declaring Homer *ridiculous*, and Madame Dacier and Boileau are wrong when they undertake to defend Homer while still adhering to the same strange manner of viewing antiquity. To understand the true sense of those exotic beauties, we must have become identified with the intellect of aggregate humanity; we must feel, live, with it in order to grasp its originality, its life, its harmony even in its most eccentric creations wherever we meet with them. Champollion wound up by perceiving beauty in the Egyptian heads; the Jews consider the Talmud replete with a morality as lofty as that of the Gospel; the lovers of the Middle Ages stand lost in admiration before grotesque statues which the profane do not deem worthy of a look. Think you that this is the mere illusion of the erudite man or the ardent amateur? No; it simply arises from the fact that in every fold and corner of the handiwork of man there is hidden a ray of divine light, the careful observer knows where to find it. The altar on which the patriarchs sacrificed to Jehovah was, materially speaking, only a heap of stones; considered in its humanitarian significance, as a symbol of the simplicity of those ancient modes of worship and of the natural and amorphous God of primeval humanity, that heap of stones was worth a temple of anthropomorphic Greece, and was certainly a thousandfold more beautiful than our temples of gold and marble raised and admired by people who do not believe in God. A little cowdung and a handful of *Kousa* herbs are enough for the Brahmin's sacrifice and for his reaching God in his own way. The rough-hewn cippus by which the Hellenes represented the Graces was more eloquent to them than beautiful allegorical statues. The value of things lies in what humanity can see in them, in the feelings it has attached to them, in the symbols it drew from them. So true is this that the imitations of primeval works, however perfect they are supposed to be, are not beautiful, while the works themselves are sublime. An exact reproduction of the pyramid of Ghizeh on the common of Saint-Denis would be mere childishness. In the latter days of Hebraic literature, the learned men composed psalms imitated from the ancient Canticles, so perfect as to puzzle every one. Well; we can only say that the old psalms are beautiful, while the modern ones are merely ingenious; and still the greatest adept can scarcely distinguish them from one another.

The beauty of a work should never be considered from an abstract point and independently of the surroundings that gave it birth. If the Ossianic poems of Macpherson were authentic they should be ranked with those of Homer. The moment it is proved that they are the work of a poet of the eighteenth century, they have only a mediocre value. For it is the true breath of humanity and not the literary merit that constitutes the beautiful. Let us suppose that a man of parts (it is almost the case with Apollonius of Rhodes) succeeded in catching a very fair imitation of the Homeric style in such a way as to produce a poem in exactly the same style, a poem that should be to Homer's what "Les Paroles d'un Croyant" are to the Bible, that poem in many people's opinion, ought to be superior to Homer's; for the author would be able to avoid what we consider blemishes, or at any rate, the wants of transition, the contradictions. I should like to know how the absolute critics would manage to prove that such a poem is not in fact superior to the "Iliad," or rather to prove effectually that the "Iliad" is worth a world's ransom while the work of the modern man is doomed to mildew and oblivion on the shelves of a library after having for an instant diverted the quidnuncs. In what then does the beauty of Homer consist, seeing that a poem absolutely like his, written in the nineteenth century would not be beautiful? It is because the Homeric poem of the nineteenth century would not be true. It is not Homer who is beautiful, it is the Homeric life, the phase of mankind's existence as described in Homer. It is not the Bible that is beautiful, it is the Biblical manners and customs, the form of existence depicted in the Bible. It is not this or that Indian poem that is beautiful, it is the Indian life. What do we admire in "Telemaque"? Is it the perfect imitation of the antique form? Is it this or that description, this or that comparison borrowed from Homer or Virgil? No, they simply elicit the cool remark, as if we were stating a mere fact, "This man has caught the antique style in a remarkably delicate way." That which arouses our admiration and our sympathy in that beautiful book is exactly the modern spirit breathing through it, the genius of Christianity which prompted Fénélon in his description of the Elysian Fields; it is that policy so moral and so rational guessed at by a miracle as it were amidst the Saturnalia of an absolute monarchy.

The true literature of an age is that which expresses and depicts that age (91). Some sacred orators of the Restoration have bequeathed us funeral orations imitated from those of Bossuet and almost entirely composed of the phrases of that great man. Well; these phrases which are beautiful in the work of the seventeenth century, because of their sincerity at that time, are insignificant later on because they are false and because they do not express the sentiments of the nineteenth century. Independently of any system, except that which dogmatically preaches utter annihilation, the tomb has its poesy, and this poesy is never more affecting perhaps than when an involuntary doubt mingles with the certainty latent in every heart, as if to moderate the too great prosaism of dogmatic affirmation. The chiaro-oscuro affords a softer and sadder tint, a less distinctly drawn horizon, more vague than and more analogous to the tomb. The few pages of M. Cousin on Santa-Rosa are more valuable, as far as our feelings go than a funeral oration imitated from those of Bossuet. A beautiful *copy* of a picture by Raphael is beautiful because it pretends nothing more than to represent Raphael. But a nineteenth century imitation of Bossuet is not beautiful, because it is a false application of forms that were true once upon a time; it does not express the humanity of its epoch.

It has been pointed out in a delicate manner how much the works of art with which our museums are crowded lose in their æsthetic value. There can be no doubt as to that, seeing that their position and their significance at the epoch when they were true contributed three-fourths of their beauty. A work has no value save in its framework and the framework of every work is its epoch. Did not the sculpture of the Parthenon possess a greater value in the place to which it belongs than stuck in little bits on the walls of a museum? I deeply admire the old religious monuments of the Middle-Ages, but face to face with our modern Gothic churches, built by an architect in a frock coat, piecing together the designs borrowed from the ancient fanes my feeling is only one of pain. Absolute admiration is always artificial; I yield to no man in my admiration of the "Pensées" of Pascal, the Sermons of Bossuet; but I admire them as works of the seventeenth century. If those works were to appear in our days, they would scarcely deserve attention. True admiration is historial. Local colour possesses an incontestable charm when real; it is insipid when copied. I like the Alhambra and Broceliande in their reality; I cannot help laughing at the romanticist who imagines that by combining these words, he can construct a beautiful work. Therein lies the error of Chateaubriand and the cause of the incredible mediocrity of his school. He is no longer himself when he leaves the domain of critical appreciation and tries to produce after the model of the works, the beauties of which he so judiciously points out.

Among the works of Voltaire those in which he has copied the forms of the past are unmistakably forgotten. Who, outside the college, reads the "Henriade" or his tragedies. But those in which he has shown the elegant proofs of his subtle tact, of his immorality, of his witty scepticism will live, for they are true. I prefer "La Fête de Bellebat" or "La Pucelle" to "La Mort de César" or the poem on Fontenoy. It may be as infamous as you please; but it belongs essentially to the century, it is the man himself. Horace is more lyrical in *Nunc est bibendum* than in *Qualem ministrum fulminis alitem.*

Hence, true admiration of primitive works is only possible from the sole point of view of human intellect and by diving into its history not out of mere curiosity, but from a deep-rooted feeling and intimate sympathy. Every dogmatical point of view is absolute, all appreciation based on modern lines is out of place. The literature of the seventeenth century is no doubt admirable, but on the condition of it being transported to its own medium, the seventeenth century. It is only the pedant of the college who is capable of seeing in it the eternal type of beauty. In this as in everything else criticism is the condition of the greater æsthetics. The true meaning of things can only be grasped by him who takes up his position at the fountain head itself of beauty; and who from the centre of human nature contemplates with rapturous ecstasy and in every direction those eternal productions in their infinite variety; temples, statues, poems, philosophical systems, religions, social forms, passions, virtues, sufferings, love, and nature itself which would have no value without the conscient being who idealizes it.

Science, art, philosophy would no longer have any value outside the point of view of the human race. He only is capable of grasping the great beauty of things who sees in everything a form of the intellect, a step towards God. For we are bound to say humanity in this instance is but a symbol; perfect beauty dwells in God alone, that is to say, in the whole. The most sublime works are those which humanity has made collectively and to which no name can be attached. The most beautiful things are anonymous. The critics who are simply scholars and nothing else regret this and bring all the resources of their art to bear upon the penetration of that secret. All this is so much blundering. Do they imagine that they have enhanced the beauty of this or that national epic because they discovered the name of the weak mortal who indited it. What do I care for that man who stands between humanity and me? What do I care for the insignificant syllables of his name? That name itself is a lie; it is not he, it is the nation; it is humanity toiling at a point of time and space who is the real author. Anonymousness in this instance is much more expressive and true; the only name which should designate the author of those spontaneous works is the name of the nation among which they saw the light; and that name instead of being inscribed on the title, is inscribed on every page. If Homer were a real and single personage it would still be absurd to say that he is the author of the "Iliad." A like composition, evolved in one piece from an individual brain, without traditional antecedent, would have been insipid and impossible to read, we might as well suppose that it is Matthew, Mark, Luke and John who invented Christ. "Only rhetoric," said M. Cousin, "can ever suppose the plan of a grand work to belong to him who carries it out." The rhetoricians who look at everything from the *literary* side, who admire the poem, while they remain profoundly indifferent to the thing that has been perpetuated in song will never understand the part of the people in those works. It is the people that provides the material, and that material the rhetoricians are blind to and simply imagine that it is the invention of the poet. The Revolution and the Empire have produced no poem worthy of being mentioned; they did better. They left us the most marvellous epic in action. It is foolish in the extreme to admire the literary expression of the feelings and the acts of mankind and not to admire those sentiments and acts in mankind. Humanity alone is admirable. Genius is only the editor of the inspirations of the crowd. Their glory lies in being so deeply sympathetic with the ever creative soul, as to feel the throbbings of that great heart resound beneath their pen. To endeavour to exalt them by revealing their individuality is merely to lower them; it is to destroy their true glory in order to ennoble them by chimeras. True nobility does not consist in having a name of one's own, a glory of one's own, but in belonging to the noble race of the children of God, in being a soldier lost in the immense army marching onward towards the conquest of the perfect.

If transported to those open fields of humanity with what pity will the critic look upon that paltry admiration that clings to the handwriting of the writer rather than to the genius of him who has dictated. No doubt good criticism should allot a great share to great men. They are valuable among humanity and through humanity. They distinctly and eminently feel what the world feels only vaguely. They impart a language and a voice to those mute instincts, which, pent up in the crowd—a stammering being if ever there was one — aspire to find their expression and recognized themselves in their accents. "Oh, sublime poet," they say to him, "we were mute and thou hast given us a voice. We were seeking in the dark for ourselves and thou hast revealed us to ourselves." It is an admirable dialogue between the man of genius and the crowd. The crowd lends him the grand material, the man of genius gives expression to it, gives it shape and creates it; then the crowd that feels but cannot speak, recognizes itself and shouts for joy. It sounds like one of those musical choruses, arranged in dialogue where now one, then several alternate with and reply to one another. Now it is the solitary voice, thin and prolonged which resounds in sweet but penetrating notes. Then comes the grand outburst, apparently discordant, but powerful in effect, amidst which the small still voice continues, but henceforth absorbed in the grand concert which at last gets beyond, and carries it along. Great men have the faculty of guessing beforehand that which becomes ere long patent to all; they are the scouts of the great army; they in their rapid and venturesome advance, can catch sight before the others of the smiling plains and lofty peaks. But in reality it is the army that has brought them where they are and has pushed them forward; it is the army that supports them and gives them confidence; it is the army which in them advances beyond its own lines, and the conquest is not realized until the main body in its slower but surer march ruts with its millions of footsteps the path which they scarcely touched and encamps its heavy masses on the ground where they first appeared as bold adventurers.

How often, in fact, have great men been literally

made by humanity, which removing from their existence every stain and every trace of vulgarity, idealizes and consecrates them like statues erected on the various stages of its march, in order to remind itself of what it is, and to become enthusiastic over its own image. Happy those whom legend thus sequestrates from criticism. For alas, we may well believe that if we touched them we should find at their feet a greater or smaller clod of earth. That which is admirable, heavenly, divine belongs nearly always by right to humanity. As a rule, good criticism must be on its guard against individuals and not allot to them too great a share. It is the masses that create, because the masses eminently possess, and to a thousandfold superior degree of spontaneity, the moral instincts of human nature. The beauty of Beatrice belongs to Dante, and not to Beatrice; the beauty of Krishna belongs to the genius of India, and not to Krishna; the beauty of Jesus and Mary belongs to Christianity and not to Jesus and Mary. No doubt it is not merely pure chance which has marked out this or that individual for idealization. But there are cases in which the woof of humanity completely covers the primitive reality. By this powerful labour and transformed by this plastic energy the most ugly caterpillar might become the most ideal of butterflies.

This labour of the crowd is an element which has been too much neglected in the history of philosophy. The debates are supposed to be finally closed when a few proper names have been opposed to one another. But no one concerns himself as to how the people looked at life, or with the intellectual system on which the age reposed; this, however, is the great motor principle. The history of the human intellect is, a rule composed in too individual a matter. It is like the scene of a play supposed to occur in a public thoroughfare in which there are at most two or three persons moving about. This or that history of German philosophy deems itself complete in devoting separate articles to Kant, Fichte, Schelling, Hegel, Hamann, Herder, Jacobi, Herbart. But where is the grand humanity amidst which they lived? That ought to be the permanent background on which the individuals should be shown. In one word, the history of philosophy should be the history of the thoughts of mankind. The current ideas of a people and of an age contain an unwritten philosophy and literature which should be inserted in the great total. It is thought that a people has no literature until it has definite and settled monuments. But the true literary productions of a people in its intellectual childhood are the mythical ideas that are not written down (the conception of a regular system of editing and the faculties implied by such work only appear among a people at a comparatively advanced degree of thought), ideas rolling their current throughout the whole of the nation, filtering tradition through a thousand secret rivulets to which every one gives a form according to his own taste. At the first glance, one would be tempted to believe that the Breton peoples have no literature, because there would be a difficulty in giving an extensive catalogue of Breton books really and truly ancient and original. But the fact is that they have a complete traditional literature in their legends, their stories, their mythological fantasies, their superstitious worships, their poems hovering about here and there. It was the same with the greater part of our heroic legends before they were repudiated by the cultured part of the nation and got into low company in the "Bibliotheque Bleue."

On entering the Spanish galleries at the Louvre we derive no doubt a great deal of pleasure from the close examination of this or that picture of Murillo or Ribeira. But there is something still more beautiful and that is the impression we derive from the galleries as a whole, from the ordinary *pose* of the personages, from the general style of the pictures, from the dominant colouring. Not a nude figure, not a smile on a single pair of lips. It is Spain in her habit as she lived that we see before us. The principle of grand criticism should in like manner consist in grasping the physiognomy of each portion of humanity. To praise this, to blame that is to fall into a paltry method indeed. We must take the work for what it is, perfect in its order, eminently representing what it does represent, and not reproach it for what it has not. The idea that the author has made a mistake is altogether out of place in literary criticism, unless we are treating of literatures altogether artificial, like the Latin literature of the decadence. No doubt, not everything is of equal value, but as a rule a piece is what it is capable of being. We must place it on a higher or lower rung in the scale of the ideal, but we should not blame the author for having conceived the thing in this or that tone, and for having voluntarily shunned this or that order of beauty. We may criticize the point of view from which each work is conceived rather than the work itself, for all its great authors are perfect from their point of view, and the criticisms addressed to them are as a rule merely so many reproaches for not having been what they were not.

I have perhaps repeated too often, still I will repeat once more that there is a science of humanity which, I hope, will have as much right to assume the title of philosophy as the science treating of the individual; a science which is impossible of attainment save by the erudite trituration of the works of humanity. We need not look for any other motive in many studies whose object is the past. Why should the most noble intellect devote itself to the translation of the Bhagavata-Purana, to commenting the Yashua? He who has accomplished this task in so learned a manner will answer you; "To analyze the works of human thought, by assigning to each its essential character, to discover the analogies which connect them with one another, and to look for the reason of those analogies in the nature of intelligence itself, which without losing aught of its indivisible unity, gets multiplied by the very varied products of art and science; to the solution of this problem the genius of the philosophers of all times has clung from the moment that Greece bestowed upon mankind the two powerful levers of analysis and observation (92)." Therein lies the only value of erudition. No one dreams of crediting it with any practical utility, apart from the fact that curiosity pure and simple would not suffice to ennoble it. Hence, all that remains is to look upon it as the condition of the science of the human intellect; *the science of the products of the human intellect*.

Both the ordinary observer and the *savant* admire a beautiful flower to the same degree, but they do not admire the same things in it. The ordinary observer only sees bright colours and shapely form. The *savant* is so delighted, is so enraptured with the marvels and secrets of its inner life as scarcely to notice those superficial beauties. It is not exactly the flower he admires, it is life, universal force which in one of its forms manifests itself in it. Criticism has up till now admired the masterpieces of literature as we admire the beautiful forms of the human body. The critic of the future will admire them like the anatomist, who penetrating beyond those perceptible beauties, finds in the secrets of their organization an order of beautiful things a thousand times superior. A dissected body is in a certain sense horrible to a degree, and yet the eye of science discovers in it a world of marvels.

Looked at in this way, the most eccentric literatures, those which, judged according to our ideas, would possess the least value, those that take us the farthest away from the actual world are the most important. Comparative anatomy obtains many more results from the observation of inferior animals than from the observation of the superior species. Cuvier might have gone on dissecting domestic animals all his life without so much as suspect the higher problems revealed to him by the study of the molluscs and annelids. In the same way those who study only regular literatures which, in the order of products of the intellect, are as the big classical animals in the animal scale will never succeed in arriving at a larger conception of the science of the human intellect (93). They only see the literary and æsthetic side; nay more; they cannot even understand that in a grand and thorough manner. For they fail to see the divine force at work in every creation of the human intellect. For instance, what are literary works in France? Elegant and subtle bits of moral gossip, never majestic and scientific works. No problem is ever propounded; there is no perception of the higher cause. The science of literatures is treated of as botany would be treated of by an amateur florist who would be content to finger and admire the petals of each flower. The higher and grander criticism, on the contrary, does not scruple to tear away the flower in order to study its roots, to count its stamens, to analyze its tissues. But we should not infer from this that it relinquishes its higher admiration. On the contrary, the higher and grander criticism alone has the right to admire, inasmuch as it alone is certain not to admire blunders, mistakes of copyists; the higher criticism alone knows the reality, and the reality only is admirable. That will be our system, we who belong to the second half of the nineteenth century. We may not possess the subtlety of those masters of atticism, their delightful way of gossiping, their witty innuendo. But we shall have the dogmatic view of human nature, we will plunge into the ocean instead of taking a pleasant dip in shallow water, and we shall return laden with primeval pearls. All that appertains to the work of the human intellect is divine, and the more primeval, the more divine. It is said that M. Villemain called M. Fauriel "an *atheist in literature.*" He should have said a *pantheist*, which is not the same thing.

CHAPTER XI.

Hence we must look upon philology or the study of ancient literatures as a *science* having a distinct object, viz. the knowledge of the human intellect. To consider those literatures merely as a means of intellectual culture is, in my opinion, to deprive them of their true dignity. To restrict their influence merely to contemporary literary production is to take a still narrower view. In a remarkable lecture delivered before the Congress of German philologists at Bonn in 1841, M. Welcker in endeavouring to define the accepted meaning of philology (*Über die Bedeutung der Philologie*), looked at it almost exclusively in that way (94). To M. Welcker philology is the science of classic literatures, that is, of model literatures, which, offering us as they do, the general type of higher learning ought to suit all peoples and be made to serve equally in their education. M. Welcker appreciates the study of antiquity above all on account of the happy influence it may exercise on the literature and æsthetic education of the modern nations. The ancients are to him models and objects of admiration rather than objects of science. Still, M. Welcker does not altogether preach servile imitation. What he asks for is an intimate and secret influence, analogous to that of electricity, which without communicating anything of its own, develops a similar state in other bodies; what he blames is the attempt of those who pretend to find in the modern peoples sufficient material for a moral and æsthetic education. Consequently M. Welcker only looks upon philology from the point of view of the classical scholar, and not from the point of view of the *savant*. As far as we are concerned, it seems to us that we should be placing philosophy in a much more certain and higher sphere, by investing it with a scientific and philosophical value with reference to the history of the human intellect, than by reducing it to a mere means of education and literary culture. If modern nations could find in themselves a sufficiently intellectual leaven, a running and primary source of original inspirations we should be careful indeed not to trouble that vein of fresh supply by an admixture of the antique. Tone in literature is the more beautiful in proportion to its being more true and more pure; to the scholar and to the critic belong the universal use and intelligent appreciation of the most diverse forms; a foreign note, on the contrary will trouble and worry the original and creative poet. But admitting that modern times could find a poesy and a philosophy as truly representative of them as Homer and Plato are representatives of the Greece of their days, even then the study of antiquity would have its value from the point of view of science. Besides; M. Welcker's considerations would not constitute a valid apology for all philological studies. If ancient literatures are only cultivated for the sake of finding models, what would be the use of cultivating those which, though having their original beauties, do not lend themselves to our imitation? We should be obliged to confine ourselves to the study of Greek and Latin antiquity, and even within those limits, the study of masterpieces only would have its value. But the literatures of the East which M. Welcker treats with great contempt, and the second rate works of classical literature, if less suitable as models for taste, sometimes present more philosophical interest and teach us more of the history of the human intellect than the most finished monuments of the ages of perfection.

The fact of classical languages has, moreover, nothing absolute in it. The Greek and Latin literatures are classical so far as we are concerned, not because they are the most excellent of literatures, but because they have been imposed upon us by history. This fact of an ancient language being selected to serve as a basis for the education, and concentrating around it the literary efforts, of a nation which has made for itself a new idiom long ago, is not as people would too often lead us to believe the effect of an arbitrary choice, but purely and simply one of the most general laws of language, a law that owes nothing to the whim or to the literary opinions of this or that epoch. In fact, to invest this denomination of classical with an absolute sense and to restrict it to one or two idioms as if they were predestined by an essential privilege resulting from their nature to be the educational instrument of all peoples is to misapprehend greatly the role and nature of classical languages. Their existence is a universal fact in the linguistic organism, and their selection, just as it implies nothing absolute with regard to all peoples, has nothing arbitrary for any of them.

The general history of languages has long ago demonstrated the fact that in every country where there has been an intellectual movement, there has already occurred the formation of two strata of languages; not through the method of one language abruptly displacing the other, but through that of the second emerging by imperceptible transformations from the dust of the first. The ancient language has been replaced everywhere by a vulgar idiom which in reality does not constitute a different language but rather a different age from that which preceded it; the former is more scholarly, more synthetical, full of inflections expressing the most delicate connections of thought, richer in its order of ideas, albeit that this order of ideas was comparatively more restricted; it is, in short an image of primeval spontaneity, in which the mind confounded the elements in one obscure unity, and lost in the whole the analytical view of parts. On the contrary, the second dialect, corresponding with a clearer and more explicit progress of analysis, divides what the ancients united and shatters the mechanism of the ancient language in order to provide each idea, each connection with its isolated expression.

By taking the one after the other the languages of every country in which humanity has a history it would be possible to verify that progress, which is the progress of the human intellect itself. In India it is Sanskrit with its admirable wealth of grammatical forms with its eight cases, its six moods, its numerous terminations, its involved and powerfully knitted phraseology which in its modification produces Pali, Prakrit and Kawi, dialects less rich, more simple and clear which in their turn are analyzed into dialects still more popular, the Hindu, Bengali, Mahratta and other vulgar dialects of Hindustan, and in their turn become dead, learned and sacred languages; Pali in the island of Ceylon and Indo-China, Prakrit among the Djaïnas, Kawi, in the islands of Java, Bali, and Madura. In the region of India to the Caucasus, Zend with its long and complicated words, its absence of prepositions and its mode of supplying them by means of cases formed by inflection, the Persian of the cuneiform inscriptions, so perfect in its structure, are replaced by modern Persian, almost as decrepit as English which has reached its last stage of erosion. In the region of the Caucasus modern Armenian and Georgian succeeded to the ancient Armenian and Georgian. In Europe the position of the ancient Sclavonic, the Teutonic, the Gothic, the Norman, is below that of the Sclavonic and Germanic idioms. And to wind up, it is from the analysis of Greek and Latin, subjected to the process of decomposition of the barbarous centuries that the modern Greek and the neo-Latin languages have sprung.

The Semitic languages, though dead languages

to a greater degree than the Indo-Germanic ones have followed an analogous course. Hebrew, their most ancient type, disappears at a remote period, to leave the field absolutely to Chaldaic, Samaritan, Syriac, dialects more analytically constructed, longer and sometimes more lucid also, which in their turn become successively merged in Arabic. But Arabic in its turn too scholarly for the everyday use of strangers who are unable to observe its delicate and varied inflections, beholds solecism usurping the common right and in consequence by the side of the literal language, which becomes the exclusive property of the schools, there springs up the vulgar Arabic, more simple in its system and less rich in grammatical forms. The languages of the West and of Central Asia present several analogous phenomena in the "superposition" of the ancient Chinese and the modern Chinese, of the ancient Thibetan and the modern Thibetan, and the Malay languages in that ancient language to which Marsden and Crawfurd have given the name of "grand Polynesian" which was the language of Javanese civilization and which Balbi calls the Sanskrit of Oceania.

But what becomes of the ancient language ousted in that way from everyday use by the new idiom? Its rôle, though changed is, for all that, none the less remarkable. If it ceases to be the intermediary of the ordinary intercourse of life, it becomes the learned and nearly always the sacred language of the people that decomposed it. Imbedded, as a rule in an antique literature, the storehouse of religious and national traditions, it remains the patrimony of *savants*, the language appertaining to the mental and spiritual domain, and it generally requires many centuries before the modern idiom in its turn dares to emerge from vulgar existence to venture into the order of things intellectual. In one word, it becomes classical, sacred, liturgical, correlative terms according to the country where the fact is verified and signifying uses which as a rule accompany one another. For instance, among the Orientals where the antique book never fails to become sacred, religious dogmas and the liturgy are generally confided to the custody of that obscure, scarcely known language.

Hence, it may be taken as a general fact of the history of languages that each nation finds its classical language in the very conditions of its history and that its choice is not an arbitrary one. It is also a fact that, among nations not very far advanced in mental culture, all matters of the intellectual order are confided to that language, and that among the peoples whose more energetic intellectual activity has forged a new instrument better adapted to their needs, the antique tongue preserves a grave and religious rôle; that of accomplishing the education of the faculty of thought and of its initiation into the things of the intellectual world.

Modern language being, in fact, wholly made up of the remains of the ancient, it becomes impossible to master it in a scientific manner, except by bringing back those fragments to the primeval structure, where each of them had its value. Experience shows how imperfect is the knowledge of modern languages with those whose knowledge is not based on the knowledge of the antique language from which each modern idiom sprang. The secret of grammatical mechanisms, of etymologies and consequently of orthography being altogether contained in the ancient dialect, the logical reason of the rules of grammar is utterly lost to those who consider those rules in an isolated way and irrespective of their origin. Routine, in that case, becomes the only method possible as in every case where practical knowledge is aimed at to the exclusion of theoretical reason. One knows the language as the workman who employs the methods of geometry without understanding them, knows geometry. Besides, being indebted for its form to dissolution, modern language will fail to imbue with life the shreds it has endeavoured to assimilate, unless it reverts to the ancient synthesis in order to find the stamp that must invest with a new unity those scattered elements. Hence its inability to constitute itself by itself into a literary language; hence the utility of those men who at given periods, had to provide its education by the antique, to preside as it were at its "classicality." Without this necessary operation, the vulgar tongue always remains what it was at its origin, a popular jargon begotten by the incapacity for synthesis, and inapplicable to matters intellectual. Not that we should mourn for the loss of synthesis. Analysis is something much more advanced and corresponds to a more scientific condition of the human intellect. But it is incapable of creating anything by itself. Eminently fit to decompose and to lay bare the secret springs of language, it is powerless to reconstruct the *ensemble* it has destroyed, unless it resorts for this to the ancient system, and derives from its commerce with antiquity its spirit of constructing a whole, and of scholarly organization. This is the law to which all modern languages had to submit in their development. And the processes by which the vulgar tongue has risen to the dignity of literary language are the very ones by which one may attain to a perfectly intellectual grasp of it. The model of philological education is traced in each country by the training undergone by the vulgar idiom in order to gain its patent of nobility.

The historical utility of the study of ancient language is in no way inferior to its philological and literary utility. The sacred book to the antique nation was that in which were recorded all the national recollections; it was consulted by every one in search of his genealogy, the meaning of all the acts of civil, political and religious life. The classical languages are in many respects, the sacred book of the modern peoples. They contain the roots of the nation, her titles, the sense of her words, consequently of her institutions. Without it a great many things would remain unintelligible and historically unexplicable. Every modern idea is grafted upon an antique stem, all actual development is the emanation from a precedent. To consider humanity from an isolated point of its existence is to condemn one's self for ever to remain in ignorance of it; it has no sense except as a whole. There lies the prize of erudition; in the re-creation of the past, in the exploration of every part of humanity; whether it be conscious or not of its mission, erudition prepares the basis necessary to philosophy.

More modest education, compelled to set itself limits, and unable to take in the whole of the past, adheres to the portion of antiquity, which, as applicable to each nation, is classic. And this choice which can never be doubtful is still less doubtful with us than with any other people. Our civilization, our institutions, our languages have been constructed out of Greek and Latin elements. Hence whether we like it or not, Greek and Latin are forced upon us by facts. No law, no rule has given them or can deprive them of this character which they derive from history, just as education among the Chinese or Arabs will never mean the acquiring of the vulgar Arabic or Chinese, but will always mean the acquisition of literal Arabic or Chinese, just as modern Greece owes its slight revival of literary life solely to the study of ancient Greek, so will the study of our classical languages, inseparable from one another, always constitute with us, and by the force of circumstances, the basis of education. Other nations, even European ones, such as for instance the Sclavonic nations, nay the Germanic peoples themselves though they were later on so intimately connected with Latinism, may look for their education elsewhere: they would at most voluntarily deprive themselves of an admirable source of the beautiful and the true; but they would not deprive themselves of direct intercourse with their ancestors. But as for us, it would be tantamount to denying our origin, to cut off all connection with our forbears. Philological education cannot possibly consist in the study of modern language, any more than moral and political education can mean the exclusive study of actual ideas and institutions, we must go back to the primary source and take our stand on the road of the past to arrive subsequently at the full understanding of the present by the same road over which humanity travelled.

CHAPTER XII.

Consequently in my opinion the sole means of constituting the *apologia* of philological sciences and of learning in general is to group them into a whole and to bestow upon them the title of *Sciences of humanity*, in contrast to the *sciences of nature*. Without this, there is no object in science, and it exposes itself to all the objections so often directed against it.

The modesty of the means it employs to attain its end should not be argued as a reproach. Cuvier dissecting snails would have raised a smile from the frivolous minded who do not understand the processes of science. The student of chemistry manipulating his various apparatus looks very much like a navvy, and still he accomplishes the most liberal work of all, the inquiry into what is. M. de Maistre has depicted modern science somewhere as having "its arms full of books and instruments of all kinds, pale with vigils and overwork, staggering along on the road to truth, quivering and inkstained, and bending its forehead wrinkled with algebra towards the ground." A *grand seigneur* like M. de Maistre must in fact have felt greatly humiliated by such painful investigations, and truth was very irreverent indeed by putting so many difficulties in his way. He must have preferred the more easy method of "Oriental science, free, isolated, flying rather than plodding along, presenting in its whole appearance something aerial and supernatural, letting the winds toy with its hair escaping from under an Eastern mitre, its spurning foot seemingly only touching the earth in order to get an impetus for its flight." It is the characteristic and the pride of modern science to attain its most lofty results only through the most scrupulous methods of experiment and to arrive at the knowledge of the highest laws of nature, its hands resting on its apparatus. It leaves to old-fashioned *a priori* the doubtful honour of seeking its support only in itself; it prides itself upon being nothing but the mere echo of facts, upon mixing no invention of its own with its discoveries.

The most humble methods are in this way ennobled by their results. The highest laws of the physical sciences have been ascertained by manipulations differing very little from those of the artisan. If the highest truths can as it were emanate from the alembic and the crucible, why should they not equally be the result of the study of the remains of the past, covered with the dust of ages? Shall the philologist who toils on words and syllables be less honoured than the student of chemistry labouring in his laboratory?

The few results attained by certain branches of philological studies constitutes in itself no objection against them. For on embarking upon an order of researches it is impossible to guess beforehand what may result from them, any more than one can know, in digging a mine, the wealth it may contain. The veins of precious metal do not lend themselves to prognostication. We may be on our way to the discovery of a new world; the laborious investigations undertaken may also lead to the sole conclusion that nothing is to be gained from them. But do not say that he who has merely attained this altogether negative result has wasted his time. For apart from the fact of there not being any absolutely fruitless research or any which does not lead either directly or accidentally to some discovery, the investigator will save others the useless trouble he gave himself. A good many orders of researches will remain in that way like mines, exploited at some previous period, but abandoned since, because they did not sufficiently reward the workers for their pains and because they no longer afford hope to future explorers. We should, however, bear in mind that results which at a given moment may appear altogether insignificant may turn out to be most important in connection with new discoveries and new comparisons. Science always presents itself to man as an unknown country, he often enters upon it by an out-of-the-way corner which fails to give him an idea of the whole. The first navigators who discovered America were far from suspecting the exact forms and true relations to one another of those parts of this new world. Was it an isolated island, a group of islands, a vast continent or the prolongation of another continent? Only the subsequent explorers could answer these questions. The same in science; the most important discoveries have often been brought about in a roundabout way; "on the slant" if I may so express it. Very few problems have been deliberately grappled with at the outset, "taken at the core." It was through fragmentary translations that Anquetil-Duperron began the study of Zend literature, as in the Middle Ages it was through very imperfect Arabic versions that the scientific authors of Greece acquired their first knowledge of the West. The celebrated passage of Clement of Alexandria on the Egyptian writings attracted little or no notice until the day when, in consequence of other discoveries, it became the key to the study of Egyptian monuments. The accessory may in this way become the principal in consequence of a change of aspect (95). The theologians who in the Middle Ages, occupied the principal scene are very secondary personages to us. The rare *savants* and thinkers who at that period conducted their investigations by the true method, and who at the time remained unnoticed or were persecuted, occupy in our opinion the first and foremost position, for only their method has been continued; they alone had issue. No kind of research should be branded at the outset as useless or puerile; one does not know what it may bring forth, nor the value it may acquire from a more advanced standpoint.

Physical science affords a great number of instances of isolated discoveries which for years remained almost without significance and only acquired importance long afterwards through the accession of new facts. For a long while students may pursue an apparently barren track, which they abandon at last in despair, when all of a sudden there appears an unexpected light; the discovery bursts forth at two or three points at the same time, and what until then had looked as a mere isolated and insignificant fact becomes in a novel combination, the basis of a whole theory. There is nothing more difficult to foretell than the importance with which posterity will invest this or that order of facts, the researches that will be abandoned, the researches that will be continued. The attractive properties of yellow amber were merely looked upon by the ancient students of physics as a curious fact until a complete scientific theory was constructed around that first atom. We must not expect a hard and fast system of logic in the order of scientific investigations, any more than we must ask the explorer to give us beforehand a plan of his discoveries. In looking for one thing one may stumble upon another, in the pursuit of a mere vision, one may hit upon a magnificent reality. Accident, chance, on the other hand claims its share. Universal exploration, a beating-up of the game on all sides, that and that only is the sole possible method. "We must look upon the fabric of science, as we would upon that of nature;" said Cuvier. "Each fact occupies its defined position, which cannot be occupied except by that fact." That which has no value in itself may possess a great deal as a necessary *means*.

The critical consideration of an object is often more serious than the object itself. One may comment seriously on a madrigal or a frivolous novel; grave scholars have devoted their lives to commenting works the authors of which only aimed at giving pleasure. All that belongs to the past deserves serious attention. Some day Béranger will become an object of scientific comment and belong to the domain of the Académie des Inscriptions. Would not Molière, who was so apt to ridicule the *savants* whose name ended in *us*, be more or less surprised at having fallen into their hands? The profane and every now and then even they who call themselves thinkers laugh at the minute investigations of the past by archæology. Such researches, if their aim were strictly confined to their own domain, would no doubt be nothing better than more or less interesting fancies of the amateur, but they become invested with the dignity of science, nay, in a certain sense sacred if one admits their connection with the knowledge of antiquity, which it is impossible to attain save through the knowledge of monuments. There are a great many studies which possess no value except with the view of an ulterior purpose. It would be difficult perhaps to find anything philosophical in the theory of Greek accentuation, but is that a reason to vote it useless? Certainly not, for without it, the thorough knowledge of Greek would be impossible. A like system of exclusion would lead to the revival of the witty argument, by which, in Voltaire's story, the education of Jeannot is simplified with a vengeance.

Besides, how many works are there which though possessing no absolute value, were highly important in their own days, on account of their opposition to rooted prejudices. We do not learn a great deal from Naudé's *Apology for great men wrongly suspected of magic*, nevertheless it may have exercised true influence in its own days. How many books of our own century will be judged in the same way by posterity? Writings intended to combat an error disappear with the error they combated. When a result has been attained, it is difficult to realize the trouble its attainment has cost. It wanted a genius to conquer the domain that afterwards may have become a child's.

The researches relating to the cuneiform inscriptions which constitute one of the most important of Oriental studies in the actual condition of science afford one of the most curious instances of studies worthy of being pursued with the greatest zeal, notwithstanding the uncertainty of their results. I leave aside the Persian inscriptions, the explanation of which is complete; I am alluding to the Median, Assyrian and Babylonian inscriptions, which, even according to those who have arduously laboured at them, remain undeciphered. How far they will continue to resist the learned onslaughts of the *savants*, it is impossible to say. But granting the most discouraging hypothesis, supposing they will remain an enigma for ever, those who have devoted their labours to them will none the less deserve as well of science, as if, like Champollion, they had restored a whole world; for even if this happy result should not be realized, its realization did not altogether belong to the impossibilities, and there was no means of knowing until they had tried.

In the actual condition of science, there is no work more urgently needed than a critical catalogue of the manuscripts contained in the various libraries. Those who have been engaged in these researches know how utterly insufficient they all are to convey an exact idea, how for instance, those of the Bibliothèque Nationale are full of mistakes and gaps. This, at the first blush, is a very humble undertaking, for which the merest pupil of the École des Chartes * would suffice. Nothing of the kind. There is no work,

* The École des Chartes is almost equivalent to our Record Office of which the Master of the Rolls has the custody, but it admits students subsidized by the State.—Transl.

requiring a wider knowledge, and all our most eminent scientific men, each examining the manuscripts within the most restricted circle of their own specific knowledge, would scarcely suffice to carry out the task in a thoroughly sufficient manner. And still, learned research will be hampered and remain incomplete until this work be done in a final manner. Even the Jews admit that in another hundred years the Talmudic-Rabbinical literature will no longer attract students. When these books shall cease to have a religious interest, no one will have the courage to tackle this chaos. And still they contain vast treasures to the critic and the student of the history of the human intellect. Had we not better be quick and utilize the five or six men of the present generation who alone are competent to let in the light upon these precious documents? I can assure you that the few hundred thousand francs a Minister of Public Education would devote to this would be better spent than three-fourths of the money usually spent upon the advancement of literature. But this minister should at the same time and beforehand don his armour against the epigrams of the "boobies" and even of literary men who will "fail to conceive why the money of the rate and tax payer should be spent on such 'tomfoolery.'"

It is the law of science as of every human undertaking to draw its plans on a large scale and with a great deal that is superfluous around them. Mankind finally assimilates only a small number of the elements of its food. But the parts that have been eliminated, are they therefore useless, have they played no part in the act of nutrition? Certainly not; they have been useful in causing the remainder to pass, they were so closely bound up with the nutritive portion that the latter without the superfluous could have neither been taken nor digested. Open a collection of antique epigraphs, and out of a hundred only one or two perhaps will be of real interest. But if the others had not been deciphered how should we have known that among them there were not some still more important? To have published those that seem useless cannot even be deemed a work of superfluity, seeing that this or that one which appears to be utterly insignificant now may become of capital importance in a series of investigations which at present we cannot foresee.

The general design of the forms of humanity is like those colossal figures intended to be seen from a distance and of which each line does not show as distinct and clear as that of a statue or a picture. The forms are largely outlined, there is a great deal too much, and if we wished to reduce it to the strictly necessary, we should have to take away a good deal. In history the outline is coarse, each feature instead of being represented by an individual or by a small number of men is represented by large masses, by a nation, by a system of philosophy or by a form of religion. On the monuments at Persepolis the nations tributary to the King of Persia are represented by a single individual wearing the dress, and carrying in his hands the products, of his country; the latter to be offered in homage to the suzerain. Here we have a picture of humanity; each nation, each form of intellect, of religion, of morality leaves behind it a short summary, which is as it were the extract and the quintessence of it, and which is often contained in one word. This abridged and expressive type remains as the representation of the millions of men who never emerged from obscurity, who lived and who died in order to be grouped under that sign. Greece, Persia, India, Judaism, Islamism, Stoicism, Mysticism, all these forms were necessary in order to complete the grandiose figure, for in order to be represented in a manner worthy of them, not a few individuals, but enormous masses were needed. Pictorial representation by masses is the grand process of Providence. There is a marvellous grandeur and a very deep philosophy in the way in which the ancient Hebrews conceived the government of God, treating nations like individuals, establishing between all the members of a community a perfect a mutual and reciprocal responsibility and dispensing with a majestic "thereabouts" his distributive justice. God only sets Himself the large, general plan. Each created being finds subsequently in himself the instincts which make his lot as mild as possible. The thought of how few traces are left behind by men, even by those who seem to play a principal part is calculated to fill us with terrible sorrow. And when we reflect that millions upon millions of creatures were born and died in that way, without leaving the slightest memorial, one experiences the same terror as one would feel in presence of utter annihilation or the infinite. Only think for a moment of those wretched existences scarcely characterized by anything, which among the savages appear and disappear like the indistinct visions of a dream. Only think for a moment of the countless generations which have been piled upon one another in our country cemeteries. Dead, dead for ever and aye? . . . No they live in humanity; they served to build the great Babel which uprises towards the sky, and each layer of which means a people.

I am going to tell you about the most charming recollection of my early youth; the thought of it almost brings tears to my eyes. One day my mother and I in one of those short excursions in the stony byways on the coast of Brittany which leave such sweet memories with all those who wander there, came upon a small village church, surrounded as usual by the churchyard, and we sat down to rest ourselves. The walls of the church of rough-hewn granite and covered with moss, the neighbouring houses built of primitive blocks, the closely serried tombs, the mouldering and overthrown crosses, the numerous skulls ranged in tiers on the steps of the tiny house which served as an ossuary (96), all these showed that people had been buried there from the most remote days, when the Saints of Brittany had made their appearance for the first time on these waves. On that day the terror-stricken feeling at the immense oblivion and the vast silence amidst which human life is swallowed up was such as to haunt me still, and to have become one of the elements of my moral existence. Among all these simple, humble folk that lie there, in the shadow of the old trees, not one, not a single one will live in the future. Not a single one has stamped his acts on the grand movement of things, not a single one will count in the final statistics of those who have given the impulse to the ever-moving wheel. In those days I served the God of my infancy, and an upward look at the stone cross on the steps of which I was seated, a glance at the tabernacle visible through the windows of the church was sufficient to explain all this to me. And besides, the sea was but at a stone's throw, so were the rocks and the foam-crested waves, I could sniff the winds from heaven, which penetrating to the very brain, awakened a kind of undefinable and indescribable feeling of freedom and expansion. My mother also was by my side; and it seemed to me that the humblest life was capable of reflecting heaven through pure love and individual affection. I considered those who lay there happy. Since then I have shifted my tent and I account for this vast darkness in a different way. They are not dead those obscure children of the hamlet, for Brittany still lives, and they have contributed to the making of Brittany; they played no part in the great drama, but they formed part of the vast chorus, without which the drama would be cold and lifeless and destitute of sympathetic actors. And when Brittany shall be no longer there, France will still be there; and when France is gone, humanity will remain, and people will go on saying; "In days gone by, there was a noble country in sympathy with all that was beautiful, whose destiny it was to suffer for the sake of humanity and to fight in its behalf." On that day the lowliest peasant who had but a few steps to go from his hut to his tomb, shall like ourselves, live in that immortal name (97); he will have contributed his small share in the great result. And when humanity is gone, God will remain, and humanity will have contributed to the making of Him, and in his vast bosom all that lived will live again, and then it will be true to the very letter that not a glass of water, not a word that has furthered the Divine work of progress will be lost.

That is the law of humanity; an enormous and lavish expenditure of the individual, a contemptuous agglomeration of human beings (I can fancy the modeller flinging his material about anyhow, and taking little or no heed of three-fourths of it that falls to the ground); the immense majority fated to enact "the wall flowers" at the grand ball conducted by destiny, or rather to figure in one of those multiple personages which the ancient drama designated as the chorus. Are they useless? No; for they also have made their show; without them the lines would have been thin and paltry; they have contributed to the splendour of the whole, which is more original and more grand. This or that nun who vegetates unnoticed, forgotten in her convent seems altogether lost as far as the living picture of humanity is concerned. Not at all; for she contributes to the sketch of monastic life; she enters as an atom in the grand mass of black necessary to that. Humanity would not have been complete without monastic life; monastic life could only be represented by a numberless group; hence all those who have made part of the group, however completely they may be forgotten, have had their share in the representation of one of the most essential forms of humanity. In short, there are two ways of influencing the world, either by one's individual force, or by the body of which one forms a part, by the ensemble in which one occupies a place. In the latter case the action of the individual seems veiled; but on the other hand it is more powerful, and the proportional part accruing to each is much stronger than if he remained isolated. Those poor women, divided, would have been vulgar, commonplace, and would have made no figure in humanity; united, they represent energetically one of the world's most essential elements; sweet, timid and pensive piety.

No one, therefore, is useless in humanity. The savage who scarcely exists, serves at any rate as waste power. And as I have already said, it was but fit that the plan of the forms of humanity should be superabundantly provided for. The belief in immortality implies nothing else than that invincible faith in the future. No action is utterly lost. This or that insect which had no other vocation than to group under a living form a certain number of molecules and to eat a leaf accomplished something which still bears consequences in the eternal series of causes.

Science, like all the other facets of human life must be represented in that large way. Scientific results should not be arrived at in a meagre and isolated manner. The final residue which will remain in the domain of the human intellect must necessarily be extracted from a vast mass of things. Just as no man is useless in humanity, so is no labourer useless in the field of science. Here, as everywhere else, there must be an immense waste of power. When we reflect upon the enormous amount of intellectual work and activity that has been engulphed for the last three centuries and even in our days in the periodical publications, in the reviews, etc.; we experience the same feeling that comes over us at seeing the eternal round of generations swallowed up by the tomb, as it were putting one another down. But this is bound to be; for if everything that is said and discovered were assimilated there and then, it would be like a man taking absolutely nothing but what is nutritious. After the lapse of a century a genius of the first order is reduced to two or three pages. The score of volumes of his complete works remain as a necessary development of his fundamental idea. A volume for each idea. The eighteenth century is summed up, as far as we are concerned in a few pages expressing its general tendencies, its spirit, its method; all this is hidden away in thousands of books, forgotten by this time and teeming with gross errors. The biggest library could be filled with the books relating to one controversy only, such as for instance that of the Reformation, of Jansenism, of Thomaism. All this expenditure of intellectual force is not lost, provided those controversies have contributed one single atom to the fabric of modern thought. A great many literary lives, apparently wasted, have been, in fact, useful and necessary. Who, at present, bestows a single thought on this or that grammarian of Alexandria, illustrious in his own time? And still he is not dead; for he helped to sketch Alexandria, and Alexandria remains an immense fact in the history of mankind.

We can conceive no idea of the largeness of the method by which the work of science should be undertaken in a condition of humanity scientifically organized. I may suppose that it took a thousand laborious lives to collect all the local varieties of a certain legend, such as for instance, that of the Wandering Jew. It is by no means certain that such labour would lead to any serious result; but it matters not; the mere possibility of finding in it some subtle induction, which by entering as an element into a more vast ensemble should reveal a feature of the system of things, would be sufficient to venture upon such an expenditure. For nothing is too dear when it becomes a question of providing a single additional atom to truth. Are not thousands of lives lost every day, what is called absolutely lost, in the furthering of the arts of luxury, in contributing a mere scrap of nourishment for the pleasures of the idle, etc. Humanity, after all, has a great deal of strength which absolutely perishes for want of employment and guidance. May not we hope that one day all this neglected or utterly wasted strength will be applied to serious things or to the attainment of supra-sensitive results?

There are often a great many false conceptions with regard to what will be the mode of life in the future; it is thought that immortality in literature will consist in *being read* by future generations. This is an illusion we had better abandon. We shall not be read by future generations, we know it, we rejoice in it and congratulate ourselves on it. But we shall have contributed to the manner of looking at things, we shall have enabled the future to do without reading us, we shall have accelerated the day when the knowledge of the world shall equal the world, when the subject and the object having become identified, God will be complete. By accelerating progress, we accelerate our death. We are not writers who are studied for their style of exposition and their classical touch; we are thinkers, and our thought is a *scientific act*. Do people still read the works of Newton, of Lavoisier, of Euler? Their books are facts; they have had their place in the series of the development of science; after which their mission is at an end. Only the name of the author remains in the annals of the human intellect like the names of great statesmen and great captains. The real *savant* never thinks of the immortality of his book, but of the immortality of his discovery. In the same way we try to enrich the human intellect by our observations, rather than to make it read the expression itself of our thoughts. We would wish our name to remain rather than our book. Our immortality consists in the insertion of an imperishable element into the intellectual movement and in that sense we may say as of old; *Exegi monumentum œre prennius* seeing that a result, an act in connection with humanity is immortal by reason of the modification it introduces for evermore in the series of things. The results of this or that obscure book which has crumbled to dust long since still last and will last for ever. The destiny of the history of literature is to replace to a vast extent the direct perusal of the works of the human intellect. Who nowadays reads the polemical works of Voltaire? And still, are there any works that have ever exercised a greater influence? The study of the authors of the seventeenth century is no doubt eminently useful to the knowledge of the intellectual condition of that period. Nevertheless I consider as good as wasted the time devoted to such study. There is nothing to be learned from it in the way of philosophical ideas and views, nor, I am bound to confess, do I conceive the result of a complete education to consist in the knowledge by heart of La Bruyère, Massillon, Jean-Baptiste Rousseau, Boileau, all of whom have but little connection with us. On the other hand no young fellow can be said to have terminated his studies without being up in Villemain, Guizot, Thiers, Cousin, Quinet, Michelet, Lamartine, Sainte-Beuve. I yield to no one in my admiration of the seventeenth century in its proper place in the history of the human intellect but I revolt the moment this heavy style of thinking utterly devoid of critical acumen is held up as the model of absolute beauty. Heaven preserve us from such a book as "L'Histoire Universelle," the object of a kind of stereotyped admiration, the work of a belated theologian, by way of a text-book of historical philosophy for our liberal rising generation.

The revolution which has transformed literature into journalism and periodical writing, which has reduced every work of the intellect to a work "of actuality" that will be forgotten in a short time naturally compels us to look at it from this standpoint. In this way the work of intellect ceases to be a monument in order to become a fact "a lever of opinion." Every one harnesses himself to the century for the purpose of dragging it in his own direction; the moment the impulse has been given, there remains only the accomplished fact. All this suggests the conception of a state of things in which the privilege of writing will no longer be a right apart, but in which masses of individuals would only think of bringing into circulation this or that order of ideas without appending to them the label of their personality. Periodical production has already reached with us a condition of exuberance such as to entail oblivion to an immense extent and the swallowing up of beautiful as well of mediocre productions. Happy the classics who came at a period when literary individuality was so powerful. There are parliamentary speeches of to-day assuredly as good as the best discourses of Demosthenes. Many of the forensic speeches of Chaix-d'Est-Ange will stand a favourable comparison with the invectives of Cicero; still Demosthenes and Cicero will continue to be published, admired, commented upon as classics, while the speeches of M. Guizot, of M. de Lamartine, of M. de Chaix-d'Est-Ange will live and die only in the columns of the journals of the day after these speeches.

CHAPTER XIII.

It is, therefore, of great importance to understand the *rôle* of the *savant's* labours and the manner in which he exercises his influence. His aim is not to be read, but to insert a stone in the great edifice. Scientific books are a fact; the life of the *savant* may possibly be summed up in two or three results, the expression of which may consist of only a few lines or may disappear entirely in more advanced formulas. He may have recorded his researches in bulky volumes that will only be read by those who travel over the same special road that he did. His immortality does not lie there, but in the brief formula in which he has summed up his life and which in its more or less exact shape will become an element in the science of the future.

Only art in which the form is inseparable from the matter goes down in its entirety to posterity. And one is compelled to admit that our worth does not lie in our form. The authors of our own century will be read very little, but they may take comfort, they will be talked of a great deal in the history of the human intellect. The monographers will read them, and will compose on them curious theses, as we do on d'Urfé, on La Boétie, on Bodin, etc. We compose none on Racine and Corneille; for they are still read, and books are only written on books that are no longer read.

Be this as it may, scientific and philosophical progress is subject to conditions utterly different from those of art. Art is not exactly a matter of progress, but of variation of the ideal. Nearly every literature has as its origin the model of its perfection. Science on the contrary advances by utterly opposite processes. By the side of its philosophical results which are never very long in becoming current, it has its special and technical part which is only intelligible to the learned. Nay, several sciences have as yet only that part and will probably never have any other.

Scientific specialities are the scandal of men of the world, just as generalities are the scandal of the *savants.* It is the result of our deplorable habit of looking upon that which is general and philosophical as superficial, and upon that which is scholarly as heavy and impossible to be read. To preach philosophy to certain *savants* is tantamount to proclaiming one's self a smatterer and a numskull. To preach science to men of the world means numbering one's self among the pedantic schoolmen. These are no doubt very absurd prejudices but they are not without their cause; because philosophy up till now has scarcely been anything but fancy *a priori* and science has only been an insignificant display of learning. The truth is, it seems to me, that specialities mean nothing except with a view to generalities, but that again generalities are only possible by virtue of specialities; the truth is, that there is a vital science which deals with the whole of man and that this science must needs be based on all other particular sciences, which are beautiful in themselves, but above all beautiful in their *ensemble.* The specials (if I may be allowed the expression) often make the mistake of thinking that the aim of their work lies in that work itself, and on that account lay themselves open to ridicule, everything savouring of result alarms, and seems of no value to, them. No doubt, if they confined themselves to making war upon generalities advanced haphazard, upon superficial observations we could only applaud their severity.

But they often seem to set great store upon details themselves. I can perfectly well understand that a date happily ascertained, the recovery of the circumstances attending an important fact, the elucidation of an obscure history may assume greater value than whole volumes of the kind frequently boasting the title of "*history of philosophy.*" But truly, are such discoveries worth anything in themselves? Does not it lie in the degree of their contribution to the founding of the true and serious philosophy of history of the future? What does it matter to me whether Alexander died in 324 or in 325, whether the Battle of Platæa was fought on this or that hill, whether the succession of the Greek and Indo-Scythian Kings of Bactriana was effected in this or that order. Truly, how much better off am I for the fact of knowing that Asoka succeeded Bindusaro, and Kanerkes to I do not know whom. If scholarship meant nothing more than that, the Hermagoras of La Bruyère who knows the names of the architects of the Tower of Babel but who has not seen Versailles is the true scholar, and all the ridicule levelled at scholarship would be truly deserved, because vanity alone could sustain people in such researches, and only mediocre intellects could devote their lives to them.

The moment it is thoroughly agreed that learning is only valuable by virtue of its results we cannot push the division of scientific work too far. In the actual condition of science, and above all of philological science the most useful work is that which brings to light new original sources. Until all the parts of science are elucidated by special monographs, the general works must be considered premature. And monographs are only possible on the condition of specialities within very severe limits. In order to clear up a given point the whole intellectual region in which it is situated must have been gone through, we must have explored all the outskirts and be able to take our stand in the centre, and with a full knowledge, of the subject. How much would not the works on Oriental literature gain if their authors were as great specialists as the philologists who have created piecemeal the science of the classical literatures. The only works of use to science are those which may be thoroughly relied upon, and whose authors have acquired, through long habit, if not the privilege of infallibility, at least that vast knowledge which constitutes the assurance of the writer and the security of the readers. Without this nothing can be said to be definitely acquired, everything will constantly have to be done over again. One may say without exaggeration that two-thirds of the works relating to Oriental languages are not deserving of more confidence than a work on classical languages by a fair scholar in the fifth form.

I should be sorry if on this point the drift of this work were misunderstood. I have eulogized polymathy and varied knowledge as a philosophical method, but I think that in the way of special work one cannot too rigorously restrict one's self to one's sphere. I like Leibnitz who under the common term of philosophy unites mathematics, the natural sciences, history and linguistic studies, but I cannot approve of a William Jones, who, without being a philosopher fritters away his activity on numberless subjects and who in a life extending over forty-seven years writes a Greek anthology, an "Arcadia," an epic poem on the discovery of Great Britain, translates the speeches of Isæus, the Persian poems of Hafiz, the Sanskrit code of Manu, the drama of Sacontala, one of the Arabian poems called "Moallakat" at the same time that he writes "A Means for Preventing Riots during Elections," and several other pamphlets on passing events, the whole "without prejudice" to his profession as a barrister.

Still less can I forgive that culpable frittering away of a scientific existence which causes science to be looked upon as a means of business and robs the *savant's* life of its most precious moments. Did not Cuvier really waste his time when he devoted hours upon hours that might have been so fruitful, to administrative functions which others might have discharged as well as he? A man only excels in one thing; I cannot conceive how one can thus admit in one's life a principal and an accessory aim. Only the principal has its value, existence has not two aims. If I did not believe that everything is sacred, that everything is of importance in the pursuit of the beautiful and true, I should consider as wasted the time devoted to anything else but special research. I can conceive the fact of a very vast, nay of a universal scope of life. That the thinker, the philosopher, the poet should be actively concerned in the affairs of his country, not in the small details of administration but in the general direction, well and good. But that the special *savant* after having written a work or so, or after having made a few discoveries should claim as a reward to be absolved from doing any more and to be allowed to enter the political arena is the sign of a paltry nature, of a man who has never understood the nobleness of science.

Hence, the true interests of science demand more than ever specialistic work and monographs. That each paving stone should have its history is a consummation devoutly to be wished. There are as yet few branches in philology and history in which general work is possible with anything like full security. Nearly all the sciences have already "enacted" their grand histories; the history of medicine, the history of philosophy; the history of philology. Well, we may unhesitatingly affirm that, with the exception perhaps of the history of philosophy, not one of these histories is capable of adequate record, and that if the work of writing monographs does not assume more extensive proportions, will not be capable before another century. In fact, we cannot expect of him who undertakes those vast histories an equal special knowledge of all the parts of his subject. He will be obliged to trust for a great many things to the works of others. And it so happens that on many important points, monographs are as yet utterly lacking, so that the author is reduced to gather here and there some sparse and second-hand notions, frequently very inexact. Let us take for instance, the history of medicine, one of the most curious and one of the most important with regard to the history of the human intellect. Let us suppose that a *savant* should undertake to rewrite in its ensemble the very imperfect work of Sprengel. By means of his personal knowledge and works already accomplished he might perhaps treat the ancient part in a definite manner. But what of Arabic medicine, mediæval medicine, Indian medicine, Chinese medicine? Granted even that he know Arabic, Chinese or Sanskrit and that he were capable of making useful monographies in one of these languages the whole of his life would not be sufficient to go even superficially over one of those fields still unexplored. Hence, in condemning himself to be complete, he condemns himself to be superficial. His book will be valuable only in those parts where he applies a special knowledge; then why not confine himself to those parts? Why devote to worthless labour which is moreover fated to become useless the time he might employ so usefully in definite researches? Why write long volumes among which one only may perhaps possess a real value? It is pitiful to see a *savant* in order not to lose a chapter of his book, condemned to write the history of Chinese medicine under about the same conditions as a man writing the history of Greek medicine after some trashy Arabic or mediæval work. And still he would be fatally condemned to do this by the very framework of his book.

The following is a curious experiment and I would wager that it might be made without exception in connection with all general histories. Present those histories to each of the men who have a special knowledge of one of the parts of which they are composed and I am certain that each of them will find his own part execrably treated. Those who have studied Aristotle are of opinion that Ritter has badly summarized Aristotle, those who have studied stoicism that he has spoken superficially of stoicism. I once presented my learned friend Dr. Daremberg with a copy of "the History of Philology" by Graefenhan that he might examine the medical part. He found it treated without the least understanding of the subject. Is it not very probable that other specialist *savants* would have judged the parts relating to the objects of their researches in the same way? So that in wanting to do too much one satisfies no one, unless, I repeat, the author of the general history be himself a specialist in one branch of it, to which branch he would have done better to confine himself.

The work of the nineteenth century then, should be the writing of monographs on every point of science, a hard, humble and laborious task, no doubt, requiring the most disinterested devotion, but a solid and lasting work withal and immensely lifted out of the common by the loftiness of the final aim. It would certainly be more sweet and flattering to human vanity to pluck at the outset the fruit which will only be ripe in a distant future. It wants a very deep-rooted scientific virtue to check one's self on this fatal slope and to refrain from rushing on when the whole of human nature clamours for the final solution. The heroes of science are they who capable of the loftiest views have been able at the same time to resist all anticipated thought and to resign themselves to the rôle of humble monographers, when every instinct of their nature would have impelled them to scale the high summits. To many, nay to the majority, we are bound to say, this is but a small sacrifice, there is but little merit in abstaining from philosophic views to which they are not inclined by nature. The really deserving are they who while understanding in the loftiest sense the supreme aim of science, while experiencing the most urgent philosophical and religious needs devote themselves for the sake of posterity to the laborious calling of mere navvies, and condemn themselves, like the plough horse to see only the furrow it turns. This in the style of the gospel is called; losing one's soul in order to save it. *To make up one's mind to ignore so that posterity may know*, is the first and foremost condition of the scientific method. For many long years science will still stand in need of those patient researches that take, or might take, the title of "*Memoranda for the use of* . . ." With a view to the welfare of posterity lofty intellects will in that way be compelled to condemn themselves to the *ergastulum* in order to store up in learned pages materials which but a small number will be able to read. To all appearance these patient investigators waste their time and their labour. There is no public for them. They will be read by three or four people, sometimes only by him who reviews their work in some scientific periodical (98), or by him who shall take up the same kind of work, that is, if the latter care at all to know what his predecessors have done. And still monographs are after all the things that live longest. A book of generalities is generally outstripped in about ten years, a monograph being a fact in science, a stone laid in the edifice is in a sense everlasting from its results. People may neglect the name of the author, the book itself may be forgotten, but the results to which it has contributed remain. It is a sufficient reward to a whole life, if it has provided a few elements to the final creed, whatever transformations these elements may undergo. Henceforth that will be the true immortality (99).

One might cite a great number of researches which to posterity will be summed up in a few lines, which lines again will imply whole lives of patient application. The Greek Kingdoms of Bactriana and Pentapotamia have been for some years the object of researches which would already make several volumes and are far from being terminated. Is it at all likely that these studies with all their details will find a lasting place in the science of the future? Certainly not. And still they were necessary to show the character of the extent, of the importance and physiognomy of those advanced colonies of Greece; without these laborious researches we should have remained in the dark with regard to one of the most curious aspects of the history of Hellenism in the East. Those results attained, the works that have been instrumental in attaining them may disappear without much inconvenience to any one, like the scaffolding when the building is finished. And even supposing that the details remain necessary for the more intelligent understanding of the general results, the means, the machinery, if I may be permitted to call it so, by which the Prinseps and the Lassens have deciphered that page of history will almost have lost its value or will at best be preserved like bas-reliefs on the pedestal of the obelisk which they were instrumental in raising. "The scholars of the nineteenth century have proved——" the text will run. And that will be all.

Science should be represented to our minds as a building of the ages which can only be raised by the accumulation of enormous masses. A whole life of assiduous labour will only be as an obscure and nameless stone in that gigantic fabric, nay it may be nothing more than an unnoticed stone hidden in the thickness of the walls. No matter, one has one's place in the temple, one has contributed to the strength of its heavy foundations (100). The authors of monographs cannot reasonably hope to see their work endure in their proper form; the results they have put into circulation will undergo modifications, they will be digested and thoroughly assimilated. But through all metamorphoses, they will have the honour of having furnished essential elements to the life of humanity. The glory of the first explorers consists in being outstripped and in giving their successors the means by which the latter can outstrip them. "But this glory is immense and should be the less contested by him who comes second, seeing that he himself will have no other merit in the opinion of those who will busy themselves with the same subject than the merit of having preceded them (101)."

The faculty of forgetting occupies a large place in the scientific education of the individual. A mass of special data more or less painfully acquired drop of themselves from the memory; but for all that we should be careful not to assume that they have been lost. For the intellectual culture which was the result of this travail, the progress accomplished by the mind through these studies remains, and these only are worth anything. It is the same in the education of humanity. The particular elements disappear, but the accomplished progress remains. There are algebraical problems for which it is necessary to employ unknown auxiliaries and to take very wide circuits. Do we regret, when the problem has been solved, the elimination of all this baggage in order to make room for a simple and final expression?

Therefore, the specialist-*savant*, far from deserting the true arena of humanity, is the one who labours most efficaciously to the progress of the intellect, seeing that he alone can provide us with the materials for its constructions. But his researches, I repeat, cannot have an aim in themselves, for they do not contribute to make the author more perfect, they are of no value until they are introduced into the grand current. We must admit that the specialist-*savants* themselves are mainly responsible for propagating strange misunderstandings on that point. Exclusively occupied with their own studies, they consider the rest useless and all those not engaged in the same researches are so many profane ones to them. In that way their speciality becomes a small world to them, in which they obstinately and superciliously shut themselves up. And still, if the special object to which a man devotes his whole life had to be taken as possessing an absolute value, every one ought to apply himself to the same object that is, to the most excellent of all. Among the ancient literatures a man should exclusively study Greek literature; among those of the East, Sanskrit literature, and he who should devote his time and labour to a mediocre literature would only be a mere blunderer. Each of these studies has only its value in view of its place in the great whole, and of its connection with the science of the human intellect. The Oriental studies, for instance, are subdivided into three or four principal branches, to each of which a small number of *savants* devote themselves exclusively; so that the researches relating to literatures which are not the object of their studies possess no interest to them. The result is that he who writes a special work on Chinese, Persian, or Thibetic literatures may hope to have about a dozen readers in Europe. And even these, being engrossed, on their side, with their own special labours, have no time to trouble much about those of others and only glance at them superficially, so that in those studies, every one works for himself alone. A strange reversal of things. Are we to infer from this that it would be desirable for every Orientalist to apply himself to all the languages of Asia? Certainly not. But what would be desirable is that the savants who are most specialistic should have the true and inmost consciousness of their labours, and that the philosophical intellects should not disdain to apply to scholarship for the material of thought. For, I repeat, if the monographer alone reads his monograph, what is the use of writing it. It would be too odd if science had no other aim than to supply food for the curiosity of this or that man. Besides, the various sciences have problems in common with, and analogous to, one another as far as form goes which are often more easy of solution by means of one science than by that of another. I am convinced, for instance, that for the undoubtedly philosophical problem of the classification and the reality of the species the naturalists would derive great light from the study of the method of the linguists and of the natural characters that help them to constitute the families and groups according to the imperceptible degradation of the grammatical processes. Let the *savants* look to it; there is a savour of vanity in that mania of condemning any and every study not made at first hand as being of doubtful alloy. Such a system, carried to extremes would lead to the shutting up of every one within himself and to the destruction of all intellectual and scientific commerce. What would be the use of monographs if for every subsequent work we were obliged to recommence. This defect proceeds from still another vanity of the *savants*, which in its turn is closely allied to the spirit of superficiality of which they have so righteous a horror, namely, the vanity of writing books not for the sake of being read, but for the purpose of proving their learning.

It cannot be repeated too often; the true scientific work is the work at first hand. As a rule results only preserve their absolute purity in the writings of him who first discovered them. It is difficult to point out how scientific matters, in passing from hand to hand, and deviating from their primary source become altered and warped, and this without the least illwill of those who borrow them. This or that fact is looked at in a light somewhat different from that in which it was first observed; a reflection is added which the author of the original work would not have made, but which he who adds it thinks himself justified in making. A general observation is put forth which the primary investigator would not have formulated for himself in the same manner. A writer at third hand will improve in that way on his predecessor, and thus, unless it reverts constantly to its sources historical science is always inexact and open to suspicion.

The knowledge possessed by the Middle Ages of classical antiquity affords the most striking instance of those imperceptible modifications of primary facts that lead to the strangest errors or to the most absurd way of representing facts. The Middle Ages knew a great many things appertaining to Greek antiquity, but it knew nothing, absolutely nothing at first hand (102), hence the most incredible errors. The mediæval writers think that they can combine in their own way the scattered and incomplete notions they possess and in that way simply multiply inexactness which at the end of three or four centuries becomes such that when in the fourteenth century, the true Greek antiquity came to be directly known, it seemed to be the revelation of another world. The Latin encyclopædists, Martianus Capella, Boethius, Isidore of Seville only compile school books and string the traditional data together. Bede and Alcuin know even less of antiquity than Martianus Capella and Isidore. Vincent of Beauvais has left the truth still further behind him. At last in the fourteenth century (outside Italy), inexactness reaches its furthest limits, the Greek civilization is no better known than India would be, if, to reconstruct the Indian world we only had the notions left to us by the writers of classical antiquity.

Several parts of literary history which as yet, have not been sufficiently endued with life by the direct study of sources afford instances of inexactness with which those committed in the Middle Ages will bear comparison. Brucker is no doubt a scrupulous investigator, but still the books he has devoted to the philosophy of the Indians, of the Chinese or even of the Arabs must be placed on the same rank as the chapter dealing with ancient history in the "*Speculum Historiale*" of Vincent of Beauvais. What then shall we say of those who came after him and who have only copied or arbitrarily extracted from him without the least feeling as to what is essential and what accessory? When people are certain that the materials they possess are the only ones extant they may allow themselves to indulge in that kind of ingenious inlaid work in which are grouped all the spangles at their disposal, on the condition, however, of making reservations and acknowledging their inability to determine the mutual relations of parts, the proportions of the whole. But when there are original sources only waiting to be explored, there is something grotesque in this dovetailing of scattered scraps, inexact, disconnected which are made into a system according to fancy and without the least notion of the manner in which the original producers proceeded. Hence the inevitable defect of all the histories of literature and philosophy composed outside the original sources as has been the case so long with regard to the Middle Ages, as is the case still with regard to the East. Those who compose those histories only copy the same errors and aggravate them by adding their own conjectures to them. Try to read in Tennemann, in Tiedemann, in Ritter the chapters treating of Arabian philosophy. You will find nothing more than you find in Brucker, that is, nothing more than approximatives. We must definitely banish from science those works at third and fourth hand, in which the same data are simply copied without an attempt at either completing or verifying them. Whosoever, in the actual condition of science, should undertake a complete history either of Arabian philosophy or medicine would literally waste his time and his labour for he would only repeat what is already known. Such a work cannot possibly be accomplished until eight or ten of the most laborious students gifted with the most special knowledge have devoted their whole lives to the publication, translation and analysis of all the Arabian authors of which we possess the text or the rabbinical versions. Until then all the general works in that direction will be utterly baseless. And it is very probable that even then nothing very marvellous will result from all this, because I have no great faith in the Arabian philosophy, but if the result were a mere atom towards the history of the human intellect a thousand well spent lives would not be too much to pay for it. In the actual condition of science we may feel a pang of regret that distinguished intellects should devote their time and labour to objects in appearance so undeserving of their concern. But if science were as it should be, namely, cultivated by great masses of individuals and practically worked in large scientific workshops, the least interesting points would like the others be cleared up. In the actual condition one may aver that there are useless researches in the sense that they take up time that would be better spent on more serious subjects. But in the normal state in which so much power, now spent upon perfectly futile objects, would be devoted to serious things there would be no ground for despising any kind of work. For the perfect science of the whole will be impossible of attainment save by the patient and analytical exploration of the parts. Certain philologists have devoted long dissertations to the particles of the Greek language, others of the Renaissance have written works on the conjunction *quanquam*; a grammarian of Alexandria has written a book on the difference between χρὴ and δεῖ. No doubt they might have set themselves more important problems; still it would be rash to assert that such works are useless. For they do something for the knowledge of classical languages and classical languages do something for the philosophy of the human intellect. In the same way the knowledge of Sanskrit will not be perfectly grasped until plodding philologists shall have composed monographs about every part and every process of it. There exists a somewhat bulky volume of Bynæus; "*De Calecis Hebræorum.*" It is a pity, no doubt that the shoes of the Hebrews should have found a monographer before the Vedas found a publisher. Nevertheless, I am convinced that this book which I mean to read one of these days contains some valuable information and should make a useful appendix to the works of Braun, Schrœder and Hartmann on the dress of the Hebrew high priest and women. Pliny's saying is literally true. "There is no book, however bad, from which you may not learn something." It is rash to exclude no matter what, to condemn beforehand no matter what research as fruitless. What priceless results have sprung from studies apparently frivolous? Is it not the progress of grammar which has contributed to the more perfect interpretation of texts and through that to the more intelligent understanding of the ancient world? The most important questions of Biblical exegesis in particular to which the philosopher cannot remain indifferent depend as a rule on the most humble and most minute grammatical discussions (103). Nowhere has the process of making grammar and lexicography more perfect wrought a more radical reform. There are numerous other cases in which the most vital questions with regard to the human intellect depend on the most minute philological details. Far, therefore, from being the mere work of minds but little given to philosophy special labours are the most important to true science and betoken the best minds. Who is there that could write more scholarly generalities on Indian literature than M. Eugène Burnouf? Well, he only does so reluctantly and casually, because he justly considers as the more essential and most urgent work the publication of texts, the philological discussion of the same. In his preface to the "Bhagavata-Purana," M. Eugène Burnouf while apologizing to the *savants* for having given some general views, protests that he has only done so for the *French reader* and that he attaches but a secondary importance to a work which will have to be done later on, and which as it could be done to-day, would necessarily be outstripped and rendered useless by its follower. Are we to take this as the humility of the mind, or as the love for humble things for their own sake? No, it is simply the outcome of a healthy method and an upright judgment. In the actual condition of Sanskrit literature the publication and translation of texts is of more value than any possible number of dissertations whether on the history of India or on the genuineness and completeness of the works. Superficial minds may be inclined to think that it would be more meritorious and glorious on the part of a lofty intellect to write for instance a literary history of India than to devote itself to the unthankful task of editing and translating texts. This is a mistake. There is, as yet, no occasion to write dissertations on a literature all the elements of which are not to hand. Petrarch, Boccaccio, Poggio might as well have endeavoured to write the theory of Greek literature. Petrarch and Boccaccio by making Homer known; Ambrogio Traversari by translating Diogenes Laertius, Poggio by discovering Quintilian and translating Xenophon; Aurispa by bringing to the West the manuscripts of Plotinus, of Proclus, of Diodorus Siculus; Lorenzo Valla by translating Herodotus and Thucydides have rendered greater services to the classical literatures than if they had prematurely endeavoured to grapple with the higher questions of history and criticism. These early humanists, no doubt, fell fatally into literary superstitions and errors of criticism, which we, with our minds sharpened by the comparisons of other literatures, are able to avoid. We at the very outset are capable of performing on those almost unknown literatures tricks of skill which were only possible with regard to Greek and Latin literatures after the lapse of two or three centuries. The first students of Manu or the "*Mahabharata*" were enabled to discover things which it took three or four hundred years to discover in Homer or Moses. Nevertheless we are bound to maintain "that the period of dissertations and memoirs on India has not yet arrived, or rather that is already past and that the labours of a Colebrooke, of a Wilson, of a Schlegel and a Lassen have for a long while barred the career so brilliantly opened by the talent of Sir William Jones (104)." In fact, to write the history of India will only become possible after two centuries of labour such as the sixteenth and seventeenth centuries devoted to the classical literatures. Works of the latter order are the only ones which, in the actual condition of science, have a real and lasting value. Still, as it is true enough that an incomplete system, provided we do not cling to it in too narrow a spirit, is better than no system at all, one could wish that some one without professing to write a definitely scientific work, should attempt a kind of manual or introduction to that literature, based upon the actual state of Sanskrit studies. I confess that the greatest obstacle I have met with in engaging upon Indian studies is the utter absence of a summary work on Sanskrit literature, its development, its principal periods, the various ages of the language, the place and rank occupied by the various works, in short, something analogous to what Gesenius has done for the language and literature of the Jews. Such a work truly would be obsolete after the lapse of ten years; but it would have had its use and contributed to facilitate the direct study of the sources. It would assuredly be a pity for an eminent man to spend upon it moments that might be better employed in making that very work useless, and still who could do it if not the man who has a thorough survey of the ground already travelled over?

It is no doubt a drawback that those who devote their lives to works of special learning should not be imbued with the grand spirit that alone can impart life to these labours, but which very often does more harm to the moral perfection of the authors than to the work itself. Perfection would mean grappling closely with the particle while keeping at the same time to the grand centre from a force of constant habit that should penetrate the whole of the scientific life. For we may truly ask where the difference lies between certain learned researches, certain collections made by intellects too feeble to set themselves an aim and the work of the mere collector who pins on his cardboard sheets butterflies of any and every colour? Oh! when life is as short as it is and when so many serious matters crop up into it, would it not be better to listen to the numberless promptings of the heart and the imagination and to taste the delicious joys of the religious sentiment than to fritter away a life that is irretrievable, and which when lost, is lost for ever and ever?

The great obstacle which checks the progress of philological studies seems to me to lie in that dispersion of work, in that self-isolation among special studies, which render the labours of the philologist only available to himself and to a small number of friends who are engaged upon the same subject with him. Each *savant* developing in that way his part without the least concern for the other branches of science becomes narrow, egotistical and lost to the lofty sense of his mission. A whole life would scarcely suffice to exhaust the researches to be undertaken on this or that special point of a science which itself is but the infinitesimal part of a much more extensive science. The same researches are constantly begun over and over again, monographs increase and accumulate to a degree such that their very number annuls and renders them useless. It seems to me that the time will come when philological students will collect and examine all those scattered labours, and when the results are finally attained the henceforth useless monographs will only be preserved as so many mementos. When the edifice has been raised there is no harm in taking away the scaffolding that was necessary to its construction. That is how physical sciences proceed. The works approved of by competent authorities are done once for all and henceforth adopted in full confidence without the self-imposed attempt of reverting to the researches of the first experimentalists except in rare cases and at long intervals. It is thus that whole years of assiduous studies have been often summarized in a few lines or in a few figures and that the vast whole of the science of nature has been composed piece by piece and with admirable joint responsibility on the part of all the labourers. The much more delicate nature of the philological sciences would no doubt not admit of the rigorous application of a similar method. Still, I am under the impression that we shall not get out of this maze of individual and isolated work save by a great scientific organization, in which everything shall be done without stint as without waste of power and in so final a manner as to render the results attained capable of acceptation in full confidence. One cannot help thinking now and again that the mass of scientific work is crushed by its own weight and that it would be for the better if publicity were more restricted. But the real defect is the want of organization and control. In an efficiently ordered scientific state it would be desirable that the number of workers were still much more considerable. For then the work would not go a-burrowing and would not choke itself, like a fire of which the fuel is too closely piled up. It is sad to think that three-fourths of the minute things which are still being sought for are already found, while other mines in which treasures are lying awaiting discovery remain without hands in consequence of the inefficient direction of the work. Science in our days is not unlike a magnificent library turned upside down. It contains everything, but so utterly pell-mell, so thoroughly unclassified that it might just as well not be there.

A moment's reflection will convince us of the absolute necessity of supposing the future to have a grand scientific reform in store for us (105). In fact, the material claiming the attention of the scholar keeps on increasing at so rapid a rate, either by reason of fresh discoveries or by the multiplying of centuries as to exceed in the long run the capacities of investigators. A hundred years hence France will count three or four literatures virtually piled atop of one another. In five hundred years there will be two ancient histories. And, if the elucidation of the first which the times and the utter absence of the printing press invested with such great simplicity to us, sufficed to occupy so many laborious lives, what will it be with ours which will have to be extracted from such an enormous mass of documents? The same argument holds good with regard to our libraries. If the National Library continues to accumulate all the new productions, it will become absolutely impracticable in another hundred years and its very wealth will make it useless (106). Hence, there is a progressive march which cannot continue indefinitely without producing a revolution in science. It would be foolish to inquire how that revolution will be brought about. Will there be an immense simplification like that wrought by the barbarians? We cannot risk the slightest reasonable hypothesis on the subject.

Without being a partisan of literary and scientific communism, I believe, nevertheless that the dispersion of forces should be urgently opposed, and the concentration of work as urgently called for. Germany has in that respect some really useful customs. It is by no means rare to find in the literary journals or in the reports of philological conferences the notice of a *savant* informing his fellow-*savants* that he has undertaken a special work on a certain subject and consequently asking them to send him any and everything bearing upon the subject which may have come in their way in the course of their own particular studies. Without attempting to lay down hard and fast rules I have an idea that under a seriously organized system one might throw open in that way public problems to which every one should be welcome to contribute his contingent of facts. The Académies, especially the académies devoted to works that have much in common, such as for instance the Académie des Inscriptions et Belles-Lettres supply the want I am pointing out, but in order to supply it in a thoroughly efficacious manner, they would have to undergo radical transformations.

CHAPTER XIV.

The plan of my work does not admit of my suggesting any ideas of practical application. My complete ignorance of practical life would, moreover, make me utterly unfit to do so. Organization, demanding experience and the careful balancing of principles by existing facts could not possibly be the work of a young man. I shall, therefore, only lay down principles.

It cannot be gainsaid that it is the duty of the State to patronize science as it patronizes art. The State, in fact, represents society and should replace individual enterprise with regard to works where isolated effort would be insufficient. The aim of society is the realization on a vast and complete scale of all the facets of human life. As it happens some of these facets can only be realized by collective wealth. Individuals cannot build unto themselves observatories; they cannot create libraries, they cannot found large scientific institutions. The State, therefore, *owes* to science observatories, libraries, scientific institutions. Individuals cannot by themselves undertake and publish certain works. The State *owes* them subsidies. Certain branches of science (and the most important) cannot provide those that cultivate them with the necessaries of life. The State is bound to afford in some shape or other to deserving workers the necessary means to pursue peaceably their labours and to keep them from harassing want. I say that this is *a duty* of the State, and I say so unreservedly (107). I do not look upon the State as a simple institution of police and for preserving order. The State is society itself, that is, man in his normal condition. It is consequently, subject to the same duties as the individual, as regards religious things. It must not merely let things take their course, it is bound to provide man with the conditions of his striving after perfection. It is a plastic and really guiding power. For society is not merely the atomic union of individuals, wrought by the repetition of the unit; it is the constituted unit, it is primeval.

I am aware that England, like France of old in some respects provides for nearly everything by private foundations, and I can understand that in a country where private foundations are so respected one may dispense with a minster of Public Education. The State, I repeat, should only step in where the individual cannot or does not suffice, hence its part is less important in a country where private individuals can do and do a great deal. Besides, England only realizes those grand things by association, that is, by small societies within the large one, and personally I consider the French organization, which sprang from our Revolution much more in accordance with the modern spirit. It is above all in the shape of religion that the watchfulness of the State over the supra-sensitive interests of humanity has hitherto been exercised. But as soon as the religious tendency of man shall manifest itself in the purely scientific and rational form all that the State formerly granted to the budget of Public Worship will by right revert to science, the sole definite religion. There will be no budget of Public Worship, there will be a budget of science, a budget of arts. The State must provide for science as it does for religion, seeing that science like religion is an essential part of human nature. Science has even a higher claim on the State, because religion though everlasting in its psychological basis, is more or less transitory in its form; unlike science it is not wholly an essential part of human nature.

Seeing that science can only exist under the condition of the most perfect freedom, the patronage which the State owes it, does not confer upon the State the least *right* to control or to regulate it, any more than the subsidy granted to public worship gives it the right to frame articles of faith. In one sense the State can exercise even less influence on science than on religions; because it can at least impose certain regulations of police on the latter; while it can impose nothing, absolutely nothing on science. Science, in fact, proceeding through the intrinsic and objective consideration of things is not itself free to yield obedience to him who would command it; if it were free in its opinions, one might perhaps ask it for this or that opinion. But it is not free; there is nothing more fatally stubborn than reason and consequently, science. To attempt to direct it, to ask it to attain this or that result is merely a flagrant contradiction; it is acting upon the supposition that it is pliable to every sensation, upon the supposition that it is not science.

Certain religious orders which applied to study that tranquillity of mind—one of the most delightful fruits of monastic life—realized of old those *grand scientific workshops* the disappearance of which is so deeply to be regretted. No doubt it would have been much better had those workers been independent (108). They would not have brought to their labours so much patience and abnegation, but they certainly would have brought a keener spirit of criticism. Be this as it may, it cannot be denied that the abrogation of the religious orders which devoted themselves to study and of the "parliaments" which afforded scholars so much studious leisure have struck a fatal blow to learned researches. This gap will not be filled up until the State shall have instituted, in some form or other, *lay chapters, lay benefices,* where the great labours of learning will be resumed by profane and critical Benedictines. By the side of the learned labour of the architect there is in science the drudgery of the hodman which requires unnoticed patience and united labour. Dom Mabillon, Dom Ruinard, Dom Rivet, Montfauçon could not have accomplished their gigantic tasks if they had not had at command a whole community of laborious workers who "fined down" the work to which they afterwards put the finishing touch. Science will make no rapid conquests until lay Benedictines will harness themselves once more to the yoke of learned research and devote lives of labour to the elucidation of the past. Glory will not be the reward of those humble workers; but life has many gentle and tranquil natures, but little exercised by passions and desires, not harassed by philosophical needs (take care not to infer from this that they are cold and withered; on the contrary, they often possess intense faculties of concentration and are exceedingly sensitive) who would be content with this peaceful life and who, amidst modest comforts and happy family surroundings, would exactly find the atmosphere suited to such modest work. Truth to tell, the most natural form of patronizing science in that way is the form of sinecures. Sinecures are indispensable in the scientific world; they are the most dignified and the fittest way of pensioning the *savant;* besides affording the advantage of grouping around scientific institutions illustrious names and eminent talents. Only barbarians and people "who see no farther than their nose" will allow themselves to be taken in by superficial objections like those raised at the first blush by the employment of extensive scientific staffs. It is very evident that the work of this or that library which numbers ten or twelve assistants would be done just as well by two or three people (and in fact among the number there are only two or three who do anything worth mentioning). Certain people would conclude from this that all the others ought to be dismissed. No doubt, if the sole aim were to provide for the material wants of the work. An odd thing. Science the one truly liberal thing in this world is only largely patronized in Russia.

It is a pity, certainly that we should have to descend to such considerations. But in the actual condition of humanity money is an intellectual force and in virtue of that deserves some consideration. A million is as good as one or two men of genius, in the sense that with a million well spent one may do as much for intellectual progress as one or two first-rate men could do, if reduced to intellectual force only. With a million I will undertake to drive modern ideas deeper into the mind of the masses than it could be done by a whole generation of poor thinkers, commanding no influence. With a million I could translate the Talmud, publish the Vedas, the Nyaya with its commentaries and bring out a number of works which would contribute more to the progress of science than a whole century of metaphysical thought. How frantic it drives one to think that with the sums silly opulence scatters about to satisfy mere whims, one might move heaven and earth. It is idle to expect the *savant* to emancipate himself from the conditions of ordinary life and to do without the everyday food. It is still more idle to hope that the wealthy who are exempt from such cares, will ever suffice for the needs of science. The grand scientific instincts nearly always find their development among well educated but poor young men. The rich always import into science a tone of superficial amateurishness of very doubtful alloy (109). Religion has never been blamed for having ministers subject to material wants like other men and claiming State aid. As for those who look upon science as a mere money-making machine, we will have nothing to say to them, they are manufacturers like a good many other manufacturers, they are not *savants*. Whosoever has been able to dwell for a single moment on the hope of becoming rich, whosoever has looked upon the outward needs in any other light than a heavy and fatal chain to which unfortunately he must submit, does not deserve the name of *savant*. Large scientific stipends and plurality of appointments would in that respect have the same serious drawbacks that wealth had formerly on the clergy; they would attract mercenary characters who look upon science as a means as good as any other to make their fortune, disgraceful simoniacs who import into sacred things their grovelling habits and their worldly views. Students should feel sure beforehand that in embracing the scientific career they condemned themselves to lifelong poverty, though they should not want for the strictly necessary; in that way there would be none but noble natures driven by a powerful and irresistible instinct who would devote themselves to it, and the scum of intriguers and adventurers would carry their pretensions elsewhere. The first condition has been already fulfilled? Why is it not the same with regard to the second?

CHAPTER XV.

In order to complete my idea and to make clearly understood what I mean by a scientific philosophy I am bound to give some instances here, which, it seems to me, will make it evident that special studies can lead to results as important to the thorough knowledge of things as metaphysical and psychological speculation. I will borrow them by preference from historical or philological sciences, the only ones with which I am familiar and with which, besides, this essay specially deals. It is not because the natural sciences do not afford data quite as philosophical. I am not afraid of exaggerating when I say that our most settled ideas on the system of things are rooted more or less deeply in the physical sciences and that the most important differences that distinguish modern from antique thought are due to the revolution those studies have wrought in our manner of viewing the world. Our idea of the *laws* of nature which has upset for ever and aye the old conception of the anthromorphic world is the grand result of physical sciences, not of this or that experiment but of a very general mode of induction, the result of the general physiognomy of phenomena. It cannot be gainsaid that astronomy in revealing to man the structure of the universe, the rank and position of the earth, the order it occupies in the system of the world has done more for true science than all the conceivable speculations based upon the exclusive consideration of human nature (110). This consideration, in fact, would lead either to the ancient finalism, which made man the centre of the universe or to pure Hegelism which admits no other manifestation of the divine conscience than humanity. But the study of the world's system and of the place man occupies in it, without upsetting either of these two conceptions forbids us to take them in a too absolute or exclusive manner. The idea of the infinite is one of the most fundamental in human nature, —if it be not the whole of human nature itself; and still man would not have succeeded in understanding in its reality the infinitude of things, if the experimental study of things had not brought him to it. Of course it is not the telescope that has revealed the infinite to him, but it is the telescope that has taken him to the extreme limits beyond which there is still an infinitude of worlds. Has not geology introduced as essential an element into philosophy by teaching man the history of our globe, the period at which humanity first appeared, the conditions of that appearance and the creations which preceded him? Physics and chemistry have done more for the inmost constitution of the body than all the speculations of the ancient and modern philosophers on the abstract quality of matter, its essence, its divisible properties. Physiology and comparative anatomy, zoology, botany are in my opinion, the sciences that teach the greatest number of things on the essence of life, and it is from them that I have drawn the greatest number of elements for my manner of regarding the individuality and the mode of consciousness resulting from the organism. Mathematics themselves, though affording no lesson on reality are precious in moulding thought, and offer us, in the way of pure reason in action, the model of the most perfect logic. But I wish to insist no longer on things of which I have no special knowledge and come back to my fundamental idea of a *critical philosophy*.

To my idea the highest degree of intellectual culture is to *understand humanity*. The physical student understands nature, no doubt not in all its phenomena, but at any rate in its general laws, in its true physiognomy. The physical student is the critic of nature, the philosopher is the critic of humanity. Where the ordinary observer only sees whim and miracle, the physicist and the philosopher see laws and the manifestation of reason. And this true intuition with regard to humanity, which is after all only criticism, the historical and philological sciences alone are able to give it. The first step in the science of humanity is to distinguish two phases in human thought; the primeval age, an age of spontaneity in which the faculties of their creative fruitfulness, unconscious of themselves, as it were and by their inmost tension attained an object at which they did not aim; and the age of reflection in which man becomes conscious and master of himself, an age of combination and painful process, of antithetical and much-debated knowledge. One of the services rendered to philosophy by M. Cousin has been to introduce among us this distinction and to demonstrate it with his admirable clearness of mind. But it will be the mission of science to demonstrate it finally and to apply it to the solution of the highest problems. Primitive history, the epics and poetry of the spontaneous ages, their religions, their languages will have no meaning until this grand distinction shall have become current coin. The enormous errors of criticism generally to be found in the essays on the works of the primitive epochs arise from the ignorance of this principle and from the habit of judging all the epochs of the human intellect by the same standard. Let us take for instance, the origin of language. Why do people advance such absurd arguments on that important philosophical question? Because they apply to the primitive epochs views that have a meaning only in our age of reflection. People say; "Seeing that the greatest philosophers are powerless to analyze language how could the first human beings have created it?" The objection holds good only with regard to deliberate invention. Spontaneous action has no need to be preceded by the analytical view. The mechanism of the intellect is much more difficult still to analyze, and still while knowing nothing of such analysis the simplest man sets in motion all its springs. The fact is that the words *easy* and *difficult* have no meaning whatsoever when applied to spontaneous action. The child that learns its language, humanity that creates science experience no more difficulty than the plant that germinates, than the organized body that arrives at its complete development. It is the hidden God, the universal force acting everywhere, producing either during sleep or in the absence of the individual soul, those marvellous effects, which are as much above human skill, as the infinite power is beyond limited strength.

It is because this creative force of spontaneous reason has not been understood that the strangest hypotheses on the origin of the human intellect have been allowed to gain ground. When the Catholic Condillac, M. de Bonald, conceives primeval man after the model of a powerless statue, devoid of originality or initiation on which God "enamels" or "lays on" if I may be permitted the expression both language, moral feeling and thought, he only continues the reasoning of the eighteenth century and denies the innate originality of the intellect; (irrespective of the absurdity of making a stump or stock utterly deprived of intelligence, speak and understand by simply speaking to it, and as if such a revelation did not imply the inherent faculty to understand; as if the receptive faculty were not correlative with the productive one). It is as untrue to say that man deliberately and with forethought created language, religion, morality as it is to say that those divine attributes of his nature have been revealed to him. All this is the work of spontaneous reason and of that hidden and inward activity which while concealing from us the motor power only shows us the effects. When we get as far down as this, it matters little whether we attribute the primary cause to God or to man, the spontaneous being both divine and human at the same time. There lies the point of conciliation between apparently contradictory opinions, but which are in fact only partial in their expression, according to their connection with one aspect of the phenomenon rather than with the other.

The false reasoning on the history of religions and their origins springs from the same cause. The great religious apparitions present a mass of inexplicable facts to him who fails to look for the cause beyond ordinary experience. The formation of the legend of Jesus and all the primitive facts of Christianity would be incapable of being explained in surroundings like ours. Let those who make themselves a narrow and paltry idea of the laws of the human intellect, who understand nothing beyond the commonplace of a drawing-room or the restricted limits of ordinary common sense; let those who have not grasped the proud originality of the spontaneous creations of human nature, let all those beware of grappling with such a problem or let them content themselves with timidly casting the solution of the supernatural upon it. In order to understand those extraordinary apparitions we must be hardened against miracles; we must lift ourselves above our age of reflection and slow combination to be able to contemplate the human faculties in their creative originality; when spurning our painful processes, they evolved from their plenitude the sublime and the divine. Then was the age of psychological miracles. To have recourse to the supposition of the supernatural in order to explain those marvellous effects is simply to insult human nature; it is to admit one's ignorance of the hidden forces of the soul; it is to imitate the commonplace man who looks upon *extraordinary* effects of which science explains the mystery as so many miracles. In every order of things, the miracle is only apparent, the miracle means that which has not been explained. The further we penetrate into the higher psychology of primordial humanity, the deeper we pierce the origins of the human intellect the more miracles we are likely to find, miracles the more admirable in that they do not require for their production a "God-machine," who is always meddling with the course of things, the more shall we become aware that they are the regular development of immutable laws like reason itself and the attainment of the perfect. The spontaneous man looks upon nature and history with the eyes of childhood; the child casts upon everything the halo of the marvellous which he finds in his own soul. His curiosity, the lively interest he shows in every new combination spring from his faith in the marvellous. *Blasés* with experiments we do not expect anything very wonderful; but the child does not know what will come next. Knowing the reality less, he believes more in the possible. That delightful intoxication of life which is contained within himself makes his head spin; he only sees the world through a delicately tinted mist; casting a joyous and inquisitive look at all things, he smiles at everything, everything smiles on him. Hence spring his joys but also his terrors; he makes unto himself a fantastic world that delights or frightens him; he has not the faculty of distinguishing, which in the age of reflection so clearly divides the *ego* and the *non-ego;* and makes us such coldblooded observers of the reality. He makes himself part and parcel of all his stories; the simple and objective relation of the fact is impossible with him; he does not know how to isolate himself from the judgment he has delivered upon it, from the personal impression of it that has remained with him. *He does not relate things, he relates the fancies he has conceived in connection with things, or rather he relates himself.* The child in his turn creates for himself all the myths that humanity created for itself; every fable that strikes his imagination is accepted by him; he himself improvises strange ones, and then affirms their truth to himself (111). Such is the process of the human intellect at the mythical epochs. The dream is taken for the reality and affirmed as such. Without the least mendacious premeditation, the fable is born of itself; accepted the moment it is born, it goes on gathering like the snowball; there is no criticism to stop it. And it is not only at the origin of the human intellect that the soul allows herself to be tricked by this delightful deception; the love of the marvellous goes on bearing ample fruit until the final advent of the scientific age, but less spontaneously and accompanied by the assimilation of many more historical elements.

Here, then, we have a principle capable of becoming the basis of a complete philosophy of the human intellect, and around which are grouped the most important results of modern criticism. Chronology counts for very little in the history of mankind. A combination of causes may once more obscure thought and revive the instincts of the primeval days. That is what did occur on the eve of modern times, and subsequent to the grand civilization of antiquity, when the Middle Ages recalled the Homeric ones, and that of the childhood of humanity. The theory of the primitive state of the human intellect, so indispensable to the knowledge of the human intellect itself, is our great discovery and has introduced thoroughly new data into philosophical science. The old Cartesian school took man in an abstract, general, uniform manner. It constructed the history of the individual as some Germans still construct the history of humanity, *a priori*, and without troubling itself about gradations which facts alone can reveal. And when I say history, I only do so for form's sake, for there was no history for that creature without any connection with its kind, who like the angels, saw everything in God. And when he had asked himself whether his faculty to think is constant, whether his senses deceived him, whether matter exists, whether the animals have a soul, everything was said and done. And what in the name of all that is sensible could they know of the living, breathing man, those long-robed and stern personages of the parliaments, of Port-Royal, of the Oratoire, the personages that cut man into two parts, the *body*, the *soul*, without any space or passage between them, and by this mere arrangement prohibited themselves from studying life in its perfect simplicity (112)? There are strange things told of the want of sensibility and sternness of Malebranche, and this could not be otherwise. It is not in the abstract world of pure reason that people beget a sympathy with life, all that touches and moves us belongs always more or less to the body. As for us, we have shifted the field of the science of man. It is his life we want to know, and life means both the body and the soul, not placed facing one another like clocks that tick in time, not soldered together like two different metals, but united into one two-fronted phenomenon which cannot be divided, without destroying it.

Our science of man, then, is no longer an abstract thing that may be built up *a priori* and from general views; it is the universal experimental method applied to human life, and consequently the study of all the products within the sphere of its activity, above all of its spontaneous activity. I prefer to the most beautiful Cartesian disquisitions the theory of primitive poesy and the national *epic* as Wolf viewed them, as they have been definitely settled by the comparative study of literatures. If aught can make us understand the aim of criticism and the importance of the discoveries we may expect from it, it is assuredly the fact of having explained by the same laws Homer and the Ramayana, the Niebelungen and the Shah-Nameh, the romance of the Cid, our own mediæval legendary poems (*Chansons de Gestes*) the heroic songs of Scotland and Scandinavia (113). There are traits of humanity capable of being fixed once for all, and for which the most ancient pictures are the best. Homer, the Bible, and the Vedas will last for ever. They will be read when the intermediary works shall have been forgotten; they will ever be the sacred books of humanity. In fact, to the two phases of human thought there are two corresponding literatures; primitive literatures, the ingenuous outpourings of the peoples in their still spontaneous condition, rustic but withal natural flowers, direct manifestations of the national genius and traditions; literatures that are the outcome of reflected thought, possessing much greater individuality than the others, and with regard to which questions of authenticity and integrity—impertinent in connection with primitive literatures—are of the greatest significance. Thus, two poems, like the "Iliad" and the "Æneid" placed side by side of old, are placed at the two poles of the idea.

The general theory of the mythologies as established by Heyne, Niebuhr, Ottfried Müller, Bauer, Strauss, comes within the same order of researches and implies the same principle. The mythologies are no longer to us a series of absurd and often ridiculous fables, but grandiose, divine poems, in which the primeval nations have poured out their dreams with regard to the supra-sensitive world. They are, in a certain sense, more valuable than history, for in history there is a necessary and fortuitous part which is not the work of humanity; while in the fables everything is its own; it is its portrait painted by its own hand. Fable is unshackled, history is not. "The Book of Kings" of Firdousi is decidedly a very inferior *history* of Persia; and still that lovely poem represents to us the genius of Persia much better than would the most exact history; it gives us her legends and epic traditions, that is, her soul. Scholars often regret deeply that India has left us no history of any sort. But in reality we have something better than her history;

we have her sacred books, her philosophy. That history would be no doubt, like all other histories of the East, a mere dry nomenclature of her kings, a series of insignificant facts. Is it not better to be in direct possession of that which at great pains we should have to extract from the history, that which solely constitutes its value; the spirit of the nation?

The races that are most philosophical are also the most mythological. India presents the most astonishing phenomenon of the richest mythology side by side with a metaphysical development much superior to that of Greece; perhaps even to that of Germany. The three characteristics that distinguish the Indo-Germanic peoples from the Semitic peoples are, that the Semitic peoples have *neither philosophy,—nor mythology,—nor epic* (114); three things in reality very closely connected and due to an entirely different mode of looking at the world. The Semites never conceived "sex" in God; the feminine of the word "God" in Hebrew would be the strangest barbarism possible (115). Because of this they have deprived themselves of the possibility of "making" either a mythology or a divine epic; a variety of complications being out of the question under an only and absolutely ruling God. Under such a *régime* the struggle becomes impossible. The God of Job replying to man with nothing but thunder-claps, is very poetical, but in noway epic. He is too strong, he crushes at the first blow. The angels do not offer any individual variety; and all the subsequent efforts to invest them with a kind of physiognomy (archangels, seraphim, etc.) have led to nothing characteristic. And besides what interest could we take in *messengers*, in *ministers* without initiative, without passion? Under the *régime* of Jehovah mythological creation could only result in depicting the executors of his orders. Hence the rôles assumed by the angels are as a rule cold and monotonous, like those of the messengers and confidants.* Variety is the element most radically wanting in the peoples of Semitic origin; their original poems would not make more than one volume. The themes are few in number and quickly exhausted. This God, isolated from nature; this nature made by God, do not lend themselves to the conception of incident and to historical composition. What an enormous distance indeed between that vast deification of the forces of nature which is the foundation of the great mythologies and the narrow conception of a world fashioned like a bowl in the potter's hands. And it is thither we strayed in search of our theology. Doubtless, this mode of conceiving things is simple and majestic, but how colourless in comparison to those grandiose evolutions of Pan which the Indo-Germanic race at its poetic beginnings as well as at its end, understood so well.

Among the secondary sciences which must aid in constituting the science of humanity, there is no one more important than the philosophical and comparative theory of languages. When we remember that this admirable science counts as yet but one generation of labour, and the precious discoveries to which it has already led, we cannot help wondering at its being so little cultivated, so little understood. Is it credible that there is not in the whole of Europe a single chair of philology, and that the Collége de France which boasts of representing in its *curriculum* the *ensemble* of the human intellect has no chair for one of the most important branches of human knowledge, created by the nineteenth century? What historical results may we not expect from the classification of languages into families, and above all from the formation of that group of which we are a part and the branches of which extend from the island of Ceylon to the deepest recesses of Brittany? What lights to be thrown on ethnography, on primi-

* To the French "messengers and confidants" is expressive enough. M. Rénan alludes to what in theatrical parlance we call "the feeders" of Corneille and Racine's tragedies.

tive history, on the origins of mankind! What philosophical results in the ascertaining of the laws that have presided at the development of language, at the transformation of its mechanism, at the perpetual decompositions and recompositions which constitute its history. Would the analytical progress of thought have been discovered if the various languages had not shown us as in mirror the ceaseless march of the human intellect from synthesis and from primitive complexity to analysis and clearness? Is it not the study of the primitive languages that has revealed to us the primitive characteristics of the exercise of the thinking faculties, the predominance of sensation and that deep sympathy which then bound man to nature? What synopsis of the human intellect, in short, can be compared to that afforded by the comparative study of the processes by which the various races have expressed the different nexus of thought? I do not know of a more beautiful chapter of psychology than the dissertations of von Humboldt on the dual number, on the adverbs of locality, or those that might be written on the comparison of Semitic and Indo-Germanic conjugations, on the general theory of pronouns, on the formation of roots, on the imperceptible deterioration and the rudimentary existence of the grammatical process in the various families, etc. What we cannot point out too often is the fact that through languages we get in touch with the primeval condition of man. Languages, in fact, do not create for themselves new processes, any more than they create new roots. Their whole progress consists in developing this or that process, in diverting the meaning of roots, but never in adding new ones. The populace and children alone enjoy the privilege of coining new words and turns of phrases that have no antecedents for their individual use. The thinking man never attempts to combine in an arbitrary way sounds to express a new idea, or to create a grammatical form to express a new nexus. We may conclude from this that all the roots of the various families are the outcome of the way the primitive peoples felt and that all the grammatical processes proceed directly from the manner in which each race treated ideas; that, in short, language in the whole of its construction dates from the first days of man and brings us in touch with his origin. Personally, I am convinced that the language spoken by the first thinking beings of the Semitic race differed very little from the common type of all those languages, such as it presents itself to us in the Hebrew or Syriac. There can, at any rate, be no doubt that the roots of these idioms, the roots that still constitute the foundation of a language spoken on a large part of the globe were the first that rang in the deep and vigorous chests of the fathers of that race. And though it may seem paradoxical to maintain the same thing with regard to our metaphysical languages, battered about by so many revolutions, one may fearlessly affirm that they do not contain a single word, a single process we may not connect by a direct affiliation to the first impressions of the first children of God. Let us, therefore, remember in Heaven's name what we have got in hand and let us labour to decipher that medal of days gone long ago.

As a rule, people imagine the laws of the evolution of the human intellect to be much more simple than they are. It is very dangerous indeed to invest with a historical and chronological value the evolutions supposed to have been necessarily successive, to suppose, for instance, that at his origin man was a cannibal because that condition is considered as the most degraded. The reality presents a very much greater variation. There is not a thinker who pondering the history of humanity does not succeed in constructing a formula; those formulas do not coincide with, but they are not contradictory of, one another. The fact is that there are no two absolutely identical developments in humanity (116). There are laws, and very deeply rooted laws, the simple action of which is never perceived, the result being always complicated by accidental circumstances. The general names by which we designate the various phases of the mind never apply in a perfectly *univocal* manner—as the schoolmen said—to two different states. "The line of humanity," says Herder, "is neither straight, nor uniform; it deviates in every direction, and presents all the curves and all the angles imaginable. Neither the asymptote, nor the ellipsis, nor the cycloid can give us an idea of its law." The relations between things are not on a plane, but in space. There are dimensions in thought as there are dimensions in the expanse. Just as a classification only explains one lineal series of beings, and necessarily neglects several as real which cross the first and would require a classification apart so do all the laws express only one system of relations and necessarily omit a thousand others. It is like a body of three dimensions projected on one plane. Certain traits will be preserved, others will be altered, others again altogether omitted. The Middle Ages are in certain aspects like the Homeric times, and yet who could care to apply the same denomination to conditions so different? In this vast picture every one lays hold of a trait, a physiognomy, a ray of light; no one grasps the *ensemble* and the significance of the whole. Let us take a traveller who has crossed France from north to south; another from east to west; a third following a different line; each of them gives his account as the complete description of France; that is the exact image of what up till now those have done who have attempted to present a system of the philosophy of history (117). A geographical map cannot possibly be drawn unless the country intended to be represented has been explored in all directions. And let us bear in mind that history is the true philosophy of the nineteenth century. Our century is not metaphysical. It cares little about the intrinsic discussion of questions. Its great concern is history, and above all the history of the human intellect. Here lies the dividing point of the schools; a man is a philosopher or a believer according to his manner of looking at history; a man believes in humanity or does not believe in it according to the system he has conceived of its history. If the history of the human mind be only a succession of systems that upset one another, all we can do is to throw ourselves into the arms of scepticism or into those of faith. If the history of the human intellect be the onward march towards the truth between two oscillations which have the effect of restricting more and more the domain of error, we are justified in still putting our trust in reason. Every one in our days is what he is according to the way in which he understands history.

The comparative study of religions when once definitely established on the solid basis of criticism will constitute the noblest chapter in the history of the human intellect, finding its place between the history of mythologies and the history of philosophies. Religion, like philosophy supplies a speculative want of humanity. Like mythology, it contains a large part of the records of spontaneous and non-premeditated exertion of the human faculties. Hence its priceless value from the philosopher's point of view. Just as a Gothic cathedral is the best piece of evidence of the Middle Ages, because the generations have dwelt there in the spirit; so is religion the best means of understanding humanity; for humanity has dwelt there; it is the deserted tent in which everything attests the traces of those whom it sheltered. Woe to him who passes by those venerable tenements with indifference, those venerable tenements in the shadow of which humanity has lingered so long and where so many noble souls still find comfort and awe. Even if the roof lets in the light of heaven and the torrents from the sky drench the upturned face of the believer on his knees, science would wish to study those ruins, to describe all the statuettes that adorn them, to lift the stained window panes which only admit a mysterious semi-glow, in order to introduce the radiant sun, and to study at leisure those admirable petrifactions of human thought.

The history of religions has, as yet, almost entirely to be created. Numberless causes of respect and timidity operate on that point against thorough frankness, without which rational discussion becomes impossible, and in reality, render the position of those grand systems more unfavourable than advantageous from the point of view of science. Religions seem to have been tabooed by humanity, it takes them a long while to obtain the recognition of their value, the value which is theirs from the standpoint of criticism; and the silence concerning them may breed an illusion as to the importance of the part they have enacted in the development of ideas. A history of philosophy (118), which should devote a volume to Plato, ought, it seems, to devote two to Jesus; and still the chances are that His name will not be mentioned once. It is not the historian's fault; it is the consequence of Jesus' position. Such is the fate of everything that attains religious consecration. How much, for instance, has not *Hebraic literature* suffered from the standpoint of science and taste by becoming the Bible? Whether from mere bad temper, or from the remains of superstition, scientific and literary criticism shrinks from considering as its own works which have been sequestrated in that way from profane and natural influences, by which we mean, that which is. Nevertheless are the books themselves to blame? Could the author of that delightful little poem called "The Song of Songs" have foreseen that one day he would have to part company with Anacreon and Hafiz to be made into an inspired singer who only sang of divine love? It is really time that criticism should become accustomed to take its material wherever it finds it, and not to make distinctions between the works of the human intellect when it becomes a question of leading opinion, of admiring. It is time for reason to cease to criticize religions as alien works, set up against it by a rival power and to finally recognize its own concern in all the productions of humanity without distinction or antithesis. It is time to proclaim the fact that one sole cause has wrought everything in the domain of intellect, the human mind, operating according to identical laws, but among different surroundings. To hear certain rationalists, we should be tempted to believe that religions came from heaven to confront reason for the pleasure of thwarting it; as if human nature had not done everything by different aspects of itself. No doubt, we may oppose religion and philosophy as we oppose two systems, but in recognizing that they have the same origin and occupy the same ground. The old method of polemics seemed to concede that religions have a different origin, and from this very fact was induced to insult it. By being bolder, we shall be more respectful.

The lofty serenity of science becomes possible only on the condition of impartial criticism, which without regard for the beliefs of a certain portion of humanity, handles its imperturbable instrument with the inflexibility of the geometrician, without anger and without pity. The critic never insults. When we shall have reached the point at which the history of Jesus shall be as open to discussion as the history of Buddha and of Mahomed, people will no longer dream of addressing harsh reproaches to those whom circumstances have deprived of the light of criticism. I am certain that M. Eugène Burnouf has never been angry with the authors of the fabulous life of Buddha, and that those among the Europeans who have written the life of Mahomed have never felt any violent spite against Abulféda and the Mussulman authors who have written the life of their prophet as true believers. The apologists maintain that it is religion which has wrought all the great things of humanity, and they are right. The philosophers believe that they are striving for the honour of philosophy by depreciating religion, and they are wrong. As for ourselves who advocate but one sole cause, the cause of the human intellect, our admiration is much more unfettered. We should fancy that we were wronging ourselves by withholding our admiration from anything wrought by the human intellect. We ought to criticize religions in the same way that we criticize primitive poems. Do we show any temper with Homer or Valmiki, because their manner is not that of our own epoch?

Heaven be praised, no one, nowadays, feels tempted to enter upon the discussion of religion with that spirit of disdainful criticism of the eighteenth century which flattered itself that it was capable of explaining everything by the use of words of superficial clearness, superstition, credulity, fanaticism. To a more advanced critic, religions are the philosophies of the spontaneous, philosophies, amalgamated with heterogeneous elements, like food that is not solely made up of nutritious parts. Apparently, the very finest would be preferable but the stomach would not be able to digest it. Exclusively scientific formulas would afford but a dry food, and so true is this, that with every great philosophical thought there is mixed up a little mysticism; that is, a compound of individual fantasy and religion.

Religions, therefore, are the purest and most complete expressions of human nature, the shell in which its forms are moulded, the bed on which it lies, on which it leaves the impression of the curves of its outlines. Religions and languages should be the first studies of the psychologist. For humanity is much more easily recognized in its products than in its abstract essence, and in its spontaneous products than in its premeditated ones. Science being wholly objective, has nothing that is personal or individual in it, religions, on the other hand, are individual, national by their very essence; they are in one word, subjective. Religions were made at a time when man put himself into all his works. Take up a work of modern science, "*l'Astronomie Physique*" of M. Biot or "*la Chimie*" of M. Regnault; you will find it the most perfect specimen of objective treatment; there is a complete absence of the author himself, the work bears neither an individual nor a national stamp; it is an intellectual, not a human work. Popular science, and in many respects, ancient science, only saw man through the prism of man and dyed him with colours altogether human. For a long while after modern men of science had created for themselves more perfect means of observation, there remained numerous causes of aberration which disfigured, and impaired with false colours, the outlines of objects. On the other hand, the telescope with which the moderns take their observations of the world is perfectly achromatic. If there are any other intellects than that of man we cannot well conceive that they can see otherwise. Scientific works, therefore, can in no way convey an idea of the originality of human nature nor of its proper character, while in a work in which fancy and the feelings have borne a large part is much more human and consequently more adapted to the experimental study of the instincts of psychological nature.

Hence the immense interest of everything that appertains to religion, to the popular instinct, of primitive narrative, of fable, of superstitious beliefs. Each nation spends her very soul over it, creates it out of her own substance. Tacitus, whatever may be his talents for painting human nature, contains less true psychology than the artless and credulous narrative of the Gospels. It is because the narrative of Tacitus is objective; he narrates or endeavours to narrate things and their causes as they really were; the narrative of the Evangelists, on the contrary, is subjective; they do not recount things, but the views they conceived of things, the manner in which they appreciated them. I may be allowed to give an instance in point. While passing by a churchyard, at night, I have been pursued by a will o' the wisp. In relating my adventure, I should express myself thus; "One evening in passing by a churchyard, I was pursued by a will o' the wisp." On the other hand, a peasant woman who happens to have lost her brother a few days before, and to whom a similar adventure had occurred, would express herself as follows; "While I was passing by the churchyard at night I was pursued by the soul of my brother." Here we have two accounts of the same fact, both perfectly veracious. What then constitutes their difference? The first recounts the fact in its naked reality; the second mingles a subjective element with it, an appreciation, a judgment, a view of the narrator herself. The one narrative was simple, the other is complex and mingles with the affirmation of the fact a judgment of cause (119). All the narratives of the primitive ages are subjective; those of the thinking ages are objective. Criticism consists in recovering, as far as possible the real colour of facts, from the colours as refracted through the prism of the nationality or individuality of the narrators.

Hence, the history of religions is the true history of philosophy. The work most urgently wanted for the advancement of the sciences of humanity would be, therefore, a philosophical theory of religions. But how could we possibly get such a theory without erudition? Islamism is certainly very well known by the students of Arabian literature; there is no religion which offers fewer obstacles to inquiry; nevertheless, in the ordinary books Islamism is the object of the most absurd fables, of the most erroneous judgments. And yet, Islamism, although the weakest of all religions from the standpoint of creative originality (the sap had already run dry), is of major importance in that comparative study, because we have authentic documents on its origins, which is not the case with any other religion; the primitive facts of the apparitions of religions, occurring in the spontaneous conditions of humanity leaving no trace. Religion does not become conscious of itself until it has reached the adult and developed state, that is, when the primitive facts have disappeared for ever. Religions, no more than individual man, remember their infancy, and it is very rare that extraneous documents are found to dispel the darkness that surrounds their cradle. Islamism affords the only exception in that respect, it is born when history is already in full swing; the traces of the disputes it provoked and of the incredulity it had to fight against still exist. The Koran is from beginning to end nothing but a mass of sophistical argumentation. There was a great deal of reflection in Mahomed and even a little of what—if driven to it—we might call *imposture* (120). The facts that succeeded the establishment of Islamism and which are eminently calculated to show the manner in which religions are consolidated, do also, every one of them, belong to the domain of history.

Buddhism does not share this advantage. Induction and conjectures will necessarily have to play a large part in the history of its origins. But what inestimable lights will not that vast development afford for the discovery of the laws that presided at the formation of a religious system; that vast development so analogous to Christianity, which starting from India has invaded half of Asia and despatched missionaries far and wide from the territories of the Seleucidæ to the uttermost corners of China. The problem of primitive Christianity will not be ripe until the day when M. Eugène Burnouf shall have finished his "Introduction to the History of Indian Buddhism."

And, the most important book of the nineteenth century should bear the title of; "A Critical History of the Origins of Christianity." Oh, the admirable book, the author of which I envy and which will be the work of my ripe age, if I am not prevented by death and the many outwardly fatal incidents which so often and forcedly cause a life to turn aside from its original purpose. People do so obstinately persist in repeating such utterly erroneous platitudes on the subject. They think that the subject is exhausted when they have mentioned the fusion of Judaism, Platonism and Orientalism, without having any notion of what is Orientalism, without their being able to say how Jesus and the apostles came by any traditions of Plato. Because as yet no one has dreamt of looking for the origins of Christianity there where they really exist, in the Deutero-Canonical books, in the apocryphal writings of Jewish origin, in the Mishna, in the *Pirké-Avoth*, in the works of the Judæo-Christians. People look for Christianity in the works of the Platonist Fathers who only represent a second moment of its existence. Christianity is primarily a Jewish fact, just as Buddhism is an Indian fact, albeit that Christianity like Buddhism was almost exterminated from the countries that gave it birth and that the admixture with foreign ingredients may have cast a doubt upon its origin.

As for me, if ever I undertook that great work, I should begin by an exact catalogue of the sources, that is of everything that has been written in the East from the captivity of the Jews in Babylon until the moment Christianity is finally constituted, without overlooking the very important aid to be derived from monuments, engraved stones, etc. Then I should devote a volume to the criticism of those sources. I should take one after the other the fragments of Daniel written in the time of the Maccabees, the Book of Wisdom, the Chaldaic paraphrases, the Testament of the Twelve Patriarchs, the Books of the New Testament, the Mishna, the Apocryphas, etc., and I would try to ascertain by the most scrupulous criticism, the exact period, the locality, the intellectual surroundings in which those works were composed. That done, I should be solely guided by those data in the formation of my ideas and completely discard all the fancies in which others have indulged either by the process of induction or on nothing more substantial than vague analogies. No doubt the universal knowledge of the human intellect would be necessary for that history. But one should be careful not to transform analogies into reciprocal loans, if history says nothing about the reality of such loans. Our French critics who have only studied the Greek and Latin worlds would have great difficulty in understanding that Christianity was at the outset an exclusively Jewish fact. Christianity in their opinion is the work of collective humanity; Socrates composed, as it were, the prelude to it, Plato laboured at it, Terence and Virgil are already Christians, Seneca still more so. This is true, perfectly true, provided one can grasp the meaning of it. Christianity in reality only became what it is when humanity *adopted* it as the expression of the wants and tendencies by which it had been stirred for ever so long. Christianity, such as it prevails amongst us, contains in fact, elements of every date, of every country. But the important point to bring to light, which is not sufficiently noticed, is that the primitive germ is wholly Jewish, that the apparition of Jesus is simply simultaneous with the Christianity anticipated by the Greco-Latin world; that the Gospel and Saint-Paul must be explained by the Talmud and not by Plato (121). The soil whence Christianity drew its sap, in which it spread its roots, is humanity, and above all the Greco-Latin world; but the kernel from which it sprang is wholly Jewish. What I wish to indicate here is the history of that curious formation and development, the history of the roots of Christianity up to the moment that the tree appears above ground, while as yet it is only a Jewish sect, up to the moment when it is adopted, or rather absorbed by the nations. It is still wholly a matter of conjecture; neither Christians, nor Jews, nor Pagans having left us anything *historical* on that first apparition, or on the principal hero. But criticism may recover history from legend or at any rate trace back the characteristic aspect of the period and its works. Scholastic precision, here as always, excludes criticism. We may set ourselves a number of questions on the Resurrection, on the miracles of the Gospel, on the character of Jesus and of the Apostles which it is impossible to answer by judging the first century after our own. If Jesus did not really rise from the dead, how is it that the belief in such resurrection spread? Are we to conclude then that the Apostles were impostors, the Evangelists liars? How is it that the Jews did not protest? How is it that . . . etc., etc. All these are questions that would have a meaning in our century of reflection and publicity, but which had none in a period of credulity when no critical thought raised its voice (122).

The first step in the comparative study of religion will be, it seems to me, to establish two very distinct classes among those curious products of the human intellect; organized religions, having sacred books precise dogmas; non-organized religions, possessing neither sacred books, nor dogmas, being only more or less pure forms of the worship of nature, and having no pretensions whatsoever to spring from revelations. The first class would comprise; Judaism, Christianity, Islamism, Parseeism, Brahminism, Buddhism, to which may be added Manichæism, which is not merely a sect or a Christian heresy, as people often imagine, but a religious apparition, grafted like Christianity, Islamism and Buddhism on an anterior religion. The second class should comprise the mythological polytheisms of the Greeks, the Scandinavians, the Gauls and in general all the mythologies of the peoples who had no sacred books. Candidly speaking, these forms of worship scarcely deserve the name of religions; the idea of revelation is utterly foreign to them; it is pure naturalism expressed by a poetical symbolism. It would perhaps be more becoming to restrict the name of religions to the dogmatic compositions of Western and Southern Asia. It is certain, though, that the existence of the sacred book is the *criterion* that should determine the classification of religions, because it is the mark of a more profound character; the dogmatic organization. It is also certain that the East presents itself as the soil whence sprang the great organized religions. The East has always lived in that psychological condition which is favourable to the birth of myths. It has never arrived at that perfect clearness of consciousness which is rationalism. The East has never understood the true philosophical grandeur which can dispense with miracles. It sets little store by a sage who is not also a thaumaturge (123). The sacred book is an exclusively Asiatic production. Europe has not created a single one (124).

Another characteristic, not less essential, and which may serve as well as the sacred book to distinguish religions is tolerance or exclusivism. The old mythological forms of worship, not claiming to be the absolute form of religion, but merely professing to be local forms, did not exclude the other forms.

"I have my God whom I serve; you shall serve yours. They are two powerful Gods."

That is the pure expression of that religious form. Each nation, each city has its gods, more or less powerful; it is perfectly natural that the one city should not serve those of another. Jehovah himself is often only the God of Jacob, cherishing his people with the same feelings of national partiality as the other local deities. Hence those challenges with regard to the respective powers of the gods, each nation insisting that her own are strongest, but which challenges in no way imply that they are the only gods. It is altogether different in the Judaism of the times of the prophets, and in general in all the great organized religions. Jehovah alone is God, all the rest are so many idols. Hence the idea of a *true* religion which had no meaning in the mythological forms of worship. And seeing that in those epochs, the truth is conceived as a revelation from the Divinity, that characteristic manifests itself in *revealed religion* (125).

Last of all, the organized religions are distinguished from the forms of mythological worship by a greater character of stability and duration. It is literally true that up till now, not a single great religion is positively dead and that the most ill-treated ones, Parseeism, Samaritanism, etc., still live in the faith of some tribe or other, or relegated to some distant nook of the globe.

Thus, on the one side; organized religions, claiming to be based upon revelation, absolute, exclusively true and possessing a sacred book.—On the other; non-organized religions, local, not exclusive, not having a sacred book.

The great Asiatic religions would group themselves as it were in three families, or rather would claim connection with three sources: (1) The Semitic family (Judaism, Christianity, Islamism); (2) The Aryan family (Parseeism, Manichæism); (3) The Indian family (Brahmanism, Buddhism). Within each family, the successive reforms have only been the developments of the self-same foundation (126). Strictly speaking we could not say that religions are a question of racial idiosyncracy, seeing that the Indo-Germanic peoples have created religions as well as the Semitic peoples. Still it would be idle to deny that the Indo-Germanic religions have a stamp apart. They are very nearly systems of pure philosophy. Buddha was only a philosopher; Brahmanism has little in common with the other organized religions save the sacred book and is in reality nothing more than the most simple expression of naturalism. A more noticeable difference still; all the Semitic religions are essentially monotheistic; the race has never had a developed mythology. All the Indo-Germanic religions are, on the contrary, either pantheistic or dualistic and boast a vast mythological or symbolical development (127). It would seem that among the peoples the creative faculties with regard to religions were in inverse ratio to the philosophical faculties. The premeditated, independent, severe, courageous — in one word — philosophical search after the truth seems to have been the inheritance of that Indo-Germanic race which from the uttermost confines of India to the farthest extremes of the West and the North, from the most distant centuries to modern times, has endeavoured to explain God, man and the world to the rationalistic sense, and has left behind it, posted as it were, at the different stages of its history, those systems, those philosophical creations, subject always and everywhere to the unvarying and necessary laws of a logical development. The Semites, on the contrary, who do not show us a single attempt at analysis, who have not produced a single school of native philosophy, are *par excellence* the race of religions, destined to give them birth and to propagate them. Theirs is the privilege of those bold and spontaneous flights of natures, still in the flush of youth, penetrating without an effort and by a most natural movement as it were into the very bosom of the infinite, and descending from it thoroughly drenched with divine dew, then letting their enthusiasm exhale in a form of worship, in a mystic doctrine, in a revealed book. The philosophical school has its cradle under the skies of Greece and India; the temple and priestly science explaining themselves in enigmas and in creeds, veiling the truth beneath mystery, often soaring higher because it is less afraid of looking backward and making sure of its onward march, such is the character of the religious and theocratic race of the Semites. They are "God's people" *par excellence*, and the exceptional atheist is to them a being devoid of meaning, an enigma, a monster in the universe. They have that moral instinct, that sound sense, practical and not capable of analyzing very deeply, but popular and easy-going which constitutes the genius of religions, and added to this the prophetic gift which often succeeds in speaking of God more eloquently and above all more exuberantly than science and rationalism (128). And in fact is it not a thing worthy of remark that the three religions which up till now have played the greatest part in the history of civilization, the three religions stamped with a special character of stability, fruitfulness and proselytism, and moreover bound to one another by such close relations as to make them seem three branches of a same trunk, three versions—not equally beautiful and pure—of a self-same idea, is it not worthy of remark that all three should have been born on Semitic ground, and from there should have started forth to the conquest of high destinies. The distance between Jerusalem and Sinai, and between Sinai and Mecca is but a few miles (128*a*).

Still, seeing that races do not differ in virtue of the possession of different faculties, but through the different extension of the same faculties, seeing that what constitutes the dominant characteristic in some is found among others in a rudimentary condition, Greece presents the unmistakable germs of the processes which in the East created the revealer, the man-god and the prophet. But they (the processes) always miscarried before they could become a genuine religious tradition. The system of Pythagoras with its degrees, its initiations, its novitiates, its distinct tinge of asceticism reminds one of the grand organized systems of Asia. Pythagoras himself is very like a theurge. He is infallible (*αὐτος ἔφα*); a disciple who has incurred his blame kills himself. He has visited the nether regions, and remembers his transmigrations. He willingly lends himself to, or even supplies the occasion for, such beliefs; he recognizes in a temple of Greece the arms he bore at the siege of Troy. In the East Pythagoras would have been Buddha. This colouring is even more striking in Empedocles, who represents the Oriental theurge in every trait. A priest and a poet, like Orpheus, a physician and a thaumaturge the whole of Sicily rings with his miracles. He raised the dead, stayed the winds, averted the plague. He only appeared in public amidst a train of servitors, the sacred crown on his head, the tinkling brass sandals on his feet, with flowing locks, his hand holding the laurel branch. His divinity was admitted throughout Sicily, he himself proclaimed it. "Ye friends, who dwell on the heights of the great city washed by the yellow Acragas;" he writes at the beginning of one of his poems; "ye zealous observers of justice, hail! I am not a man, I am a God. On my entering the flourishing cities men and women prostrate themselves. The multitude follow my footsteps. Some ask me for oracles, others the cure for cruel maladies by which they are tormented." The processes by which his miraculous legend is constituted remind one in every particular of those of the East. A trance from which he has roused some one becomes a resurrection. He arrests the Etesian gales that are devastating Agrigentum by closing a gap between two rocks; hence the surname of κολυσανέμας. He drains a marsh close to Selinus, which is sufficient to make him an equal of Apollo. Here we have analogies very characteristic of the founders of religion in the East. But alas, Greece was too flighty to dwell for very long on those beliefs and to constitute them into religious traditions; the divinity of Empedocles came to grief against the scepticism of the scoffers, and spiteful legend made merry over his sandals that were found on Etna. Asia has never known how to laugh, and it is because of this that she is religious.

As for the mythological forms of worship that were not organized, that had no sacred book, their variety is much greater, or to speak correctly all classification here becomes impossible. They are all so much pure fantasy, it is human imagination constantly broidering on a self-same canvas, which is natural religion. Comparing poem with poem, creed with creed, the variety in this instance becomes every now and then almost individual, a simple family affair. The most one can do is to indicate the diverse degrees and ages of those curious processes. At the lowest degree stands fetichism, that is, the individual or family mythologies, fables evolved from dreams and affirmed in the most arbitrary way, fables without the slightest traditional antecedent, without the idea of their *truth* ever presenting itself to the mind, any more than it did in the dream, the fable for the fable's sake. Then came the more premeditated myths, in which the instincts of human nature express themselves in a more distinct manner, that is, they already show a certain analysis, but without reflection or perception of allegorical symbolism. At last comes the thought-out symbolism, the allegory created with the distinct consciousness of the double meaning which utterly escapes the first creators of myths.

In reality every mythological creation like every religious development goes through two very distinct phases; the creative age when the grand traits of legend are traced deep down the popular conscience, and the age of remodelling, of adjustment, of verbose amplification when the grand poetical vein is lost and when the whole thing consists in the re-dishing of the old poetical fables after a stereotyped process beyond which they no longer go. Hesiod on the one side, the Alexandrian mythologists on the other; the Vedas on the one side, the Puranas on the other; the canonical Gospels on the one side, the Apocryphas on the other afford so many examples of this transformation of mythologies. It is simply a way of taking the myths of olden times and amplifying them by fusing all the original traits in the new narrative, and to a certain extent of composing the monograph of what was but a mere detail in the grand primitive fable; the whole of it requiring but little power of invention, because there is never a deviation from the given theme. That which must have probably happened is added, the situation is amplified, certain links are forged. It is, in short, a carefully considered composition and in one sense, a literary one, having for its bases a spontaneous creation. That age is necessarily insipid and wearisome. For the spontaneous production, so full of life, so graceful, does not admit of being remodelled. What chance is there for the naïve thoughts of a child, ponderously commented on by pedants, delicate flowers that wither by passing from hand to hand? Do not you think that to the primitive men that created them, Venus, Pan, the Graces had a meaning different from that which they convey in the park at Versailles, reduced as they were to a chilling allegorism by a thinking age, which from mere fantasy goes in search of a mythology of the past, in order to coin from it a conventional language for its own use (129)?

Those two phases in legendary creation correspond with the two ages of every religion; the primitive age in which religion is evolved beautiful and pure from man's conscience, like the sun's rays; the age of naïve and simple faith without afterthought, without objection or refutation; and the thinking age, marked by the rise of objection and apology; the subtle age in which reflection becomes exacting, without being able to obtain satisfaction; in which the marvellous, formerly "so easily accepted," so beautiful in its imaginings, so gentle in its conceptions, the so eminently pure reflex of the moral instincts of humanity, becomes timid and paltry, sometimes immoral, pettily supernatural, a series of miracles arrogated by *coteries* and brotherhoods, etc. Everything shrinks and dwindles, the practice of religion becomes devoid of meaning and materialized; prayer becomes a mere mechanical performance, worship a mere ceremonial, formulas become a kind of Cabbalism, in which words operate not as formerly by their moral sense, but by their sound and articulation; legal prescriptions, originally stamped with a deep sense of morality become mere irksome prohibitions which people seek to elude until the day when they shall find some subtle argument to rid themselves altogether of them (130). In the primeval age, religion has no need of creeds; it is a new spirit, a fire that goes on consuming everything before it, it is free and boundless. Then, when the enthusiasm has subsided, when the original and native force is extinguished, man begins to *define*, to combine, to speculate upon that which the first believers had embraced in faith and love. That day beholds the birth of scholasticism, on that day the first germ of incredulity is sown. I cannot state all my perceptions on this fruitful subject nor the wealth of psychological knowledge that might accrue from the study of those admirable works of human nature. I am aware that our position face to face with those strange works is a singular one. Replete with life and truth to the peoples who created them, they are to us merely an object of analysis and dissection. This, in one sense, constitutes an inferior position which will always debar us from arriving at a perfect understanding of them. How often, in pondering the mythology of India for instance, have I been struck with the absolute impossibility, as far as we are concerned of understanding the life and soul of it. We, in this instance are confronted with works, deeply expressive, brimful of significance to a portion of humanity, we the sceptics, the analysts. How could they possibly tell us what they tell them? They who believed in Christ can understand Him. In the same way, to understand in all their import those sublime creations, one must have believed in them, or rather (for the word *to believe* has no meaning in the world of fantasy) one must have lived with them. Would it not be possible to realize that wonder by such a vast progress of the scientific spirit as would make people profoundly sympathetic with everything humanity has accomplished? It is difficult to say, but it is certain at any rate, that those systems containing as they do, more or less precious atoms of human nature, that is, of truth, he who succeeds in understanding them, would find solid nourishment there. It may be taken as a general rule that when a work of the human intellect seems too absurd or too odd, it is because we do not understand it, or take it in the wrong sense. If we placed ourselves in the true light we should see the sense of it.

I wanted to show by a few instances the philosophical results to which the sciences of pure scholarship may lead and how unjust is the contempt for those studies felt by certain intellects, themselves gifted with the philosophical sense. What if I were to show, in the case of the philosophy of history, that this marvellous science, which one day will be the ruling science, will never succeed in constituting itself in a serious and dignified manner except by the aid of the most scrupulous learning, that until then it will remain in the same condition as were the physical sciences before Bacon, that is, wandering from one hypothesis to another, without a settled "way-bill," not knowing what form to give to its laws and never going beyond the sphere of artificial and fantastic creations?

What, if I were to show that literary criticism which is our domain proper, and of which we are deservedly proud can never become serious and profound except by the aid of learning? How can we grasp the physiognomy and originality of primitive literatures, if we do not penetrate into the moral and intimate life of the nation, if we do not place ourselves on the same standpoint of humanity which it occupied, in order to see and to feel as it did; if we do not watch its life, or rather if we do not share its life, if only for a moment? Besides, there is as a rule nothing more silly than the admiration bestowed upon antiquity. People do not admire its original and really admirable features; but in a paltry spirit pick out in the works of antiquity the traits which approach our own; they endeavour to point out the beauties which with us—one is bound to admit it—would be considered as of a second rate order. The way superficial intellects stand embarrassed when face to face with the grand productions of classical literature is very amusing indeed. They start from the principle that those works must be beautiful under any circumstances, seeing that the *connoisseurs* have decided as much. But as they are incapable—for lack of scholarship—to grasp their intense originality, their truth, their value in the history of the human intellect, they wind themselves up to a pitch of admiration, to an enthusiastic appreciation of the beauties of antiquity, which in fact, is only admiration of their own silliness. Utterly conventional admiration, "turned on" for form's sake and in order to escape from the self-reproach of barbarism which the non-admiration of works admired by *connoisseurs* would imply. Hence the self-inflicted tortures in order to rise to the occasion when face to face with works which absolutely must be voted beautiful, in order to discover a small detail, an epithet, a brilliant trait, a sentence which translated into French would sound startling. If they were honest about it, they would prefer Seneca to Demosthenes (131). Certain people who have been told that Rollin is beautiful are astonished at only finding very simple phrases, and do not know where to lay hold of a point for admiration, incapable as they are of conceiving the beauty resulting from that naïve and delightfully upright nature. It is the man himself, who is beautiful, it is the things themselves that are beautiful, and not the way they are expressed. But there are so few people capable of judging æsthetically. They admire on trust and in order not to remain behind. How many people, standing before a picture by Raffaele know what constitutes its beauty? How many if they frankly stated their opinion would prefer a modern picture, with a clearer style and a more brilliant colouring? One of the keenest pleasures is to see mediocre intellects floundering about with regard to works they have been told beforehand are beautiful. Fréron admires Sophocles for having respected certain conventionalities which, assuredly were the last things to enter that poet's mind. As a rule the Greeks knew nothing of the beauties of design, and the credit we bestow upon them in that respect is purely gratuitous. I have heard people express their admiration of the entrance upon the scene of "Oedipus Tyrannus," because his first line contains a pretty antithesis for which there is an equivalent in a line by Racine.

Ever since people have gone on repeating (and justly) that the Bible is admirable, every one professes great admiration of the Bible. The result of this favourable disposition is precisely the admiration of that which is not there. Bossuet, who is supposed to be such a very Biblical scholar, and who is scarcely anything of the kind, goes into ecstacies over the blunders and the solecisms of the Vulgate, and professes to discover beauties in them of which there is no trace in the original (132). That worthy Rollin is more ingenuous still and points out in "the Song of Moses and Miriam" the *exordium*, the connection of ideas, the plan, even the *style*. To wind up, Lowth, more insipid than all the others, gives us a treatise of Aristotelian rhetoric on the poesy of the Hebrews, in which we find a chapter "On the Metaphors of the Bible," another on the *Comparisons*, a third on the *prosopopoeia*, a fourth on the sublime in diction, etc.; without suspecting for a moment that which really does constitute the beauty of those ancient poems, namely; the spontaneous inspiration —independent of artificial and premeditated forms— of the human intellect, still young and fresh to the world, carrying with it everywhere God of whom it still preserves the recent impression. Admiration, in order to be to the purpose and useful must, therefore, be historical and scholarly. Each work is beautiful when considered amidst its own surroundings, and not because it fits one of the pigeon holes that have been established more or less arbitrarily. To establish absolute divisions in literature, to declare that a work shall be an epic, or an ode, or a novel, and to criticize the works of the past according to the rules that have been laid down for each of these kinds, to blame Dante for having written a work which is neither an epic, nor a drama, nor a didactic poem, to blame Klopstock for having taken too perfect a hero, to do all this, is simply to ignore the liberty of inspiration, and the right of the intellect to blow from the quarter whence it chooses. Every mode of realizing the beautiful is legitimate and genius has ever the same right to create. The beautiful is that which represents in finished and individual traits, the eternal and infinite beauty of human nature.

The *savant* alone has the right to admire. Not only do criticism and æstheticism, considered as opposed to one another, not exclude one another, but they are inseparable. Everything is at the same time admirable and susceptible of being criticized, and he only who knows how to admire, knows how to criticize. How, for instance can a man understand the beauties of Homer unless he be a *savant*, unless he know antiquity, unless he have the sense of primeval things? What do we generally admire in those ancient poems? Some small instances of great simplicity, some traits that raise a smile; not that which is truly admirable, the picture of an age of humanity in its inimitable truth. The admiration of Chateaubriand is only therefore so often at fault because the æsthetic sense with which he was so eminently gifted was not based upon solid learning (133).

Hence, in the actual condition of the human intellect it is by labours in the direction of scientific philosophy that we may hope to add to the domain of already acquired ideas. When we reflect upon the rôles in the history of human intellect enacted by men like Erasmus, Bayle, Wolf, Niebuhr, Strauss, when we reflect upon the ideas they have put into circulation, and the advent of which they accelerated one feels surprised that the name of philosopher, so lavishly bestowed upon obscure pedants, cannot be applied to such men. I am aware that the results of the higher sciences are a long while in becoming current. Of the immense labours already accomplished by modern students of Indian literature, only a very few have as yet become common property. A numberless hive of learned philologists has produced a complete reformation in biblical exegesis in Germany without France being the wiser, as yet, with regard to a single word of their works. Still there are secret canals for science as well as for philosophy by means of which results do filter through. The ideas of Wolf on the epic or rather those to which he has led have become public property. The grand pantheistic poesy of Goethe, of Victor Hugo, of Lamartine implies an acquaintance with the whole of the labours of modern criticism, the final upshot of which is literary pantheism. I can scarcely imagine that M. Hugo has read Heyne, Wolf, William Jones, still his poesy would breed that supposition. The day comes when the results of science fill the air, if I may be allowed the expression, and affect the general formation of literature. M. Fauriel was only a critical *savant;* the gift of artistic production was almost denied to him; nevertheless, there are few men who have wielded so profound an influence on productive literature.

But the mines of the past are still very far from having yielded all the treasures they contain. The task of modern scholarship will only be accomplished then when all the facets of humanity, that is, all the nations shall have been explored definitely, when China, Judæa, Egypt shall have been restored to us in their primeval aspects, when we shall have finally arrived at the perfect understanding of the whole of human development. Then, and then only the reign of criticism will be inaugurated. For criticism will only proceed with perfect surety when the field of universal comparison shall be thrown open to it. Comparison is the great instrument of criticism. The seventeenth century was ignorant of criticism, because the comparison of the different facets of the human intellect was impossible to it. Herodotus and Livy must have been considered serious historians; Homer must have passed muster as an individual poet, previous to the comparative study of literatures having revealed the very delicate facts connected with mythism, primitive legend and apocryphism. If the seventeenth century had been acquainted, like ourselves, with India, Persia and ancient Germania, it would not have admitted in so ponderous a fashion the fables of the Greek and Roman origins. Bossuet, whose claim to eminence rests upon the fact of having represented in a marvellous abridgment the whole of the seventeenth century, its grandeur as well as its weakness, would he have brought to bear so detestably critical a method on his exegesis, if instead of having derived his Biblical education from Saint-Augustine, he had got it from Eichhorn and Von Wette (134)?

People are not inoculated with the critical sense in an hour; he who has not cultivated it by means of a protracted scientific and intellectual training will always find arguments to oppose to the most delicate inductions. The theses of refined criticism are not of the kind that are demonstrated in a few minutes and on which one can force the ignorant adversary or the one determined not to fall in with the views proposed to him. If there be myths on the face of them among the works of the human intellect, they are assuredly the first pages of Roman history, the story of the tower of Babel, of Lot's wife, of Samson; if there be a thoroughly characteristic historical romance, it is that of Xenophon; if there be a historian story-teller, it is Herodotus. Still, it would be so much time wasted to endeavour to persuade to that effect those who decline to take that view. To elevate and to cultivate the minds of the majority, to popularize the great results of science, these are the only means of spreading the understanding and acceptance of the new ideas of criticism. It is science, it is philology, it is the vast perception and comparison of things, it is, in short, the modern spirit, that converts. We may leave to mediocre intellects the satisfaction of believing themselves invincible in their ponderous arguments. We should not as much as try to refute them. The results of criticism cannot be *proved*, they must be perceived, to understand them requires long training and a thorough culture of the perception of the *finesse* of things. It is impossible to convince the man who obstinately rejects them, just as it is impossible to prove the existence of microscopic animalculæ to the man who refuses to make use of a microscope. Determined to shut their eyes to delicate considerations, to take count of no shades whatsoever, they fling their everlasting "prove to us that it is impossible," at your head. (There are so few things that are impossible.) The critic will leave them to enjoy their triumph by themselves and abstain from discussing with narrow intellects, determined to remain such, he will pursue his road, supported by the thousands of inductions which the universal study of things will cause to spring forth from all parts and which so powerfully converge towards the rationalistic point of view. Obstinate denial cannot be grappled with ; in no matter what order of things is it possible to make a man see who is determined not to see. It is moreover inflicting a wrong upon the results of criticism to clothe them in that heavy syllogistic form in which the mediocre intellects excel, and which delicate considerations could never assume.

An instance in point. The four canonical Gospels often report the same fact with very considerable variations of circumstances. These are easily accounted for by natural hypothesis for we have no right to be more exacting with regard to the Gospels than with regard to other historical or legendary narratives which often present more startling contradictions. But it would appear that this constitutes an altogether unanswerable objection against those who think it incumbent to see in each of these narratives a history literally true and exact in the most minute details. This is, however, not the case. For if the circumstances are only different and not absolutely irreconcilable, they will say that one of the texts has preserved details omitted by the others and they will endeavour to make the various circumstances meet at the risk of concocting the most grotesque narrative out of them. If the circumstances are decidedly contradictory, they will aver that the fact narrated occurred two or three times, albeit that in the eyes of sound criticism the narrators had the same event in view. Hence, the narratives of John and the synoptics (by which collective name are meant Matthew, Mark, and Luke) of the last entry of Jesus into Jerusalem being irreconcilable, the sticklers for harmony suppose that he came thither twice, with scarcely an interval between the two occurrences. Similarly, the three denials of Peter being told differently by the four Evangelists constitute in the opinion of those critics eight or nine different denials, while Jesus predicted that he would deny him only thrice. The circumstances attending the Resurrection give rise to analogous difficulties, to which they oppose similar solutions. What is there to be said about such an explanation? That it is metaphysically impossible? No. It will always be impossible to silence the man who will obstinately maintain it; but it will always be rejected by whosoever is gifted with more or less critical education as contrary to all the laws of logical hermeneutics, especially if it is often repeated. There is no difficulty of which one cannot get rid by some subtle argument, and in reality, a subtle argument may sometimes be a true one. But there is no possibility of a hundred subtleties being true at the same time. We may say the same with regard to the "non-suit" opposed by certain exegetists to what they term *negative argument*, that is, to the inductions drawn from the silence or absence of texts. For instance; because the most ancient of histories of the Jews settled in Palestine offers no trace of the accomplishment of the Mosaic prescriptions, the rationalistic critic concludes from it that these prescriptions did, as yet, not exist. "How do you know," says the orthodox student, "but what they may have existed without their being mentioned?" The romance of Antar and the Moallacats does not lead us to suppose the existence of any judiciary institution, of any penalty whatsoever among the Arabs, before the advent of Islamism. "How do you know but what they may have had a jury without it being mentioned?" To cope with such criticism, we should want a text conceived as follows; "The Arabs at that period had no jury;" which text, I admit, it would be difficult to find. You may as well require a similar text to prove that artillery was not known in Homeric times, and in general, for all the results of criticism expressed in negative form.

This impossibility of imposing his results and of silencing his adversaries may at first make the critic impatient and goad him into entering that coarse arena. That would be an unpardonable error. For many long years to come still the critic will be a solitary creature and ought to confine himself to merely regretting that the necessary education to understand him should be confined to so small a circle. How can it be otherwise, when the first instruction received in childhood and which is often the only philosophical doctrine of life, is the very negation of criticism? Poetical and vague superstition is gross and nothing else. If the critical spirit is much more wide-spread in Northern Germany than in France, the cause lies no doubt in the difference of religious instruction, with us positive and hard, with them not hard and fast and purely human.

CHAPTER XVI.

I WONDER whether I have succeeded in making people understand the possibility of a *scientific philosophy* which would no longer be a system of vain and empty speculation, aiming at no real object, of a science which would no longer be dry, barren, exclusive, but which in becoming complete, would become religious and poetical? There is no word to express that intellectual condition in which all the elements of nature would be blended into one superior harmony, and which, when realized in a human being, would constitute the perfect man. I am inclined to call it *synthesis*, in the special sense which I will explain.

Just as the simplest fact of the human understanding as applied to a complex object is composed of three actions; (1) a general and confused view of the whole; (2) a distinct and analytical view of the parts; (3) a synthetical recomposition of the whole with the knowledge one possesses of the parts; so the human intellect, in its march, traverses three conditions which may be defined under the three names of syncretism, analysis, synthesis, and which correspond with these three phases of the understanding.

The first age of the human intellect which we too often fancy to have been that of simplicity, was that of complexity and confusion. We are too easily persuaded that the simplicity which we conceive to be logically anterior to complexity is also chronologically anterior to it; as if that which, in regard to our analytical processes is the simpler, ought to have come into existence before the whole of which it is a part. The language of the child, apparently more simple, is in fact more comprehensive and more closely knit together that that which explains word for word the most carefully analyzed thought of ripe age. The most profound linguists have been surprised to find at the origin and among the peoples we call children rich and complicated languages. The primitive man has not the faculty of division, he sees things in their natural state; that is, organically and in the plenitude of life (135). Nothing is abstract to him, for abstraction means life piecemeal; everything is concrete and alive. The power of distinguishing is not given to man at his origin, the first view is a general one, comprehensive, but obscure and incorrect; everything is huddled together without distinction. Like the beings who are meant to be perpetuated, the human intellect was, from its initial moments, complete, though not developed; nothing has been added since, but everything has been unfolded in its natural proportions, everything has been put in its proper place. Hence, the extreme complexity of the primitive works of the human intellect. Everything was contained in a single work, all the elements of humanity were gathered into one unity, which was undoubtedly very far removed from modern clearness, but which, it must be admitted, was incomparably majestic. The *sacred book* is the expression of that first condition of the human intellect. Let us take the sacred books of ancient peoples and what do we find in them? The whole of the supra-sensitive life, the whole of a nation's soul; her poesy; her heroic recollections; her legislation, her politics, her code of morals; her history, her philosophy and science; in short, her *religion*. For the whole of this first development is wrought in a religious form. The religion, the sacred book of primitive peoples, is the syncretic accumulation of all the human elements of the nation. Everything is there in a confused but withal beautiful unity. Hence arises the lofty placidity of those admirable works; antithesis, contention, distinctions being banished from them, harmony and peace reign supreme without ever being disturbed. To struggle is the characteristic of the state of analysis. How could religion and philosophy, poesy and science, morality and politics have contended in those grand primitive works, seeing that they repose side by side on the same page, often in the same line? The religion was the philosophy, the poesy was the science, the legislation was the morality; the whole of humanity was condensed in each of its acts, or rather the power of humanity was emitted complete in each of its efforts.

Therein lies the secret of the incomparable beauty of these primitive books, which are still the most adequate representations of complete humanity. It is idle to look specially for science in them; the science of our time is unquestionably better than that which they contain. It is idle, too, to look for philosophy in them, for we are beyond doubt better analysts. It is, again, equally idle to look in them for legislation and public law; for the rulers of our day are better informed, and that is not saying much. But what we do find in them is "simultaneous humanity," the grand harmony of human nature, the likeness of our glowing youth. Hence, also, is derived the superb poesy of those primitive types in which was incarnated the doctrine of those demigods who stand as religious ancestors to all nations—Orpheus, Thoth, Moses, Zoroaster, Vyasa and Fohi, who are at once savants, poets, legislators, social organizers, and, summing up all these qualities, priests and mystagogues. These admirable types still lasted for some time in the early ages of analytical reflection, producing those primitive "sages" who, though something more than mystagogues, are not quite philosophers, and who have also their legend (a fabulous biography), though one not so well formed as that of the initiators (a pure myth). Such are Confucius, Lao-Tseu, Solomon, Locman, Pythagoras. Empedocles, who form a connecting link with the early philosophers by means of the still more refined types of Solon, Zaleucus, Numa, etc.

Such is human intelligence in the primitive ages. It has its beauty, which our timid analysis cannot rival. It is the divine life, of childhood, during which God reveals Himself so closely to those who know how to adore. I admire not less than M. de Maistre does this ancient wisdom, wearing the crown of the sage and the priestly robe. I regret its loss, but I do not, on that account, blame the ages devoted to the toilsome work of analysis, which, inferior as they are in certain respects, represent, after all, a necessary progress of the human mind.

The human mind, in fact, cannot remain steadfast in this primitive unity. Thought, when applied more closely to objects, recognizes their complexity and the necessity of studying them piecemeal. Primitive thought could see only one world; thought in its second stage of years perceives a thousand worlds, or rather sees a world in all things. Its gaze, instead of becoming more extensive, pierces and plunges downward; instead of taking a horizontal direction, it extends vertically; instead of losing itself in a boundless horizon, it settles earthwards and upon itself. It is the age of partial sight, of exactitude, of precision, of the making of distinctions; instead of creation we have analysis. Thought becomes broken up and divided. The primitive style possessed neither the division of phrases nor the division of words. Analytical style calls to its assistance a complicated mode of punctuation, intended to dissect the different members. There are poets, savants, moralists, philosophers, politicians; there are even still theologians and priests (136). This is singular, for as theology and sacerdotalism are the complete form of primitive development, one would imagine that they would have disappeared with it. So they would have done if humanity advanced with perfect harmony and precision. As such is not the case, theology and sacerdotalism survive that which should have been mortal to them; they remain one specialty among many others. This is a contradiction in terms, for how are you to make a specialty of that which is only something upon condition of being everything? But as analytical science imposes itself as a necessity, the timid endeavour to conciliate this necessity with the remains of institutions which are opposed to analysis, and they fancy that they succeed in maintaining the two things face to face. I repeat that if theology was worth preserving, it would necessarily take the precedence of all else, and all other things would only be of importance by comparison with their bearing on it. The theological point of view is in contradiction with the analytical point of view; the analytical age ought to be atheistic and irreligious. But humanity, fortunately, prefers contradicting itself to leaving without sustenance one of the essential cravings of its being.

It is not of his own choice, it is by the fatality of his nature, that man thus quits the delights of his primitive garden, so smiling and so full of romance, to plunge into the quicksands of critique and science. One may regret these early attractions, just as, in the prime of life, one often regrets the dreams and joys of childhood; but one must go manfully onward, and, instead of casting longing looks behind, follow the hard path which will lead, no doubt, to a state of things a thousand times better. Even if the analytical condition which we are traversing were distinctly inferior to the primitive state (and it is only so in certain particulars), analysis would, nevertheless, represent a more advanced stage than syncretism, because it is an intermediate stage which must be gone through in order to reach a higher state. True progress sometimes seems a retrogression and then a return. The introgradations of humanity are like those of the planets. Viewed from earth, they are retrogradations; but in reality they are not so. The retrogradation appears as such only to the gaze of those whose sight embraces but a limited portion of the curve. Whether circular or spiral, as Goethe would have it, the march of humanity is along a line the two extremities of which meet. A vessel navigating along the western and wild coast of the United States in order to reach the eastern and civilized coast, would, to all appearances, be nearer its destination when starting than when it was being assailed by the storms and snows of Cape Horn. And yet, looking at the reality, this vessel would be nearer its destination when off Cape Horn than it was upon the banks of the Oregon. This great circuit was unavoidable. In the same way, the human mind will have had to traverse deserts in order to reach the promised land.

Analysis is war. In the primitive synthesis, with men's minds scarcely differing, harmony was a very simple matter. But in the state of individualism, liberty is ready to take umbrage; each person is bent upon saying what he thinks fit and cannot see why he should subject his will and thought to those of others. Analysis is the revolution and the negation of the one, absolute law. Those who dream of peace in this state dream of death. Resolution is a necessary element in it, and, whatever one may do, it will take its course. Peace is not the lot of the analytical state, and the analytical state is necessary for the progress of human intelligence. Peace will only reappear with the great synthesis, upon the day when men shall once again meet in the fond embrace of reason and of human nature properly cultivated. While this necessary transition is in progress, anything like a general association is impossible. The individual existence of each is too strong; individualities so marked in their characteristics will not allow themselves to be gathered up in sheaves. It would be impossible in the present day to create those great religious unities, those vast agglomerations of souls into the one doctrine called religious, or those military orders of the Middle Ages in which so many individualities quite insignificant in themselves became fused in view of one common aim. It is easy to bind up the ears when they have been cut or knocked off by the storm, but not so long as they are growing. To allow oneself to become thus absorbed in a great corporation, through which one lives, and with the fame or prosperity of which one becomes identified, one must have little individuality, few views of one's own, simply a great fund of unreflecting energy ready to place at the service of a great common idea. Reflection would be insufficient to bring about the unity; diversity is the essential characteristic of the philosophical epochs; any great dogmatic foundation is impossible in them. The primitive state was the age of solidarity. Crime even was not regarded as individual; the substitution of the innocent for the guilty appeared quite natural; a misdeed was transmitted and became hereditary. In the age of reflection, on the contrary, such dogmas seem absurd; each man answers only for himself, each man is his own artificer. With us, all knowledge is antithetical; in face of good we see evil; in face of the beautiful the ugly; when we make an affirmation, we deny, we see the objection, we harden ourselves, we argue. In the primitive age, on the contrary, an affirmation was plain and simple, with no going back upon it.

Assuredly, if analysis had no ulterior aim, it would be distinctly inferior to the primitive syncretism. For the latter seized the whole life, whereas analysis does not grasp it. But analysis is the necessary condition of the true synthesis; this diversity will anew dissolve itself into unity; perfect science is only possible upon condition of its being first of all based upon analysis and a clear view of the parts.

The conditions of science are for humanity the same as they are for the individual; the individual only knows thoroughly the whole of which he also knows the separate elements one by one, as also the part which these separate elements play on the whole. Humanity will not be learned until the day that science has explored every corner and put the living being together again after having taken him to pieces. Do not, therefore, sneer at the savant who sinks deeper and deeper into this slough. No doubt, if this toilsome operation was an end in itself, science would be only an ungrateful and degrading pursuit. But all is noble in view of the grand definite science, wherein poetry, religion, science and morality will find their lost harmony in complete reflection. The primitive age was religious but not scientific; the later age will be at once religious and scientific. Then there will be once more an Orpheus and a Trismegistos, not to sing to peoples in a state of childhood their fanciful dreams, but to teach a humanity grown wise the marvels of reality. Then there will be once more sages, poets and organizers, legislators and priests, not to govern humanity in the name of a vague instinct, but to lead it rationally into its paths, which are those of perfection. Then will appear, once more, superb types of human character, which will recall the marvels of the early ages. Such a state will seem a return to the primitive age; but between the two there will be the abyss of analysis; there will have been centuries of patient and attentive study; there will be the possibility, in embracing the whole, to gain simultaneously consciousness of the parts. No two things can be more like each other than syncretism and synthesis; nothing can in reality be more diverse, for synthesis virtually preserves within itself all analytical process; it assumes it and builds itself thereupon. All the phases of humanity are, therefore, good, seeing that they tend to what is perfect; the only thing is that they may perhaps be incomplete, because humanity accomplishes its work bit by bit, and draws its sketches one after the other, all in view of the grand final tableau and of the ulterior epoch, in which, after having traversed syncretism and analysis, it will complete the circle with synthesis. A little reflection may have rendered impossible the marvellous creations of instinct; but complete reflection will bring to life again the same works with a superior degree of clearness and of determination.

Analysis is powerless to create. A simple synthesist, devoid of critique, has more power to change the face of the world and to make proselytes than the stern and inaccessible philosopher. It is a great misfortune to have discovered within oneself the springs of the soul; for one is always fearful of becoming self-duped; one regards with suspicion one's feelings, joys and instincts. The simple-minded marches straight ahead, with a determined step and with unconquered energy. The age in which criticism is the most advanced is by no means the one best able to realize the beautiful. Germany is the only country in which literature allows itself to be influenced by the preconceived theories of criticism. Each new growth of literary production is brought about in that country by a new system of æsthetics; to which may be attributed so much of what is artificial and affected in its literature. The defect of the intellectual development of Germany is the abuse of reflection: I mean to say, the deliberate and intentional application to spontaneous production of the laws which are recognized in the anterior phases of thought. The great result of the historical critique of the nineteenth century, as applied to the history of the human intelligence, is to have recognized the necessary ebb and flow of the systems, to have got a dim idea of some of the laws according to which they are piled the one upon the other, and the way in which they incessantly oscillate towards the truth, when they follow their natural course. This is a speculative truth of the first order, but one which becomes very dangerous when put into application. For to conclude from this principle "The ulterior system is always the best," that any superficial or frivolous person who may happen to utter some drivel after a man of genius is preferable to the latter because he is chronologically posterior to him, this is, in sooth, to give mediocrity too much the best of it. And yet this is what too often happens in Germany. After a great philosophical or critical work has appeared, it is certain that a whole swarm of "advanced thinkers," so-called, will spring up, with the pretension of outvying it, when in reality they often merely contradict it. I cannot too strongly insist that the law of the progress of systems is only applicable when their production is perfectly spontaneous, and when their authors, without thinking of anticipating each other, concentrate their attention only on the intrinsic and objective consideration of things. To neglect this important condition is to hand over the development of the human intelligence to chance or to the grotesque claims of a few vain and presumptuous individuals (137).

Criticism does not understand the art of assimilation. Dogmatic eclecticism is only possible upon the condition of not being too closely tied down. Our attempts to effect a fusion between doctrines fail because we know them too well. The early Christians, the Alexandrians, the Arabs, the men of the Middle Ages and Mahomet were able to practise an eclecticism far more powerful than ours, because it was much more in the rough. They did not possess such exact knowledge as we do, and they had less of criticism; these elements they mixed up, without knowing whence they came. They amalgamated without scruple, mixing up the whole without any particular regard to details, putting in their originality without being aware they were doing so. Criticism, upon the contrary, cannot digest; the morsels remain whole; so that it is easy to see the difference. The dogma of the Divinity would never have been formed if the Christian doctors had taken into account the thousand variations of detail which we see. Modern eclecticism is excellent as a principle of criticism, but barren as an attempt at dogmatic fusion; it will never be anything more than a piece of marquetery work; a juxtaposition of distinct pieces. In former times, a new spirit or new institutions were formed by a thorough mixture of different principles, just as our coarsest aliments are transformed by cooking. The institutions or dogmas of the past were taken as they came, and arranged according to individual fancy. The Middle Ages made for themselves an empire out of ancient and very inaccurate recollections. If the Middle Ages had been as familiar with history as we are, they would not have indulged in this pretty fancy. The counter-sense (*contresens*) had much to do with these strange creations, and I hope some day to show the part it played in the formation of our most essential dogmas; or rather the uncritical tendency sought to discover its aim in the past, and, in order to do so, fashioned the past at its fancy. This, assuredly, is a rude kind of science, if ever there was one. Nevertheless, it created more than ours has, thanks to its very rudeness. Clear and delicate vision serves only to draw distinctions; analysis can never be more than mere analysis.

And yet analysis is, in its way, a progress. In syncretism, all the elements were jumbled together without that precise distinction which characterizes analysis; without that splendid unity which results from perfect synthesis. It is only in the second degree that the parts begin to outline themselves clearly, and this is done, it must be confessed, at the cost of that unity of which the primitive state offered at all events the outward show. Then it is the multiplicity, the division which predominate until synthesis, taking possession of these isolated parts—which having had a separate existence, are henceforth conscious of the fact—fuses them anew in a superior unity.

In reality, this great law is not solely the law of human intelligence (138). Evolution from a primitive and syncretic germ by the analysis of its members, and fresh unity resulting from this analysis, such is the law of all that has life. A germ is deposited, containing *in posse*, without distinction, all that the being will some day become; the germ develops, the forms become constituted in their regular proportions, that which was potential becomes a fact; but nothing is created, nothing is added. I have often successfully used the following comparison to make this view understood. Let us imagine a mass of homogeneous hemp, drawn out in separate strands, the mass will represent syncretism, in which all instincts have a confused existence; the strands will represent analysis. If we imagine the strands, while remaining distinct, to be afterwards interlaced so as to form a rope, we have the synthesis, which differs from the primitive syncretism, insomuch as that the individualities, while knotted together in unity, remain distinct. In an hypothesis which I am far from assuming as dogmatically certain, but merely as a striking illustration of the system of things, the law of God would not be very different from this. Primitive unity was without life, for life can only exist upon the condition of analysis and of the opposition of the parts. The being was as if he did not exist; for nothing was distinct in him; the whole was without individualization or separate existence. Life only began when the obscure and confused unity was developed in multiplicity and became the universe. But the universe, again, is not the complete form; the unity is not sufficiently distinct in it. The return to unity operates in it by the intelligence, for the intelligence is merely the unique outcome of a certain number of multiple elements. The history of the being will only be complete when multiplicity is entirely converted into unity, and when from everything that exists shall issue an unique resultant which will be

God, just as in man, the soul is the resultant of all the elements which compose it. God will then be the soul of the universe, and the universe will be the body of God, and life will be complete; for all the parts of that which is will have lived apart and will be ripe for unity. The circle will then be closed, and the being, after having traversed the multiple, will anew rest in the unity. But why, it may be asked, emerge therefrom to re-enter it again? What good will the voyage athwart the multiple have done? The good will be that all will have lived its own life, and that thus analysis will have been introduced into the unity. For life is not absolute unity or multiplicity; it is multiplicity in unity, or rather multiplicity resolving itself into unity (139).

The perfection of life in the animal is in direct ratio to the distinctiveness of its organs. The lower animal, in appearance more homogeneous, is, in reality, inferior to the vertebrated animal, because a grand central existence is the outcome, in the latter, of several perfectly distinct elements. France is the first among nations, because she is the unique concert resulting from an infinity of different sounds. The perfection of humanity will not be the extinction, but the harmony of nationalities; nationalities continue to increase rather than diminish in strength; to destroy a nationality is to destroy a sound in humanity. "Genius," writes Michelet, "is only genius in that it is at once simple and analytical, at once child and full-grown man and woman, barbarian and civilized (140)." In the same way, science will only be perfect when it is at once analytical and synthetical; when exclusively analytical, it is narrow, dry and scanty; when exclusively synthetical, it is chimerical and gratuitous. Man will only really have knowledge when, while affirming the general law, he has a clear view of all the detailed facts which it implies.

All the special sciences start by the affirmation of

unity, and only begin to distinguish when analysis has revealed numerous differences where before had been visible nothing but uniformity. Read the Scottish psychologists, and you will find at each page that the primary rule of the philosophical method is to maintain distinct that which is distinct, not to anticipate facts by a hurried reduction to unity, not to recoil before the multiplicity of causes. Nothing can be better, but upon the condition that, by an ulterior outlook, one makes sure that this reduction to unity, which is not yet ripe, will one day be effected. It would, assuredly, be very strange that there should be in nature sixty-one simple bodies, neither more nor less; that there should be in man eight or ten faculties, neither more nor less. Unity is at the foundation of things, but science must wait for its appearance, while still feeling assured that it will appear. It is an error to reproach science with thus reposing in diversity, but science, upon the other hand, would be wrong if it did not make its reservations and recognize this temporary diversity as being destined to disappear some day after a deeper investigation of nature.

The present state is open to criticism and is incomplete. True science, the complete and felt science, will be for the future, if civilization is not once again arrested in its march by blind superstition and the invasion of barbarism, in one form or another. But, whatever happens, even should a Renaissance become necessary, it is unquestionable that it would take place, that the barbarians would look to us as to the ancients to advance further than we have done, and to obtain in their turn new points of view. Then pity will be felt for us, the men of the age of analysis, reduced to see nothing but a corner of things; but we shall be honoured for having preferred humanity to ourselves, for having deprived ourselves of the pleasure of general results, in order to put the future in the position of being able to deduce them with certainty, very different from those egotistical thinkers of the early ages, who endeavoured to improvise for themselves a system of things rather than to gather for the future the elements of the solution. Our method is *par excellence* the disinterested one; we do not labour for ourselves; we are willing to be ignorant in order that the future may have knowledge; we labour for humanity.

This patient and severe method seems to me to be suited to France, which, of all countries, has practised with the most firmness the positive method, but which is also the one in which abstruse speculation has been the most barren. Without accepting in all its fulness the reproach which Germany levels at us, of understanding absolutely nothing in religion or in metaphysics, I admit that the religious sense is a very weak one in France, and it is precisely on that account that we attach more importance than others, in religion, to narrow formulæ which exclude altogether the ideal. This is why there will never in France be any medium between the strictest Catholicism and incredulity; this is why it is so difficult to make people understand that, if you are not a Catholic, you are not necessarily a Voltairian. The metaphysical speculations of the French school (excepting, perhaps, Malebranche) have always been paltry and timid. The true French philosophy is the scientific philosophy of a Dalembert, a Cuvier, or a Geoffroy St. Hilaire. Theological development has been quite null in France; there is no country in Europe where religious thought has been less active. Strange to say, the very men who have been so quiet, so delicate, so swift to note the slightest shades of difference in real life are regular simpletons in metaphysics, and accept without question enormities which to the critical sense are simply revolting. They feel this, and do not concern themselves with them. But as the need for a religion is one common to humanity, they find it convenient to take ready made the system which lies handy, without stopping to consider whether it is acceptable (141). Religion has always in France been a sort of separate wheel, a stereotyped form, like "Louis by the grace of God," having no connection with the remainder and which is not read; a dead letter. Our religious wars are in reality only civil or party wars. If France had possessed a stronger religious sentiment, she would have become Protestant, like Germany. But not having the sentiment of the theological movement, she saw no half-way house between a given system and the disdainful repudiation of this system. France is in religion what the East is in politics. The East can imagine no other government than an absolute one. Only, when absolutism becomes intolerable, the sovereign is stabbed. This is the only political tempering understood there. France is the most orthodox country in the world, for it is the least religious and the most positive country. The Franklin type of man, the man of the present day who is as atheistic as possible, is often the one most closely attached to formulæ. If clever people look into the matter at all closely, they either fall back with characteristic facility upon our incompetency to judge matters of this kind, or else they fairly laugh at us. There is among unbelievers even, in France, a certain fund of Catholicism. The pure ideal religion, which, in Germany, has so many proselytes, is quite unknown with us (142). A system ready made, which it is not necessary to understand and which spares us the pains of searching, that is what France requires in religion, because she quite understands that she has not the delicate perception of things of this kind. France pre-eminently represents the analytical, revolutionary, profane and irreligious period of humanity; and it is because of her very powerlessness in religion that she clings with this sceptical indifference to the formulæ of the past. It may be that some day France, having accomplished her task, will become an obstacle to the progress of humanity and disappear; for the parts are quite distinct; the man who has effected the analysis does not do the synthesis. To each his work, such is the law of history. France will have been the great revolutionary instrument, will she be equally powerful for religious re-edification? The future will show, but, be this as it may, she has done enough for fame in having delineated one side of humanity.

CHAPTER XVII.

WOULD to God that I could have succeeded in making clear to a few lofty minds that there is in the pure cultivation of human faculties and of the divine objects which they attain a religion as suave, as rich in delights, as the most venerable forms of worship. I have tasted in my childhood and early youth the purest joys of the believer, and I say from the bottom of my heart that these joys are nothing by comparison with what I have felt in the pure contemplation of the beautiful and the passionate search after truth. I wish to all my brethren who have remained orthodox a peace to be compared with that in which I live since my struggle is over and since the appeased storm has left me in the midst of this great pacific ocean, a sea without wind or shore, upon which one has no star but reason, no compass but one's own heart.

A scruple, however, sometimes rises within me, and the idea which I have endeavoured to express in these pages would be incomplete if I did not offer the solution of it here. It is, in fact, the grand objection incessantly raised against rationalism ; and I am anxious to express my feelings upon this point. Science and humanism, I may be told, offer you sufficient religious sustenance. But can the religion be that of all the world ? Can the man of the people, bending beneath a never-ending toil, the limited intelligence, to which the secrets of a higher life must ever be a closed book, hope to have his part in this worship of the perfect? That if your religion is for a small minority, that if it excludes the poor and the humble, it is not the true one; more than that, it is barbarous and immoral, inasmuch as it banishes from the kingdom of Heaven those who are already dispossessed of the joys of earth.

These objections are all the stronger because I am the first to admit that science, to arrive at this degree in which it offers a religious and moral aliment to the soul, must elevate itself above the vulgar level, that ordinary scientific education is here quite insufficient, that you must have, in order to realize this idea, a life entirely devoted to study, a scientific asceticism which never falters and the most absolute renunciation of the pleasures, the business and the affairs of this world; that not only the ignorant man is radically incapable of understanding a word of this system of life, but that even the immense majority of those who are looked upon as educated and cultivated are absolutely incapable of attaining to it.

Yes, I admit that rational and pure religion is only accessible to the small minority. The number of philosophers has been, as it were, imperceptible in humanity. The most modest of religions has had a thousand times as many followers and has had more influence upon the destinies of humanity than all the schools put together. Philosophy after our fashion presupposes a long period of culture and habits of thought which very few are capable of. I do not know whether it is possible anywhere in France outside Paris to place oneself properly at this point of view, and I should be afraid of going too far if I said that there are now in the world two or three thousand persons capable of worshipping after this fashion. But the humble are not, on that account, excluded from the ideal. Their formulæ, although inferior, suffice to make them lead a noble life, and the people more especially have in their grand instincts and powerful spontaneousness an ample compensation for what is denied them in the way of science and reflection. Is he who can understand the preaching of a village Jocelyn, and these parables,

> Où le maître, abaissé jusqu'au sens des humains,
> Faisait toucher le ciel aux plus petites mains,

disinherited from the heavenly life? All men, by the sole fact of their participation in human nature, have their right to the ideal; but it would be running counter to evidence to pretend that all are equally apt to taste the delights of it. While repeating with Michelet "Oh! who will deliver me from the bond of inequality!" while admitting that, in respect to intelligence, inequality is harder to bear for the privileged man than for the inferior, it must be said that this inequality is in the nature of things, and that the theological formula is in this particular perfectly true; all men have sufficient grace to work out their own salvation; but all are not called to the same state of perfection. Mary has the better part which shall not be taken from her. What may be regarded as certain is that if humanity were as highly cultivated as we are, it would have the same religion as we have.

If, therefore, you blame the philosopher for the exceptional excellence of his religion, you must also blame the man who seeks in the ascetic life a higher perfection for being called to an exceptional state. You must also reproach the man who cultivates his mind for breaking the vulgar line of humanity. It must be allowed, painful as the admission may be, that perfection, in the present state of society, is possible only to very few. Are we to conclude from this that perfection is bad and injurious to humanity? Assuredly not; we need only regret that it is subject to such narrow conditions. It is intolerable pride on the part of the philosopher to imagine that he has the monopoly of the higher life; it would be a very blameworthy piece of egotism for him to rejoice in his isolation and designedly to prolong the degradation of his fellow-men in order that he might have no equals. But it cannot be imputed as a crime to him that he should raise himself above the common level, and exclaim with St. Paul: *Cupio omnes fieri qualis et ego cum.* Do not therefore contend that the inferiority of philosophy consists in its being accessible to the small minority; for this is, on the contrary, its chief title to glory. The only practical conclusion to be drawn from this melancholy truth is that one should labour to hasten the advent of the blessed day in which all men will have their place in the sunshine of intelligence and will be called to the true light of the children of God.

It would be a very pleasant but very chimerical optimism to hope that this day is at hand. But it is the property of faith to hope against hope, and there is nothing, after all, which the past does not justify us in hoping from the future of humanity. For how different were the conditions of intellectual culture in the antiquity of Greece from what they are to-day. In the present day, science and philosophy are a profession. "One does not get credit with the world," says Pascal, "for understanding poetry if one has not hung out the poet's sign, nor for being clever at mathematics, if one has not displayed that of the mathematician." In the noble ages of antiquity, a man was philosopher or poet, as one is an honest man in all situations of life. No practical interest, no official institution were necessary to stimulate the zeal for research or for the production of poetry. Spontaneous curiosity, the instinct of what was beautiful, sufficed. Ammonius Saccas, the founder of the highest and most learned philosophical school of antiquity, was a porter. Imagine a market porter of the present day creating in France an order of speculation analogous to the philosophy of Schelling or of Hegel. When I think of the noble people of Athens, where every one felt and lived the life of the nation, of this people which applauded Sophocles' plays, of this people which criticized Isocrates, of this people where the women said: "This, then, is Demosthenes," of this people where a female vendor of herbs detected Theophrastus to be a stranger, where every one had been taught in the same gymnasia and had learnt the same songs, where every one knew Homer and understood him in the same sense, I cannot help feeling rather sore at our society being so profoundly divided into men of culture and barbarians. With the Greeks, all men had their share in the same souvenirs, all gloried in the same trophies (143), all had contemplated the same Minerva and the same Jupiter. What are Racine, Bossuet, Buffon and Fléchier to our fellow-countrymen? What do they know of the heroes of Louis XIV., of Condé and Turenne? What meaning do Nordlingen and Fontenoy (144) convey to them? The people in our day are disinherited from the intellectual life; there is no literature for them. There was only one kind of taste at Athens—the taste of the people, good taste. With us there is the popular taste and the taste of men of refinement; the distinguished kind and the petty kind. In order to appreciate our literature, a man must be well-read, a critic and more or less of a wit. The vulgar herd admires without knowing why, and does not venture upon a judgment of its own upon works which exceed the limits of its intelligence. Germany does not know what provincial taste is because she has no taste of the capital; antiquity had nothing of the weak and popular style because it had no aristocratic literature.

I cannot conceive an elevated mind remaining indifferent to such a spectacle and not suffering at the sight of the greater part of humanity being excluded from the domain which it possesses and which might so readily be divided. There are some people who do not understand happiness except as an exceptional favour to themselves, and who would not appreciate wealth, education and intelligence if they were the possession of all the world. These latter do not love perfection in itself, but relative superiority; they are full of vanity and egotism. For my own part, perfect happiness is, as I understand it, that all men should be perfect. I cannot understand how the opulent man can fully enjoy his opulence, while he is obliged to veil his face in presence of the misery of a portion of his fellow-creatures. My greatest sorrow is to reflect that all cannot share my happiness. There will only be happiness when all are equal, but there will only be equality when all are perfect. What pain for the savant and the thinker to find themselves, through their very excellence, isolated from humanity, having their world apart and their belief apart! And yet you wonder that with this they are sometimes sad and solitary! But even if they were in possession of the infinite, of absolute truth, they could not but be pained at possessing it alone, or fail to regret the commonplace dreams which they at least had in common with their fellow-men. There are souls which cannot endure this isolation, and which prefer attaching themselves to the weak rather than to stand out by themselves in humanity. I admire and like these men. . . . At the same time, the savant cannot follow this course, even if he wished to do so, for what has been proved to him to be false is henceforward incapable of being accepted. The spectacle of the physical sufferings of the poor is, no doubt, a lamentable one, but I confess it does not come home to me so keenly as to see the immense majority of my fellow-men condemned to intellectual helotism, to see men similar to myself, possessing perhaps intellectual and moral faculties superior to my own, reduced to a state of brutal degradation, unfortunate passengers through life who are born, live and die without having for a moment lifted their eyes from the servile instrument which gives them their daily bread, without having for a single moment breathed in God.

One of the commonplaces most frequently repeated by vulgar minds is this: To initiate the masses devoid of fortune into the intellectual culture usually reserved for the higher classes of society is too often for them a source of pain and suffering. Their education will merely serve to make them feel the social want of proportion, and to render their condition intolerable. That is, I repeat, just the bourgeois view, which only envisages intellectual culture as a complement to worldly fortune, and not as a moral good. Yes, I admit that the simple are the happiest; but is that a reason for not raising oneself? Yes, these poor creatures will be more unhappy when their eyes are opened. But it is not a question of being happy; it is a question of being perfect. They are as well entitled as others to the nobility of suffering. Remember that the question at issue is true religion, the only thing which is serious and sacred.

I can understand the most radical divergencies as to the best means of operating for the greatest good of humanity, but I do not understand that honest minds should differ as to the aim and substitute egotistical ends for the great divine end: perfection and life for all. Upon this first question, there are only two classes of men: the honest men who subordinate themselves to the great social end, and the immoral men who are resolved to get enjoyment, and who care little whether they do so at the cost of others. If it were true that humanity was so constituted that there was nothing to be done for the general good, if it were true that politics consisted in stifling the cries of the wretched and in looking with folded arms upon evils as if they were without a remedy, nothing could induce noble minds to endure life. If the world were thus constituted, we should have no alternative but to curse God and then commit suicide.

It is not enough for the progress of human intelligence that a few isolated thinkers should reach very advanced posts, and that a few heads should shoot up like wild oats above the common level. Of what service is the most magnificent discovery if only a hundred persons or so are to profit by it? How is humanity served if seven or eight persons have been able to perceive the true reason of things? A result can only be regarded as acquired when it has entered into general circulation. Now the results of abstruse science are not of the kind which have simply to be enunciated. Mens' minds have to be raised up to them. It would be all very well for Kant and Hegel to be right; their science, in the present condition of things, would be incapable of being communicated. Would this be their fault? Not at all, but that of the barbarians who cannot understand them, or rather the fault of the society responsible for the existence of barbarians. A civilization is only really strong when it has a widely extended basis. Antiquity had thinkers almost as advanced as our own, and yet ancient civilization perished owing to its paucity, buried beneath the multitude of barbarians. It did not rest upon a sufficiently large number of men; it disappeared not for want of intensity, but for lack of extension. It is a matter of great urgency, I take it, to enlarge the whirl of humanity; otherwise a few individuals might reach heaven while the mass is still dragging along upon the earth. A progress of that kind would not be a genuine one and would be of no effect.

If intellectual culture was merely a form of enjoyment, it would not be a cause of complaint that only a minority had a share in it, for man has no right to enjoyment. But from the moment that it is a religion, and the most perfect of religions, it becomes barbarous to deprive a single soul of it. Formerly, in the age of Christianity, that was not so revolting; upon the contrary, the lot of the unfortunate and the simple was in one sense to be envied, inasmuch as they were nearer to the kingdom of God. But the charm has been broken and cannot be restored. The outcome of this is very shocking, for we see men condemned to suffer, without a single moral thought, without an elevated idea, without a noble sentiment, retained only by force like brutes in a cage. That, assuredly, is intolerable.

But what is to be done? Are these beasts to be let loose upon men? No, for humanity and civilization must be saved at any cost. Are the brutes to be kept under lock and key and well beaten when they offer resistance? That is a horrible alternative. No, they must be made men of, they must be given their part in the delights of the ideal, they must be elevated, ennobled, made worthy of liberty. Till that is done, to preach liberty will be equivalent to preaching destruction; it will be very much as if, out of respect for the *right* of bears and lions, one opened the bars of a menagerie. Until then, violent actions are necessary, and although to be condemned in the analytical appreciation of facts, they are, in effect, legitimate. The future will absolve them, as we absolve the great Revolution, while deploring its culpable acts and stigmatizing those who provoked them.

But it is a waste of time to vex one's mind over these problems. They are in a speculative sense insoluble; they will be solved by brute force. It is like reasoning upon the crater of a volcano, or at the foot of a dyke when the waters are rising. Humanity has many a time thus found itself arrested in its march like an army brought up short by some unfathomable precipice. The shrewdest then lose their heads, and human prudence is at its wit's end. The wisest suggest turning back and making a circuit of the precipice. But the crowd behind is ever pressing forward; those in the foremost ranks are toppled over into the yawning gulf, and when their bodies have filled up the abyss, the last comers pass over on the level. God be praised, the abyss is crossed. A cross is erected at the spot, and the tender hearted come and weep there.

Or, to take another comparison, it is as when an army has to cross a broad and deep river. The cooler heads wish to build a bridge or to construct pontoons, but the more impatient determine to let the men swim across; three-fourths perish, but anyhow the river has been crossed. Humanity, having at its disposal forces without limit, does not show itself very economical in regard to them. These terrible problems are insoluble, and one can only fold arms and look on in despair. Humanity will leap over the obstacle and do all for the best. Absolution for the living, and holy water for the dead!

Oh! how fortunate it is that passion undertakes these cruel executions. Men of delicate mind would hesitate too much and go to work too timidly. When the work to be done is to found the future while inflicting blows upon the past, you require some of those redoubtable sappers who are not affected by woman's tears, and who are not afraid to use the axe. It is only by a revolution that institutions which have been long since condemned can be destroyed. In a time of tranquillity, people cannot make up their mind to strike, even when that at which the blow is aimed has ceased to have any *raison d'être*. Those who believe that the renovation which was necessitated by all the intellectual triumph of the eighteenth century could have been effected peacefully are mistaken. Efforts would have been made to compromise, a thousand personal considerations, which in time of peace are much prized, would have been brought into "play;" no one would have dared to abolish outright either the privileges or the religious orders, or so many other abuses. The tempest took this in hand. The temporal power of the Popes is assuredly out of date. Yet if everybody had this opinion, no one would make up his mind to clear away this relic of the past. We must wait for the next earthquake to do that. Nothing is done in times of tranquillity; it is only in revolution that people show daring. One should always endeavour to lead humanity in the paths of peace and to let revolutions glide along the soft inclines of time, but if one is more or less critical, one is fain to admit to oneself that this is impossible, that the matter cannot be effected in that way. But in any case the thing will be done in one way or another. It is a waste of time to calculate and cunningly combine the means; for brutality will have its finger in the pie, and there is no calculating with brutality. We have here an antinomy and an unstable equilibrium, as in so many other questions relative to humanity, when we envisage them exclusively in the present. There are men who are necessarily detested and cursed by their age; the future will explain them and say with calm impartiality: it was necessary that there should be men of that stamp (145). Moreover, this posthumous rehabilitation is not rigorously just; for as they are nearly always immoral, they have had their reward in the satisfaction of their brutal passions. I can imagine in theory a virtuous revolutionist, who would act in a revolutionary spirit through the sense of duty and in view of the calculated good of humanity, so that circumstances alone would be to blame for his acts of violence. But as a matter of fact there has never been any individual of the sort, and it may be that such a character is outside the limits of humanity. For such acts cannot occur without passion being imported into them, and, upon the other hand, such passions cannot fail to evoke some disinterested view. The character of revolutionists is very complex, and the extremely simple explanations given of them are convicted of being false by their very simplicity.

Theophylactus relates that Philippicus, a Roman general, being on the point of giving battle, began to weep as he thought of the great number of men who were about to be killed. Montesquieu calls that bigotry, but it was, perhaps, merely the result of a large heart. It is good to weep over these terrible necessities, provided that the tears do not prevent you from marching forward. What a cruel alternative for the high minded! Either to form alliance with the wicked and draw upon one the curses of those one loves, or to sacrifice the future.

Woe to him who brings about revolutions; happy

he who enjoys their fruits, and happier still they who, born in a better age, will no longer need to have resort to the most irrational and absurd means to effect the triumph of reason. The moral point of view is too narrow to explain history. One must raise oneself up to the level of humanity, or, let it rather be said, one must soar above humanity and raise oneself to the Supreme Being, where all is reason and where all differences are reconciled. There is the great white light, which, lower down, is refracted in a thousand hues separated by undiscernible limits.

M. Pierre Leroux is right. We have destroyed paradise and hell. Whether we have done right or wrong, I cannot say, but it is certain that we have done this. One cannot replant a paradise, one cannot relight a hell. We must not remain half-way. We must bring down paradise upon earth that all may enter it. And paradise will be here below when all have their share in light, perfection, beauty, and therefore in happiness. When the priest, with a congregation of believers around him, preached resignation and submission, because, after all, it was merely a question of suffering for a short time, after which would come eternity, when all sufferings would be reckoned as merits, that was all right enough. But we have destroyed the influence of the priest, and it is not in our power to re-establish it. We decline to submit to it ourselves; it would be a strange thing that we should wish to impose it upon others. Even supposing that we still had some influence upon the people, supposing that our advice had some weight and would not rather excite their mistrust, with what sort of a face could we sceptics go and preach Christianity, which we admit that we no longer have need of, to people who require it, in order to make things comfortable for us? What would be the name to give to such a procedure? For, since the beginning of the world, has there been a single instance of such a miracle as lying and hypocritical scepticism making men believers. Conviction alone effects conviction. I have read—where, I forget—a history of some bonzes guaranteeing to an old woman paradise in another world, if she would give them her fortune in this one. But the sceptic who preaches paradise and hell, in which he does not believe, to a people which does not believe in it either, plays a much more contemptible part. "My friends, leave me the enjoyment of this world, and I promise you enjoyment in the next." This, assuredly, would be a very good comedy scene, and the people, who have a very keen sense of the humorous, would be much amused by it.

God forbid that I should say that belief in immortality is not in one sense necessary and sacred. But I maintain that when the sceptic preaches this consoling dogma to the poor without believing in it, simply to keep him quiet, this should be termed a swindle ; it is equivalent to making a payment in notes which one knows to be bad ; it is turning the simple out of the track of the real and the true by means of a chimera. It cannot be denied that too much concern for the life of the future is in some respects injurious to the welfare of humanity. When one reflects that all things will be rectified above, it is no longer worth while to pursue so eagerly order and equity here below. Our principle is that we should regulate our present life just as if the future life did not exist, that it is never justifiable to refer to what is beyond in extenuation of any social condition or action. To appeal incessantly to the future life is to deaden the spirit of reform, to relax the zeal for the rational organization of humanity. All the work of social reform accomplished by the French bourgeoisie since the eighteenth century rests upon this implicitly recognized principle, that the present life must be organized without regard to the future. It is the surest way of not letting any one be made a dupe.

But, it will be said, at all events do not interfere

with the priest, who is a believer himself, and who, therefore, may effect a conversion. True enough; but do not place too much reliance upon this apostleship improvised in a moment of panic; the people will feel that you are very pleased that they should be thus preached to, and will notice that you remain incredulous. You may pay missionaries to preach in all the villages; but your incredulity will be a more effective sermon than any they can preach.

Well, then, let us be converted ourselves. To make the people believe, we must ourselves believe. Of all courses, this is the most impracticable; religions are not to be resuscitated. A man cannot be converted at will. You will believe in a moment of panic; you will try to believe. But what strange Christians are the Christians of fear! As soon as the sun comes out, you will revert to your incredulity. You may have driven Voltaire out of your library; you will not drive him out of your memory, for Voltaire is yourself.

The idea, then, of containing the people by means of ancient ideas must be given up. There remains brute force, but be on your guard. Do not place much reliance upon that; helots in a minority are still the stronger. One false step, one maladroit act will be sufficient for them to push you down and trample upon you. Are you quite certain of not making one false step in twenty years? Remember that they are there, behind you, waiting their opportunity. And besides, this is immoral and intolerable when one comes to think upon it. The happiness which I enjoy is only to be had at the cost of a portion of my fellow-creatures. If, for an instant, the mastiffs which keep watch at the door of the *ergastulum* relax their vigil, all is over. I never have been able to realize a sense of security in a country constantly threatened with the invasion of the waters, nor moral happiness in a society which presupposes the degradation of a part of the human race.

Nor must you fail to remark the fatality which has brought things to this point, and which has riveted each link in the chain, and do not think that you have said all there is to be said when you have declaimed against this or that. It was by sheer force of things that cultured humanity broke off the yoke of ancient creeds, and was led to find them unworthy of acceptation. Can it be blamed for this, and can people believe what they please? There is nothing more fateful than reason. It was by the fatal working of things, and without any impulse from the philosophers, that the people in turn became incredulous. Who is to be blamed for this, seeing that it did not rest with the first sceptics to remain believers, and that they would have been hypocrites had they simulated a belief which they did not possess, while it would have been of very little effect, seeing that falsehood is powerless in the history of humanity. It is, lastly, by the force of things, that the unbelieving people has risen up against its masters in unbelief and has said to them: "Give me my share here below, seeing that you take from me my share in heaven." Thus there is nothing but what harmonizes in this development of the modern spirit; the whole march of Europe for four centuries is summed up in this practical conclusion: to elevate and ennoble the people, and to let all men have a share in the delights of the intelligence. Turn the problem about as you will, that is what it amounts to. In my opinion, this is the capital question of the nineteenth century; all the other reforms are secondary and premature, for they presuppose that one. To maintain a portion of humanity in a state of brutality is immoral and dangerous, to give back to it the chain of the ancient religious beliefs, which had a fairly moralizing effect upon it, is impossible. Only one course, therefore, remains, and that is to widen the basis of the family and to find room for all at the banqueting-table of light. Rome only escaped from social wars by opening its ranks to allies, after having vanquished them. Thank God, we also have conquered. Let us, therefore, open our ranks.

Society is not, in my opinion, a merely conventional tie, an external institution and a simple matter of police. Society has the charge of souls, it has duties towards the individual; it does not owe him life, but the possibility of life; that is to say, the first fund which, fertilized and multiplied by each man's labour, will in due course become the aliment of his physical, intellectual and moral life. Society is not the atom-like and fortuitous assemblage of individuals, as is, for instance, the tie which brings together passengers by the same vessel. It is primitive (146). If the individual were anterior to society, his acceptance would be necessary for him to be considered as a member of society and subject to its laws, and one might conceive, for instance, the possibility of his refusing to participate in its liabilities and advantages. But seeing that man is born into society, as he is born to reason, he is no more free to repudiate the laws of society than those of reason. Man is not born free, with full liberty to afterwards embrace voluntary servitude. He is born part of society, he is born under the law. He is no more entitled to complain of being subject to a law which he has not accepted than he is to complain of being born a man. The old societies had their sacred books, their epics, their national rites, and their traditions, which were, so to speak, the depôt of education and national culture. Each individual, upon coming into the world, found, in addition to the family, which does not suffice to make man, the nation, which is the depository of another and higher life. Christianity, which has destroyed the ancient conception of the nation and of country, has taken the place with modern peoples of this great national culture, and for a long time has quite sufficed. Thus, man has always had open before him a grand school of the higher life. Man, like the plant, is wild by nature; to be a man does not mean merely to have the human face or to reason upon a few plain subjects after the fashion of other people. To be a man, you must have intellectual and moral culture.

I believe, with the Catholics, that our profane and irreligious society, paying heed solely to order and discipline, caring little about the immorality and degradation of the masses, so long as they continue to turn the mill in silence, rests upon an impossibility. The State owes the people religion; that is to say intellectual and moral culture; it owes them the school even more than the temple. The individual is only completely responsible for his acts if he has received his share of the education which makes the man. By what right do you punish this wretch who has been shut off since his youth from moral ideas, having barely the power of discerning between good and evil, impelled by coarse appetites which are his sole law, and perhaps also by pressing needs? You punish him for being a brute, but is it his fault if no one took him at his birth to cause him to be born to the moral life? Is it his fault if the only escape he has received has been that of vice? And to remedy these crimes which you have been unable to prevent, you have only the galleys and the scaffold. The true culprit in all this is the society which has not elevated and ennobled this poor wretch. What a strange coincidence that nearly all criminals should spring from the same class! Nature, I would say with Pascal, is not so uniform. Is it not evident that if nineteen-twentieths of the crimes punished by society are committed by persons deprived of all education and prompted by want, the cause lies in this lack of education and in this want? God forbid that I should ever seek to excuse crime or disarm society against its enemies. But crime is only crime when it is committed with full consciousness. Do you suppose that this poor wretch would not, like you, have been honest and good if he had, like you, been cultivated by a long course of education and ameliorated by the salutary influences of the family? We must start from the principle that man is not actually born good, but with the power of becoming good, any more than he is born a savant, but with the power of becoming so; that the main thing is to develop the germs of virtue which he has in him, that man is not inclined to evil of his own choice, but by want, by the fatality of things, and especially for lack of moral culture. Assuredly, in the present state, when society cannot exercise a civilizing influence upon all its members, it is important to maintain punishment so as to deter those whom education has failed to keep from crime. But this is not the normal state of humanity, for, I repeat, you do not punish a man for being savage, though, if you have savages to govern, you may, so as to keep them in order, have recourse to the sanction of punishment. In that case, it is no longer a moral punishment; it is making an example, nothing more. I willingly admit that for a man to reach the utmost limits of want, the point where morality expires in presence of want, there must at one time or another of his life have been some fault of his own—I except of course the infirm and women—that with morality and intelligence a man can always find a way out of his difficulties and resources of some kind. But is it the fault of these poor people if they do not possess this morality and intelligence, seeing that these faculties need to be cultivated, and that no one has taken any pains to develop them?

All the evil which there is in humanity proceeds, as I think, from lack of culture, and society is not entitled to complain of this, seeing that it is, to a great extent, responsible for it. When calling the two parties which now dispute for the mastery in the world aristocrats and democrats, we may say that the one and the other are, in the present state of humanity, equally impossible. For the masses being blind and deficient in intelligence, to appeal only to them is to appeal from civilization to barbarism.

Upon the other hand, the aristocracy constitutes an odious monopoly if it does not set before it for its aim the tutelage of the masses; that is to say their gradual elevation. I was a spectator of those fatal days concerning which we may say:

> Excidat illa dies æro, nec postera credant
> Sæcula, nos etiam taceamus, et oblita multa
> Nocte tegi nostræ patiamur crimina gentis.

God knows that never for a moment did I desire the triumph of the barbarians; and yet I suffered pain to hear honest men pouring out mockery or anger upon these lamentable follies. It irritated me to hear people applauding the bloodiest acts of revenge or regretting that there were not more such. For, after all, did these senseless people know what they were doing, and was it their fault if society had left them in this state of imbecility through which they were destined, upon the first day of trial, to become the tool of the perverse and the foolish?

No one can deplore popular folly more than I do, and I am glad that it should be put down. But these acts of folly evoke in me only one regret, and that is that one half of humanity should be thus abandoned to its native bestiality, and I cannot understand any honest and clear-sighted mind failing at once to draw this conclusion. Of these beasts let us make men. People who laugh over these follies irritate me; for these follies are, in part, their work.

It used to be said in respect to hapless Italy: "Look and see if this people be worthy of liberty. Look what use they make of it and in what a way they defend it." No doubt, but whose is the fault? Is it the fault of those who are condemned to nullity and who, advanced in age, wake up children; or of those who have kept them under, and who then come and reproach a great country with the immorality of which they themselves have been guilty (147)? This indignation will ever remain among the most vivid recollections of my youth. A guardian has made his ward imbecile in order to preserve the management of his property. Chance restores to the ward for a moment the use of his fortune, and, as a matter of course, he makes ducks and drakes of it; from which fact the guardian draws a strong argument for placing his ward again under his charge.

So it is not "Away with the barbarians," but "Let there be no more barbarians," that we should exclaim, for as long as there are any, an invasion will always be on the cards. If there were, face to face with each other, two races of men, the one civilized, the other incapable of civilization, the only policy would be to stamp out the uncivilizable race, or to make it strictly subject to the other. If it was true, as Aristotle believes (148) that, just as the soul is destined to command and the body to obey, so there are in society men who have their reason within themselves, and others who, having their reason out of themselves, are only fitted to execute the will of others, the latter would naturally be slaves; it would be just and expedient that they should obey, their revolt would be as great a misfortune and crime as if the body revolted against the spirit. From this point of view, the conquests of the democracy would be the conquests of the spirit of evil, the triumph of the flesh over the spirit. But it is this very point of view which is deceptive; an incontrovertible degree of progress has put a ban upon this aristocratic theory and laid down as an axiom the inviolable rights of those who are weak in body and mind as against the strong. All men bear within them the same principles of morality. It is impossible to love the people as they are, and it is only the ill-intentioned who are desirous of keeping them in their present condition, in order to be able to make them answer their own purposes. But let them have a care; one day or other, the wild beast may likely enough turn and rend them. I am firmly convinced, for my own part, that unless we make haste and elevate the people, we are upon the eve of a terrible outbreak of barbarism. For if the people triumph in their present state, it will be worse than it was with the Franks and Vandals. They will destroy of their own accord the instrument which might have served to elevate them ; we shall then have to wait until civilization once more emerges spontaneously from the profound depths of nature. We shall have to traverse another period of the Middle Ages, to pick up the broken thread of learned tradition.

Morality, like politics, is summed up, then, in this grand saying: To elevate the people. Morality should have prescribed this course at all times; policy dictates it more imperiously than ever, now that the people have been admitted to share in political privileges. Universal suffrage will only be legitimate when all men shall possess that share of intelligence without which one does not deserve the name of man, and if, in the interim, it is to be maintained, this is solely because it is calculated to hasten the advent of that condition of things. Stupidity has no right to govern the world. How is it possible, I ask of you, to entrust the destinies of humanity to unfortunate beings, whose ignorance lays them open to all the tricks of charlatanism, who are scarcely entitled to rank as moral beings? A deplorable state of things truly, when, in order to obtain the suffrages of an omnipotent multitude, the great point is not to be true, learned, clever or virtuous, but to possess a name or to be a brazen charlatan!

I will suppose some learned and laborious searcher to have discovered, if not the definite solution, at all events the most advanced solution of the great social problem. It is undeniable that this solution would be so complicated that there would be barely twenty people in the world capable of understanding it. Let us hope that he may be gifted with patience if he intends waiting, in order to get his discovery accepted, the adhesion of universal suffrage. An empiric who proclaims loudly that he has found the solution, that it is as clear as noonday, and that only the bad faith of interested persons can refuse to recognize it, who repeats every day in the columns of a newspaper certain stale commonplaces, such a man as this will assuredly make his fortune more quickly than one who looks for success to science and reason.

Let it, therefore, be well understood that those who refuse to enlighten the people are those who want to make tools of them, and who need their blindness in order to succeed. Shame to those who, when they talk of an appeal to the people, know that they are only making an appeal to imbecility. Shame upon those who base their hopes upon stupidity, who rejoice in the multitude of fools as in the multitude of their own partisans, and who believe that they triumph when, thanks to an ignorance which they themselves have fostered, they can say: "You see that the people will have none of your modern ideas." If there were no more fools to be got over, the profession of sycophant and parasite of the people would soon be at an end. The immoral means of government, a Machiavelian police, and restrictions upon certain natural liberties and so forth, have hitherto been both necessary and legitimate. They will cease to be so when the State is composed of intelligent and cultivated men. The question of government reform is not, therefore, political; it is moral and religious; the Ministry of Public Education is the most important, I was going to say the only important one. After scrutinizing all the necessary antinomies of the present political programme, it will, I think, be admitted that the intellectual rehabilitation of the people is the remedy for them all, and that the most liberal institutions will be the most dangerous, as long as what has been so well called "the slavery of ignorance" shall last. Until then, government *à priori* will be the most detestable of all governments.

At the first dawning of modern liberalism, it was for a moment thought that absolutism was dependent for existence upon the force of a strong government.

But we have since found out that it has a much more powerful support in the stupidity and ignorance of the governed, inasmuch as we have seen peoples which have been set at liberty regret their chains and ask to have them forged afresh. It is no great thing to destroy tyranny; that has been done a thousand times in the course of history. But to get on without it. . . . In the view of some, that is the best apology which can be offered for those who govern; to my mind it is their gravest offence. Their offence is that they should have rendered themselves necessary and have maintained mankind in such a state of degradation that they themselves ask for slavery and shame. M. de Falloux has expressed his surprise that the Tiers État of '89 should have sought to avenge ancestors who were not conscious of any injury having been done them. That is true, and the most revolting part of the business, and that which cries most for vengeance, is that these forefathers did not, as a matter of fact, feel that they had been wronged.

As the greatest good of humanity should be the aim of every government, it follows that the opinion of the majority is only entitled to impose itself when that majority represents the most enlightened reason and views. What! In order to please the ignorant masses, you will do a perhaps irreparable wrong to humanity? I will never consent to recognize the sovereignty of unreason. The only sovereign by divine right is reason, the majority only has power so far as it is supposed to represent reason. In the normal state of things, the majority will, as a matter of fact, be the most direct criterion for ascertaining the party which is in the right. If there was a better means for ascertaining the truth, it would be expedient to have recourse to it and take no account of the majority.

If we were to be guided by certain politicians who dub themselves liberals, the sole duty of a government is to follow public opinion without ever attempting to direct the movement. It is, they say, an intolerable piece of tyranny that the central power should impose upon the provinces institutions, men and schools which are little in harmony with the prejudices of these provinces. They cannot admit that it is right that the administrators and teachers from these provinces should come to Paris and acquire an education which will render them superior to the people over whom they are placed. This is a singular scruple. Paris, enjoying a superiority of initiative and representing a more advanced stage of civilization, has the full right to impose herself and to carry forward towards perfection the less enlightened masses. Shame be to them who have no other prop than ignorance and stupidity, and who endeavour to preserve them as their best auxiliaries. The question of the education of humanity and the progress of civilization takes precedence of all others. No injury is done to a child in endeavouring to draw him out of his natural indifference, for the development of his intellectual and moral culture. The time is not yet near at hand when there will be no need to do good to humanity in spite of itself. To govern in the spirit of progress is to govern by right divine.

Universal suffrage presupposes two things: (1) that all men are competent to form an opinion upon questions of government; (2) that there is not, at the time of its being instituted, any absolute dogma; that humanity is at that moment without a fault and in the condition which M. Jouffroy has called practical scepticism (*scepticisme de fait*). These epochs are epochs of liberalism and toleration. When one man is not more richly endowed than his neighbour with the knowledge of the truth, the simplest way is to count heads; numbers constitute the right, or at least an external and practical right, which may very possibly not convert the minority but compels its acceptance. In reality, this is not very logical, for as numbers are not an indication of intrinsic truth, the minority might say: "You force yourselves upon us not because you are right, but because you are stronger in numbers. That would be right, if numbers represented force; for then, instead of fighting the question, it would be more reasonable to count heads and so avoid useless evils. But, although less numerous than you, we have better muscles and we are braver; so let us fight it out. We are no more or less in the right than you; you are the more numerous, we are the stronger; so let us come to." The fact is that such a state of things is not normal for humanity; and reason alone, that is to say the established dogma, confers the right of imposing one's will; numbers are, in short, of as superficial a character as force, and that nothing can be firmly established except upon the basis of reason.

I say it with due deference, and with the conviction that those who read these pages will not set me down as a sedition-monger. I say it as a pure critic, looking at the revolutions of the present day as we do at those of Rome, for instance, and just as people will look at ours five centuries hence; a triumphant insurrection is often a better criterion as to which party is right than is a numerical majority. For the majority is often composed or at all events based upon people who are very insignificant and inert, regardful only of their own repose, who do not deserve to be taken into account; whereas an opinion capable of stirring the masses and above all of causing them to triumph testifies by that very fact to its force. The vote by battle is at all events as trustworthy as any other form of vote, for with that only the living forces are counted; or rather the energy which opinion gives to its partisans is weighed; and that is an excellent criterion. People do not fight for what is dead; whatever stirs the pulse the most is that which is fullest of life and truth. Those who are attached to what is absolute and to clearly defined solutions are ready to appeal to the numerical tests, for nothing can be clearer than numbers; all you have to do is to count heads. But this would be making things too easy, and humanity does not go to work in quite so simple a way. Do what we will, we cannot find any other absolute basis than reason, and until humanity shall have reached a definitely scientific age, we shall have no other criterion of reason than the definite fact. The fact does not constitute reason, but indicates it. The best proof that the insurrection of June (1848) was unlawful was that it did not succeed.

Here we have a necessary, insoluble antinomy, and one which will endure until some great dogmatic formula has once more englobed humanity. In the periods of scepticism, when people are aspiring to a new form which has not yet taken shape, no one being quite sure as to what the true religion is, it would be intolerable that such an one should, of his own individual authority, come and impose his creed upon others. People only declare all religions to be equally good when no particular one is sufficient. If there were any religion really alive, which corresponded to the requirements of the age, we may be sure that it would not be long in establishing its claims and that the nation would not be inclined to haggle with it. Indifference in politics is what scepticism is in philosophy, a halt between two dogmatisms, one dead, the other in the germ. During this interregnum, each person is free to attach himself to any doctrine, to be, according to his fancy, Pythagorean or Platonian, stoic or peripatetic. All forms are equally inoffensive, and the only function of authority is to keep the peace between them and to prevent them from exterminating one another. It is not the same in the dogmatic states, in which there is a living and actual reason, a doctrine outside of which is no salvation. Strong in all the life of the nation, it is the prime necessity and right of that nation. It is in one sense superior to the political law, inasmuch as it finds therein its reason and its sanction. The government is then absolute and is carried on in the name of the doctrine which is universally accepted. All bends before it, and the spiritual power, which represents it, is all the more above the temporal power because the spiritual requirements of man are above his material interests, or, as it used to be said, the spirit is above the flesh. And this absolute rule is not tyranny. Tyranny only comes in when the chain is felt, when the ancient dogma has grown old and uses the same authoritative methods to maintain its supremacy. We are sometimes unjust towards the persecutions of the Church in the Middle Ages. She was bound to be intolerant at that time, for so long as a whole society accepts a dogma and proclaims this dogma to be absolute truth, and that without any opposition, it is charitable to persecute. It is neither more nor less than defending society. The wars of the Albigenses, the persecutions of the Waldenses, the Cathares, the Bogomites and the poor of Lyons do not shock me more than the crusades; they were, as a matter of fact, stray sheep, who had quitted the great flock of humanity; and as to the really advanced men of the Middle Ages, Scotus Erigena, Arnauld de Bresse, Abélard and Frederick II., they underwent the just punishment of being in advance of their time. The reason why these acts of the Inquisition in the Middle Ages excite our indignation is that we judge them from the standpoint of our own sceptical age; it is very evident that in our day, when there is no longer any dogma, such acts would be execrable. To massacre other people for an individual opinion is horrible. But when it is done for the dogma of humanity, the whole question is altered. That a man should be violent and even cruel in the defence of his disinterested faith is regrettable, but may always be excused. Persecution only becomes odious when it is the work of interested agents, who sacrifice the thoughts of others to their own ease and comfort.

This is why the persecutions of the Church in the Middle Ages and in our time must be judged quite differently. For in these modern times, persecution has ceased to be what it was in the Middle Ages; it is now no more than an antiquated form of oppression, worn out, cumbersome and illegitimate; all that it does to retain its power is odious, for it no longer has any *raison d'être*. The death of John Huss itself excites my anger, for he represented the future; the death of Vanini and Giordano Bruno revolts me, for the modern spirit had already become definitely emancipated, and as to the absurd religious persecutions of Louis XIV., no one but a narrow-minded and hard woman, Jesuits and Bossuet could have been capable of advising them to a worn-out and aged king. When the Church was the legitimate authority, she had much less cause to persecute than since she has ceased to be so. That great and odious persecution, the Inquisition, did not become positively monstrous until the sixteenth century, that is to say when the Church had been finally vanquished by the Reformation. Louis XIV. had not, so far as I can recall, a single act of severity to perform in order to maintain his absolute authority, and this was almost a matter of course; for his sovereignty was legitimate and accepted; no prince could have been more absolute and less tyrannical. The Restoration, upon the contrary, was always upon the *qui vive* to maintain a power assuredly much less extensive, and the smallest act of violence upon its part was revolting, for it was self-imposed. The measure of the violence which a power is obliged to display to maintain its position, and especially the indignation which this violence excites, is the measure of its legitimacy. We are legitimists in our own way. The legitimate government is that which is based upon the reason of the age; the illegitimate government is that which employs force or corruption to maintain itself in opposition to patent facts.

It is through not comprehending the difference between these two ages of humanity that so many sophisms are current as to the relations between Church and State. In the earliest age, that in which there is a true religion, the embodiment of society, the State and religion are one and the same thing, and so far from the State salarying religion, religion maintains itself, and it is rather the State which, upon certain occasions, appeals to the Church. It is even superior to the State, inasmuch as the State derives from it the principle of its existence. But in periods when the State, having no creed, says publicly: "I do not understand anything about theology, do as you please," it should not give a salary (it is then only that this ignoble word comes into existence) to any one form of worship, or, what amounts to the same thing, it should do so to all. What the State gives to religions is but mere alms; they may well blush in taking it, and I can quite understand the indignation of the Ultramontanes when they see God inscribed in the State budget like some public functionary. In these days, there are nothing except mere opinions, and why should the State pay a salary to opinions. I can understand the State recognizing a single creed, or not recognizing any. But I do not understand it recognizing all the creeds (149). The liberal theory of indifferentism is superficial. Humanity requires some doctrine. If Catholicism is true, the most extreme claims of the Ultramontanes are well grounded, the Inquisition is a beneficent institution. In fact, as from this point of view sound belief is the greatest good to which all the rest should be sacrificed, the sovereign does a fatherly act in separating the grain from the chaff and in burning the latter. All must give way to the one necessary fact: the saving of souls. The principle of *compelle intrare* is legitimized by its results. If in sacrificing a thousand polluted souls you may hope to save a single one, from an orthodox point of view the sacrifice is justified (150).

I am sorry that it should be so, but there is no getting away from the dogmatic question. Those who want to keep this question apart are in the impossibility of reaching a logical solution.

It shows want of shrewdness to presuppose an absolutely legal order of things, against which no objection can be raised and which imposes itself absolutely. Society is never in either a perfectly legal or totally illegal state. Every social state of things is inevitably illegal, so far as it is imperfect, and tends always to become more legal, that is to say towards being perfect. It is not less superficial to suppose that the government is merely the expression of the will of the greatest number, so that universal suffrage would be a natural right, and that, this suffrage being acquired, there would be nothing left but to let the will of the people express itself. That would be too simple and easy. Only college pedants, superficial and simple minds could be deceived by the apparent evidence of the representative theory. The mass is only entitled to govern if we suppose that it knows better than any one else what is best. The government represents reason, God, or, if that phrase be preferred, humanity in the highest sense (that is to say the lofty tendencies of human nature) not a set of figures. The representative principle was all very well to uphold in opposition to the ancient dispositions of individuals, when the sovereign considered himself entitled to command of his own right, which is much more absurd still. But, as a matter of fact, universal suffrage is only legitimate if it can hasten the march of social improvement. A despot who effected this improvement against the wishes of the greatest number would be entirely in his right. When the Napoleon we want, the great political organizer, shall come upon the scene, he will be able to do without the papal benediction or popular sanction.

The ideal government would be a scientific one, in which competent specialists would treat government questions as scientific ones and would seek for a rational solution of them. Up to the present time, it has been birth, intrigue or the privilege of first come first served which have generally conferred grades upon the governing class; and the first intriguer who succeeds in sitting down in front of a board of green cloth is dubbed a statesman. I am not at all sure whether some day we shall not have in some form or other, something equivalent to the Chinese institution of men of letters, and whether the government will not become the natural possession of competent men, of a sort of academy of moral and political sciences. Politics may be regarded as a kind of science and required as much study and knowledge as any other. In the primitive societies, the college of priests governed in the name of the gods; in the societies of the future, savants will govern in the name of rational search for what is best. In our day, such an academy would have a hard task if it had to demonstrate to ignorant and headstrong presumption the legitimacy of its action. The mania of foolish people for a reason to explain what they do not understand and to be angry when they do not understand is one of the greatest obstacles to progress. The wise men of the future will despise it.

But, it will be said, how are you to impose upon the majority that which is best, if it refuses your offer. That is the very point which requires the most delicate handling. The sages of old had some very convenient auxiliaries in the oracles, the augurs, the Egeriæ, etc. Others had armed forces at their disposal. All these means have become impossible. The religion of the future will cut the knot with its heavy sword. Let us learn at all events not to be so severe upon those who have resorted to a certain amount of ruse and what it is the fashion to call corruption, if in reality (and that is the essential condition) they have only had in view the welfare of humanity. If, upon the contrary, they have only had selfish considerations in view, they are tyrants and wretches.

It is doing a disservice to a ward to place him too soon in possession of his property. But it is a crime to keep him in a state of imbecility in order to retain perpetual control over him. Better far is a premature emancipation, for, after a brief period of dissipation, it may contribute to render him amenable to reason.

Until the people has become initiated to intellectual life, intrigue and falsehood are evidently put up for public sale. It is a question of securing the good graces of a blind old man, and, in order to do so, you must lie and cajole. The vivid scenes of Aristophanes are not in the least exaggerated. The suffrage of the unenlightened people can only lead to the rule of the demagogue or of the aristocracy of birth, never to a government based on reason. The philosophers, who are sovereigns by right divine, are unsympathetic in the eyes of the people and exercise little influence over them. Look at what was the fate of all the sages (*oï äristoï*) at Athens: Miltiades, Themistocles, Socrates and Phocion. They are not brilliant externally, they do not flatter, they are serious and severe, they do not laugh, they speak a language which is not understood of the multitude; that of reason. How can you expect that men of this kind, if they attempt to speak to the multitude, will fail to get into disfavour? Only those who appeal to the passions of the people, or who style themselves dukes or counts, speak a language which is intelligible. These two languages are easily understood.

This explains the disfavour the people has always shown against the philosophers, especially when they have had the temerity to concern themselves with public affairs. Left to choose between the charlatan and the genuine physician, the people always incline to the former. The people like to be told only things which are clear and easy of comprehension, and the unfortunate part of the matter is that in nothing is truth to be found upon the surface. The people are fond of banter. The most superficial and State views put in a grossly humorous vein which set the teeth of refined persons on edge, transport the ignorant with delight. The true interests of the people are rarely to be found where they appear to be. The wise men who go to the reality are regarded as the people's enemies, and the charlatans who confine themselves to commonplaces are as a matter of course their friends. Besides, there is, somehow or other, in the wise an indefinable degree of pride, however hard they try to be humble and condescending. It is not their fault; pride (and the word is not used in a depreciatory sense) is innate in them. The grand seigneur is proud too; but his pride does not shock the people so much. They console themselves for not possessing the gold and the ribbons of the grand seigneur; but they cannot forgive the thinker for being superior to them in intelligence, and they regard themselves as being at least as competent as he is in politics. The people are much more indulgent for the great than for the middle classes which are well educated and enlightened. The latter they regard as being upon the same level as themselves, and they look upon their superiority with great jealousy. The king and the royal family are as demigods and attract his affection. But a plain bourgeois, whose talents have carried him into power, cannot fail to be a thief and an intriguer. The great are too much above the people to excite their envy; jealousy can only exist among equals. A government composed of men with no great names is bound to be suspected and vilified. "What has this man, who is my equal, done that he should have attained this position? He must be a dishonest man, for otherwise he would be my superior, which, of course, is impossible. He has had the handling of the State funds, he must have let some of it stick to his fingers; for I know that if I had been in his place, I should have been much tempted to do so." Such is the language of coarse and vulgar envy. These suspicions never are directed upon those who are looked upon as being of another species and with whom it is hopeless to compare oneself. When in the company of some peasants, I noticed that they were very dissatisfied at the idea of their representatives in Parliament receiving a small salary even during the vacation when they were doing nothing for it; yet these same people had nothing to say against the millions of the civil list.

Assuredly, if everybody was like us, not only would government be much easier, but there would scarcely be any need of one. Governmental restrictions are in inverse proportion to the perfection of individuals. Now, all other people would be like us if they all had our culture, if all possessed like us the complete idea of humanity. Why is it that all liberty is accompanied by a corresponding danger and stands in need of a corrective? The reason is that liberty is the same for the wise as for the foolish. But when all men are wise, or when public reason is strong enough to keep the foolish in order, no restriction will be necessary.

Fichte has gone so far as to conceive a social state of things so perfect that the very thought of evil is banished from the mind of man. I believe with him that moral evil will have marked but one age of humanity, the age in which man was neglected by society and did not receive from it the moral inheritance to which he was entitled. "There are men," says M. Guizot, "who have full confidence in human nature. According to them, when left to itself, it tends in the direction of what is good. All the ills of society come from government, which corrupts mankind by violence or fraud." I am one of those who feel this confidence. But I believe that the evil is derived not from governments committing deeds of violence or fraud, but because they do not elevate. I, as a man of culture, do not find any evil in myself, and I am impelled spontaneously towards what seems to me the most noble. If all others had as much culture as myself, they would all, like myself, be incapable of doing an evil act. Then it might be said with truth: you are gods and sons of the Most High. Morality has hitherto been conceived in a very narrow spirit, as obedience to a law, as an internal struggle between contradictory laws (151). For my own part, I can declare that when I do right, I do not obey any one, that I do not engage in any struggle or win any victory, that I accomplish an act as independent and as spontaneous as that of the artist who derives from his inner sense the beauty of which he gives an external realization, that I have merely to follow with delight and in perfect acquiescence the moral inspiration which comes out from the recesses of my heart. The man of elevation has only to follow the pleasant beat of his inward impulse; he might adopt the motto of St. Augustine and of the Abbey of Thelema: "Fais ce que tu voudras," for he can only wish what is good. The virtuous man is as an artist who realizes the beautiful in human life as the sculptor does in marble, or the musician in sound. Is there anything like obedience or struggle in the act of the sculptor or the musician? This is pride you will say. But that depends. If by humility is meant the small value which man attaches to his nature, the slight esteem in which he holds his estate, I utterly decline to give to such a sentiment the title of virtue, and I reproach Christianity with having sometimes taken this view. The basis of our moral law is excellence, the perfect autonomy of human nature; the foundation of all our philosophical and literary system is the absolution of all that is human.

The ennobling and the emancipation of all men by the civilizing action of society, such is, then, the most pressing duty of government in the present state of things. Everything which is done outside of that is useless or premature. People are always talking about liberty, the right of public meeting, the right of association. Nothing could be better if the intelligence of those who are to profit by them were in a normal state; but until such is the case, nothing could be more frivolous. It will be all very well for people who are imbecile or ignorant to meet in conclave; no good can come out of their assemblage. Sectaries and party men imagine that it is only compression which prevents their ideas from prevailing, and they fret against this compression. They are mistaken. It is not the ill-will of governments which gags their ideas; it is that their ideas are not yet ripe; just as it is that peoples are kept in subjection not by the force of an absolute government but by the depression of the subjects of that government. Do you suppose that, if they were ripe for liberty, they would not secure it for themselves at once? Our French liberalism, thinking that it can explain away everything by means of despotism, thinking exclusively of liberty, regarding government and its subjects as natural enemies, is in reality very superficial. Let us get to understand that it is not a question of liberty, but of acting, of creating, of working. The true always finds enough liberty to make itself visible, and liberty can only be hurtful when it is sought for by those who have not got good sense. It serves only to favour anarchy, and is of no service for the real progress of humanity. If a commissioner of police comes into a room where a few weak and empty headed individuals are mutually exciting their instinctive passions, we declare that we are shocked at such a violation of liberty. Do you imagine that it is these poor wretches who are going to solve the problem? We use force to preserve for everybody the right of twaddling at his pleasure; would it not be much better to speak the language of reason and teach all men to speak and understand their language. Close the clubs and open schools in their stead, and you will be doing true service to the popular cause.

The liberty of saying what one pleases presupposes that those to whom one addresses oneself have the intelligence and discernment required for passing an opinion upon what is said to them; for accepting it if it is good advice, for rejecting it if it is bad. If there were a class legally to be defined incapable of showing this discernment, it would be necessary to exercise supervision over what was said to them; for liberty is only tolerable when accompanied by the corrective of public good sense which takes due account of errors. This is why liberty of teaching is an absurdity, as regards children. For as the child, accepting what is taught him, without being able to pass criticism upon it, regards his master not as a man who gives his opinion to his fellow-men that they may examine it, but as an authority, it is evident that supervision should be exercised over what is taught him, and that some other liberty should be substituted for his in order to effect this discernment. As it is impossible to trace categories between adults, liberty becomes, so far as they are concerned, the only course possible. But it is certain that until a people is educated, all liberties are dangerous and require restrictions. As a matter of fact, in questions relating to the liberty of expressing one's thoughts, we have not only to consider the right appertaining to the speaker—a right which is a natural one and only limited by that of others—but also the position of the listener, who, not always possessing the necessary discernment, is placed, as it were, under the tutelage of the State. It is from the point of view of the hearer and not from that of the speaker that restrictions are permissible and legitimate. The liberty of saying what one pleases can only be admissible when all men will have the necessary discernment, and when the best punishment for the foolish will be the contempt of the public.

I am sorry not to be able to make sufficiently clear my conviction as to the vanity and hollowness of our political and liberal agitation, which would be all very well in a State where men's minds were generally cultivated, and where many scientific ideas came into existence (for science cannot exist without liberty); but in a society composed for the most part of ignorant people open to every kind of seduction, and in which intellectual power is evidently on the decline, to do no more than defend these empty formulæ is to neglect the essential for almost insignificant legal forms, inasmuch as authority can always exclude them or interpret them in its own way.

M. Jouffroy has expressed this most clearly in his admirable discourse upon the scepticism of the time, which I might copy word for word as expressing my own views on this subject: "Each one of our liberties has appeared to us in turn as the object we have been longing for, and its absence as the cause of all our woes. And yet we have gained these liberties, and we are no better off, and the day following each revolution finds us in a hurry to draw up the programme of the next one. The fact is that we misconceive the true state of things; the truth is that each of these liberties we have so eagerly desired, that liberty itself is not and cannot be the aim for which a society like ours aspires. . . . Take them one after another, the liberties which we enjoy, and you will see that they are no more than guarantees and means; guarantees against whatever might stand in the way of the moral revolution, which alone can cure us, means of hastening on this revolution, etc." It is not saying much to declare that the public liberties are better guaranteed now than they were at the dawn of Christianity; and yet I maintain that a great idea would find in our day more obstacles in the way of its propagation than Christianity did at its birth. If Jesus were to appear now, He would be brought up before the magistrates; which is worse than being crucified. Imagine the life of Jesus being crowned by a commonplace death; what a difference! People are too ready to believe that liberty is favourable to the development of really original ideas: As it has been remarked that, in the past, every new system has been born and has grown outside the law, until it has in time become law in its turn, it was not unnatural to expect that in recognizing and legalizing the right of new ideas to develop, the state of things would improve. But it is the very contrary which has happened. There has never been so little originality in thought as since thought has been free. A true and original idea does not require permission to come into existence, and it matters little whether its right is recognized or not; it always finds enough liberty, for it makes for itself all the liberty which it requires. Christianity did not require liberty of the press or the right of public meeting in order to conquer the world. A liberty which is officially recognized requires to be regulated. Now a regulated, or restricted liberty constitutes a tighter chain than the absence of the law. In Judæa, under Pontius Pilate, the right of meeting was not recognized, and, as a matter of fact, people were all the more free to meet; for by the very fact of the right not being recognized, it was not in any sense restricted. Originality, I repeat, is better served by arbitrary methods and the drawbacks it entails than by the tangled web in which we are enfolded by thousands of laws and rules, which are an arsenal for weapons of every kind. Our formalistic liberalism is only really of service to agitators and to the petty originality which is so injurious to great originality and which is of so little use to the true progress of the human mind. We wear out our strength in defending our liberties, forgetting that these liberties are only a means, that they are only of value in so far as they may facilitate the advent of new ideas. We are above all things anxious to be free to produce, and as a matter of fact we produce nothing.

We shudder at the idea of the outward bond, and we understand nothing of the great boldness of thought. The very shadow of the Inquisition terrifies even the Catholics, and inwardly we are timid and devoid of "go;" we are only too ready to resign ourselves to public opinion and to habit; we sacrifice our originality to it, and whatever travels at all outside the commonplace routine is declared to be absurd. Germany, no doubt, at the end of the last century and the beginning of this had less outward liberty than we have, and yet I defy any one to deny that all the free thinkers in our Republic had not a quarter of the boldness and liberty which are breathed in the writings of Lessing, of Herder, of Goethe and of Kant. In reality, thought was more free half a century ago at the Court of Weimar, under an absolute Government, than in our country which has fought so many battles for liberty. Goethe, the friend of a Grand Duke, might have been prosecuted if he had written in France; the translator of Feurbach could not find a publisher who ventured to bring out his book. This is our peculiarity; we are an external and superficial nation, more concerned for the form than for the reality. The great and broad ideas about God have been and still are in Germany the doctrine of all philosophically cultivated minds; in France, no one has yet had the courage to profess them, and if he did he would encounter greater obstacles than he would have done at Tubingen or Jena under absolute governments. If it be asked whence the obstacle would come, the answer is from the intellectual timidity which closes our minds against any idea and traces around us the narrow horizon of the finite. France, I repeat, has only understood external liberty, but not in any respect liberty of thought. Spain, in reality as free and as philosophical as any other nation, has not felt the need of an external emancipation, and can you imagine that if she had seriously desired it, she would not have secured it? Liberty there is entirely internal; Spaniards have preferred to think freely in the dungeons and at the stake. Mystics like St. Theresa of Avila, indefatigable theologians like

Soto, Bañez and Suarez were in reality as bold speculators as Descartes or Diderot.

Let us, therefore, set ourselves to think more freely and scientifically, rather than to be more free to express our thoughts. The man who is right is always sufficiently free. Is it not very probable that those who protest against the violation of liberty are not so much the people who, possessed of the truth, suffer pain because they cannot divulge it, as those who, having no ideas of their own, work to their own advantage that liberty which ought only to serve for the rational progress of human intelligence? The innovators who have been right in the eyes of the future may, perhaps, have been persecuted; but persecution has never retarded for much more than a year the triumph of their ideas, and has in other ways been of more service to them than if they had secured immediate acceptance.

No doubt, it is our duty carefully to preserve the liberties which we have acquired with so many efforts, but, what is of greater importance still, is to convince ourselves that this is merely a primary condition which is advantageous if one has ideas and fatal if one has not. For what boots it to be free to meet if one has nothing of value to communicate to one another? What boots it to be free to write and speak, if one has nothing new or true to say? Each side has its part to play; persecutors and persecuted alike have their shoulders to the everlasting wheel; and after all the persecuted owe a debt of gratitude to the persecutors, as but for them they would not have the same title to admiration.

Persecution has the great advantage of getting rid of the petty originality which endeavours to gain kudos by a paltry kind of opposition. When men risk their lives for their ideas, it is only those who are possessed of God, those who are carried away by a powerful conviction and by the invincible need to utter their thoughts who put themselves in the forefront. Our guaranteed semi-liberties give too fine a field for intrigue; for one does not incur much risk, and the annoyances to which one exposes oneself are, at most, but a good investment for the future. This makes things too easy, whereas in former times, out of ten innovators nine were put to a violent death, so that the tenth was really original. The pruning hook which lops off the weak branches does but give more strength to the others. In these days, there is no pruning hook, but there is no more sap. In short, all this is a matter of little importance, and humanity will go on its way without the liberals and despite the retrogrades. The mind is never bolder and more self-reliant than when it just feels the hand checking it. Leave it quite unfettered, and it will be lacking in ballast, so contented with its liberty that it will think only of preserving it without any idea of profiting by it. The history of the human mind shows us how all ideas are born outside the law and grow in secret. If we go back to the origin of all reforms, they will appear in the regular course of things to be incapable of being put into force. Let us take our stand, for instance, in 1520, and ask ourselves how the new idea will succeed in piercing this sea of ice. It is impossible; the chain is too strong, what with the Pope, the Emperor, kings, religious orders and universities; and against all these only a poor monk. It is quite impossible. Or let us take the origin of modern rationalism. The age is in the meshes of the Jesuits, the oratory, priests, and kings. The Jesuits have made of education a machine to shrink the heads and to depress the minds, according to Michelet's expression. And against all this what have we except a few poor and obscure savants, with no backing in the masses, such as Galileo and Descartes. What can they expect to do, and how are they to lift such a weight of authority? Yet a century and a half later it was done.

Thus all reforms would have been prevented if the law had been rigorously observed, but the law never provides for every contingency, and the mind is so subtle that a very small outlet suffices for it. So that it matters little whether the law grants or refuses liberty to new ideas, for they make their way all the same; they come into existence without the law and despite the law, and they are all the better for this than if they had grown in full legality. When a river which has overflown its banks pours onward, you may erect dykes to arrest its progress, but the flood continues to rise; you may work with eager energy and employ skilful labourers to make good all the fissures, but the flood will continue to rise until the torrent has surmounted the obstacle, or until, by making a circuit of the dyke, it comes back by some other way to inundate the land which you have attempted to protect from it.

CHAPTER XVIII.

The end of humanity, and therefore the aim which political conduct should keep before it, is to realize the highest human culture possible, that is to say the most perfect religion, by science, philosophy, art and morality: in a word by all the means of attaining the ideal which are in the nature of man.

This high culture of humanity could only be solid in so far as it was realized by the individual. Consequently, the end would not be attained if a civilization, however elevated, was only accessible to a small number, and especially if it constituted a personal enjoyment and one without any tradition. The end will only be attained when all men shall have access to this true religion, and when all humanity shall possess culture.

Every man has a right to the true religion, to what makes man perfect; that is to say every man ought to find in the society into which he is born the means of attaining the perfection of his nature, according to the formula of the day; in other words, every man ought to find in society, as regards the intelligence, what the mother furnishes him as regards the body, the milk, the primordial element, the primary foundation which he cannot procure for himself. This perfection needs a certain degree of material well-being. So that in a normal state of society, man would also be entitled to the primary funds necessary for procuring this livelihood.

In a word, society owes to man the possibility of

life, of the life which man, in his turn, is bound, if necessary, to sacrifice to society.

If socialism were the logical consequence of the modern intelligence, one would have to be a socialist; for the distinguishing feature of modern intelligence is the unquestionable. Many persons, indeed, with opposite intentions, maintain that socialism is the direct filiation of modern philosophy. Whence the one side concludes that socialism must be admitted as a necessity, while the other side maintains that modern philosophy should be rejected.

Nothing causes more misunderstandings in the moral sciences than the absolute use of names for designating systems. Wise men never accept any of these names, for a name is a limit. They criticize doctrines, but never take them just as they are. What man of any mark in our day would take for himself the names of pantheist, materialist, sceptic, etc.? Give me ten lines of any author, and I will prove to you that he is pantheist, and with ten more I will prove that he is not. These words do not designate a unique and constant shade of ideas; they vary according to the aspects.

It is the same with socialism. For my own part I would willingly adopt as my formula of opinion in this respect what M. Guizot says: "Socialism derives its ambition and its strength from sources which no one can dry up. But dominated by the forces of unity and order in society, it will always be combated and vanquished in so far as it is absurd and perverse, while gradually taking its place and its share in the immense and imposing development of humanity which is going on in our days."

What constitutes the force of socialism is that it corresponds to a perfectly legitimate tendency of the human mind, and in this sense it is the genuine, natural development of it. One must be blind not to see that the work begun four centuries ago in the literary, scientific and political order is the successive exaltation of the whole human race, the realization of that inmost craving of our nature for "more light."

At the present juncture, the problem is set forth in particularly difficult terms. For, upon the one hand, it is necessary to preserve the conquests already secured for civilization, while upon the other all must have their share in the blessings of this civilization. This will appear contradictory, for it must seem at first sight as if the abjection of a certain number and even of the majority was a necessary condition of society as it has been moulded by modern epochs, and especially by the eighteenth century.

I do not hesitate to assert that never, since the origin of things, has the human intelligence set itself so terrible a problem. That of slavery in antiquity was much less so, and yet it took centuries to conceive the possibility of a society without slavery.

In proportion as humanity advances, the problem of its destiny becomes more complicated; for it has to combine more data, to weigh more motives, to conciliate more contradictions. So humanity marches on, with one hand clutching within the folds of its robe the conquests of the past, and in the other holding the sword which is to effect fresh conquests. Formerly, the question was very simple; the most advanced views, merely because they were the most advanced, might be regarded as the best. It is no longer so. No doubt it is always well to take the shortest route, and I do not at all approve those who maintain that we should walk, but not run. We should always do the best we can, and do it as quickly as possible. But the essential thing is to discover what is the best, and that is not an easy matter. It is barely fifty years ago that humanity saw clearly the object which it had hitherto been unconsciously pursuing. This is an immense progress, but it is also an undeniable danger. The traveller who looks only at the horizon of the plain risks not seeing the precipice or the quagmire at his feet. In the same way, humanity, when looking only to the distant object, is tempted to make a jump for it, without regard to the intermediate objects against which it may not improbably dash itself to pieces. The most remarkable characteristic of the Utopists is not to be historical, not to take into account of what we have been brought to by accomplished facts. Supposing that the society of which they dream were possible, supposing even that it were absolutely the best, it would still not be the true society, that which has been created by all the antecedents of humanity. The problem is, therefore, more complicated than it may appear; the solution can only be obtained by balancing two orders of consideration; upon the one side, the object to be attained, upon the other the present state of things, the ground we are treading. When humanity went instinctively forward, one might put confidence in the divine genius which directed its course, but one shudders at the thought of the dread alternatives it holds in its hands since it has reached an age of consciousness, and of the incalculable consequences which a blunder or an act of caprice might have.

With these great problems confronting them, the philosophers reflect and wait; among those who are not philosophers, some deny the problem and maintain that the present state of things must at any cost be preserved, while others hope to meet what is wanted by solutions too simple and too self-evident. It is needless to say that each side has good arguments to adduce, for the reformers throw in the teeth of the conservatives undeniable wrongs and evils which call loudly for a remedy, while the conservatives have no difficulty in demonstrating that with the system proposed by the reformers there would be no society possible. And better by far is a defective society than one which does not exist.

I have often reflected that a pagan of the time of Augustus might have urged in favour of the maintenance of ancient society all that is said in our days to prove that nothing ought to be changed in the present state of society. What is the aim of this sombre and melancholy creed? What strange people these Christians are, people who avoid the light, unsociable, the very residuum of the people (152). I should be very much surprised if some of the self-satisfied people of those days did not say as in our own: We must not refute Christianity, but suppress it. Society in the presence of Christianity is as it were in the presence of an implacable enemy. Society must crush it, or it will itself be crushed. In these conditions, all discussion is reduced to a struggle, and all reasoning to a weapon. What do we do in presence of an irreconcilable enemy? Do we enter into controversy with him? No, we go to war with him. In the same way, society must defend itself against Christianity, not by reasonings but by force. It must not discuss or refute its doctrines, but suppress them. I can fancy Seneca coming by chance across those words of St. Paul: "Non est Judæus neque Græcus; non est servus neque liber; non est masculus neque femina; omnes enim vos unum estis in Christo." "Surely," he would have said, "Here we have a Utopian. How can a society exist without slaves? Would you have me cultivate my land with my own hands? This is subversive of public order and then what is this Christus who is playing so singular a part? These people are dangerous. I will speak to Nero about them." No doubt if the slaves, taking literally and as being immediately applicable the words of St. Paul, had established their dominion upon the smoking ruins of Rome and of Italy and had deprived the world of the benefits it was to derive from the dominion of Rome, Seneca would have been to a certain extent right. But if a Christian slave had said to the philosopher: "Oh! Annæus, I know the man who wrote these words, he preaches only submission and patience. What he has written will be accomplished, without rebellion and by the masters themselves. A day will come when society will be possible without slaves, although you, as a philosopher, cannot imagine this," Seneca would not, in all probability, have believed him; but perhaps he would not have had the simple-minded dreamer flogged.

Socialism is, therefore, right to the extent of discerning the problem, but solves it badly; or rather socialism is not yet possible of solution. Individual liberty, in fine, is the primary cause of the evil. Now, the emancipation of the individual is secured, finally secured, and must for ever be preserved. "Society," said Enfantin, "does not consist solely of idlers and workers; politics should have for aim the moral, physical and intellectual improvement of the lot of the workers and the progressive effacement of the idlers." Here we have a clearly defined problem. Now let us listen to the solution. "The means are, as regards the idlers, the destruction of all the privileges of birth, and, as regards the workers, a classification according to capacities and a scale of rewards according to the work done." This is a remedy worse than the disease. It is a necessity of the human mind that when a problem is thus stated for the first time, certain ingenuous and generous minds, not possessed of enough rational critique nor a sufficient experience of history, nor of any idea of the extreme complexity of human nature, should dream of the formation of a society too simple to be possible, and should imagine that they had found the solution in some obvious or superficial idea, which, if it could be realized, would go directly counter to their object. No social problem can be attacked from the front; whenever a solution appears to be clear and easy, we must be on our guard. The truth, in this order of things is deep and hidden. But the duller minds, which do not seize these delicate shades of difference, blunder straight on through morasses and quagmires. This is an inevitable and irremediable error. Persuaded that they hold the key of the enigma, these worthy souls are importunate and very eager to be doing; they are anxious to be given a free course, and they are convinced that only selfishness and ill-will stand in the way of their system being adopted. Those who laugh at these simple enthusiasts or who insult them are still less excusable; for they are not any better informed, and they are perhaps still more backward, for they have not grasped the problem. My conviction is that the day will come when it will be said of socialism as of all reforms: It has attained its object, not according to the aim of the sectaries, but for the good of humanity. Reforms never triumph directly; they triumph by compelling their adversaries, in order to overcome them, to partially adopt them. It is like a storm which draws back into its vortex those who attempt to face it (153), a stream which carries with it those who seek to withstand it, a knot which tightens when you attempt to untie it, a fire which kindles when you blow to put it out. Humanity, like God of the Old Testament, accomplishes the will by the efforts of its enemies. Examine the history of all great reforms, and it will seem at first sight as if they were defeated. But in reality the reaction which has resisted them has only triumphed over them by conceding what was just and legitimate in their demands. It might be said of the reforms as of the crusades: Not one succeeded; all succeeded. Their defeat is their victory, or rather no one gains an absolute triumph in these great struggles unless it be humanity, which profits by the energetic initiative of the innovators, and by the reaction which undesignedly corrects and improves that which it sought to suppress.

We should, in my opinion, feel grateful to those who attempt to solve a problem, even when they are fated not to succeed. For, before reaching the true solution, many erroneous ones must be tried, and the panacea and the philosopher's stone must be allowed to have their chance. I cannot profess much respect for that negative wisdom, so much in favour with us, which consists of making light of those who seek the truth and of remaining motionless and inactive so as not to risk being regarded as subversive. It is a poor merit not to fall when one does not move. The first people to enter upon a new order of ideas are bound to be charlatans more or less in earnest. It is easy for us nowadays to sneer at Paracelsus, Agrippa, Cardan, and Van Helmont, and yet without them we should not be what we are. Humanity only reaches the truth through a series of successive errors. It is the aged Balaam who falls and whose eyes are opened (154). As one sees the tide bringing the ever collapsing waves upon the shore, the feeling aroused is one of powerlessness. The wave arrived so proudly, and yet it is dashed to pieces against the sand, and it expires in a feeble career of the shore which it seemed as though about to devour. But upon reflection, one finds that this process is not so idle as it seems, for each wave, as it dies away, has its effect, and all the waves combined make the rising tide against which heaven and hell would be powerless.

Foreign nations often laugh at the waste of time and strength which a revolution entails in France, and at the disappointments which cause her to revert to the point from which she started, after having paid very dearly for her excursion. It is very easy for them, seeing that they do not attempt to do anything, leaving us to make experiments at our own cost, to laugh when we make a false step upon this unexplored ground. But let them try what they can do, and we shall see. England, for instance, is obstinately attached to the most flagrant contradictions. Her religious system is the most absurd of any, and she is not to be moved from it. She refuses to open her eyes. Her quietism and her prosperity are a shame to her, and testify to her nullity.

Such, then, is the situation of the human mind.

A vast problem lies before it, the solution of which is urgent; but the solution is impossible and perhaps will not be ripe for a century. Then come the empirics with their deplorable *naïveté;* each one of them has discovered at a glance what has so perplexed and baffled the wise and the experienced; each of them undertakes to effect a general pacification, the only condition he asks for the salvation of society being that he should be left a free hand. The wise men who know how difficult the problem is shrug their shoulders. But the people have not the sentiment of the difficulty of a problem, and the reason is very simple; they imagine it to be too simple and do not take account of all its elements.

To seek a perfect equilibrium and repose at such an epoch as this is to seek the impossible; for we are by necessity in the midst of what is provisional and unstable. The calm is but an armistice and a breathing space. Humanity, when it is fatigued, is willing to pause, but to pause is not to rest. It is impossible for society to find calm in a state when it is suffering from an open wound such as that of to-day. The mere consciousness of the malady prevents repose, and one can but doze between one attack and another. At such a period, no one can be right unless it is the critic who does not offer a decided opinion. For the age is oppressed by a problem at once inevitable and insoluble. At these epochs, doubt and indecision are the truth; the man who is not in doubt is either a simpleton or a charlatan. The life of humanity, like the life of the individual, rests upon necessary contradictions. Life is but a transition, an intolerable burden long endured. There is not a moment in which one can say that one rests upon any solid foundation; one is always hoping to reach that solid basis, but only hoping.

We must not, therefore, be surprised at these insoluble contradictions. Only narrow minds can construct for themselves at any given moment a well-defined and rounded-off system, and imagine that this infinite void (155) can be filled up with an *à priori* constitution. The party man feels it necessary to think that he is entirely in the right, that he is fighting for a sacred cause, that those opposed to him are criminal and perverse. The party man seeks to force his antipathies upon the future, not reflecting that the future is devoid of passion against any one, that Spartacus and John of Leyden are merely objects of interest to us. Strange to say, we are only impartial and critical for the fanaticisms of the past, and we are fanatical ourselves. We barricade ourselves in our party in order not to see the reasons of the other side. The wise man does not feel anger against any one, for he knows that human nature only has its passions moved for the incomplete truth. He knows that all parties are at once wrong and right. The conservatives are wrong, for the state of things which they uphold as good and which they do right to uphold, is bad and intolerable. The revolutionists are wrong; for, if they discern the evil, they do not possess, any more than the conservatives, the idea of organization. But it is absurd to destroy when you have nothing to put in place of what you destroy. Revolution will be sacred and legitimate when the regenerating idea, that is to say the new religion, having been discovered, all that will be needed will be to upset the worn-out state of things to give it its legitimate place; or rather there will no longer be any need to effect a revolution; it will come of itself. Any constitution would be at once abrogated by it; for it would be absolute sovereign. So it was in 1789. The revolution was ripe at that period; it had already been effected in the public mind; for every one saw what a flagrant contradiction there was between the new ideas created by the eighteenth century, and existing institutions. It was the same in 1830. The liberal revolution had preceded it, the liberal principles were accepted in advance. Was it so in 1848? The future will show, but it is a remarkable fact that the victorious side was the one most embarrassed on the morrow of the victory. The revolution of '48 was not at all a political revolution; compare the politicians and the politics of to-day with those of the epoch preceding the 24th of February, and you will find them to be absolutely identical. Its signification was that of social revolution, and as such it was certainly premature, inasmuch as it proved abortive. Revolutions must be made for well-ascertained principles, and not for tendencies which have not yet been formulated in a practical manner.

Herein, then, lies the secret of our situation. The present state of things being defective and felt to be so, whoever comes forward with a proposed remedy is welcome. Upon the morrow of one revolution, the germ of another begins to form. This accounts for the favourable consideration which may be reckoned upon by any party which has not yet been put to the test. But no sooner has it triumphed than it is in as great a difficulty as the rest, for it does not know any more than they do. Hence arises the inevitable unpopularity of all authority, and the fatal position in which every Government finds itself. For it is expected to provide in a moment what it cannot give and what no one possesses, the solution of the problem of the hour. Every government thus becomes, by the force of things, a target for every weapon, and is condemned to be unable to fulfil its task. It is an unfair piece of tactics to remind the governing body of what they have said and promised during the time they were in opposition and to make them appear inconsistent; for this inconsistency is necessary, and those who declare so stoutly that they would do differently if they were in power either lie or deceive themselves. If they were in power, they would be subject to the same necessities and would act in the same way. For the last sixty years there has not been a single Head of the State who has not died on the scaffold or in exile, and it was necessary that such should be the case. Any future one will have the same fate, unless a periodic law, more favourable to him in reality than he may think, comes in time to deliver him from office. How is it possible to avoid succumbing beneath an impossible task? In reality, this does honour to France; it proves that she has a high idea of perfection. It is to our credit that we are hard to please and dissatisfied. Mediocrity is easily satisfied; lofty minds are always full of disquiet and agitated, for they are constantly aspiring after something better. Only the infinite could thoroughly satisfy them.

Thus humanity is in the position of a sick person, who suffers in whatever position he may be, and yet who allows himself to be constantly lured by the hope that he will be better if he turns round. Revolutions are the upheavals of the everlasting Enceladus turning over when Ætna weighs too heavily upon him. It is superficial to envisage history as being composed of periods of stability and periods of transition. It is transition which is the customary state. No doubt humanity remains fixed for a more or less lengthened period upon the same ideas; but it is like the bird of paradise in the legend, which broods as it flies. All is the end, everything is the means. In human life, mature years are not the aim of youth, old age is not the aim of maturity. The aim is life taken in its unity.

There is an optical illusion to which we who were born between 1815 and 1830 are subject. We have not been the witnesses of great events, so we go back to the Revolution for our estimate of everything; that is our horizon, the hill of our childhood, our world's end. Now it so happens that this horizon is a mountain; we measure everything by that. This is deceptive, and cannot form any induction for the future, as since the invasion which constitutes the limit of ancient and of modern history, there has been no fact like that, and perhaps will not be again for centuries. But whenever there is any question of revolution, even if the reference be to mere child's play, we at once carry our minds back to this gigantic cataract, and never to the much slower changes recorded by earlier history, say that of the sixteenth and seventeenth centuries.

I will take care not to follow political economy in its deductions, for the economists would no doubt attribute to my incompetence the suspicion which these deductions arouse in me; but I am competent in matters of morality and of the philosophy of humanity. I do not concern myself with the means; I speak of what should be and consequently of what will be. Well, I am convinced that humanity will succeed within a century in realizing that towards which it is now tending, with the exception, of course, that it will then be guided by new requirements. Then one will be in a position to criticize both sides: those who resisted, and those who fancied that society could be reconstructed like a house of cards. Each man will have his part to play; we, the critics, like the rest. What may be taken for granted is that nobody will be altogether right or absolutely wrong. Barbès himself, the unreasonable revolutionist, will then become an exponent of legitimacy, and the mutual explanations which must then ensue will be interesting. The common error of the socialists and of their opponents is to suppose that the question of humanity is a question of ease and enjoyment. If that were so, Fourier and Cabet would be perfectly right. It is horrible that one man should be sacrificed to the enjoyment of another. Inequality is only conceivable and just from the standpoint of moral society. If it were merely a question of self-indulgence, it would be better that all should have Spartan fare than that some should have luxuries and others go hungry. Would it, in fact, be worth while to sacrifice one's life and happiness for the good of society if the sole result was to procure a little insipid enjoyment for a few insignificant idlers who have put themselves beyond the pale of humanity, in order to live more at their ease? Let me repeat that if the object of life was but self-indulgence, it would not be unreasonable that each one should claim his share, and from this point of view any enjoyment which one might procure at the expense of others would be in reality an injustice and a robbery. The communistic follies are, therefore, the consequence of the hideous hedonism of the last few years. When the socialists say: The aim of society is the happiness of all; when their adversaries say: The aim of society is the happiness of a few, both are alike wrong, but the former less so than the latter. What should be said is: The aim of society is the greatest possible perfecting of all, and material ease is only of value in so far as it is to a certain extent the indispensable condition of intellectual perfection. The State is neither an institution of police, as Smith would have it, nor a charity bureau and a hospital, as the socialists would have it. It is a machine for making progress. Every sacrifice of the individual which is not an injustice, that is to say the spoliation of a natural right, is permissible in order to reach this end; for in this case the sacrifice is not effected for the enjoyment of another individual, it is made upon behalf of society as a whole. It is the idea of the ancient sacrifice, the man for the nation: *expedit unum hominem mori pro populi.*

Inequality is legitimate whenever inequality is necessary for the good of humanity. A society is entitled to what is necessary for its existence, however great may be the apparent injustice resulting for the individual.

The principle that there are no such things as individuals is true as a physical fact, but not as a teleological proposition. In the plane of things, the individual disappears; the large shape mapped out by individuals generally is alone of any account. Socialists are not really consistent when they preach equality, for equality is derived mainly from the consideration of the individual, and inequality is only conceivable from the point of view of society The possibility and the requirements of society, the interests of civilization, take precedence of all the rest. Thus, individual liberty, emulation and competition being conditional to all civilization, the present iniquity is better than the final servitude of socialism. Thus, learned and lettered culture being absolutely indispensable in the scheme of humanity, even when it can only fall to the share of a small minority, this flagrant privilege would be excused by necessity. For there is not, as a matter of fact, any tradition for happiness, but there is a tradition for science. I will go so far as to say that if at any time slavery was necessary for the existence of society, slavery was legitimate, for in such a case the slaves were slaves of humanity (156), slaves of the divine scheme—a thing which is no more repugnant than the existence of so many beings inexorably attached to the yoke of an idea which is above them and which they do not understand (157). If a day arrived when humanity once more needed to be governed in the old way, to be subject to a Lycurgus-like code, that would be quite justifiable (158). In the same way, the day may come when international rights will reach such a point that each nation will be sensitive like the limb of a body to what is going on in others. With a more perfect code of morality, rights which are now false and dangerous will be unquestioned, for the condition of these rights will be laid down, which has not yet been done (159). This may be conceived when once you attribute to humanity an objective aim (that is to say independent of the well-being of individuals) the realization of the perfect, the great deification. The subordination of animals to man, that of the sexes the one to the other, does not shock any one, because it is the work of nature and of the inevitable organization of things. At bottom, the hierarchy of men according to their degree of perfection does not shock one's ideas of fitness a whit more. What is shocking is that the individual, of his own right and for his personal enjoyment, should enslave his fellow-man in order to have self-indulgence at his expense. The inequality is revolting when we consider solely the personal and egotistical advantage which the superior derives from the inferior; it is natural and right if considered as an inevitable law of society, the transitory condition at all events of its perfection. Those who envisage rights, like the rest, as being always rigorously the same, launch anathemas against the most necessary facts of history. But this way of looking at things has grown obsolete; the human mind has passed from the absolute to the historical, envisaging everything from the point of view of *becoming*. Rights create themselves like other things; they are created, not, of course, by positive laws, but by the successive exaltation of humanity, which manifests itself in the conquest which it effects of these rights. The fact does not constitute the right, but manifests the right. All rights must be conquered, and those who cannot conquer them prove that they are not ripe for these rights, that these rights do not exist for them, unless it be potentially. The freeing of the negroes was neither achieved nor deserved by the negroes, but by the progress in civilization of their masters. It is not because you have proved to a nation that it is entitled to its independence that it rises to claim it; the young lion goes off to hunt for his food when he feels strong enough, without any one telling him. The wishes of humanity do not constitute the right, as Jurieu would have it; but it is, in its general tendency and its main results, the indication of the right. The champions of the absolute right, like the jurists, and of blind facts, like Callicles, are both wrong. The fact is the criterion of right. The French Revolution is not legitimate because it has taken place; but it took place because it was legitimate. Right is the progress of humanity; there is no right in opposition to this progress, and, *vice versâ*, progress is sufficient to legitimize everything. Whatever serves to advance the work of God is permissible.

We Frenchmen, who are gifted with an absolute and exclusive spirit, fall into strange illusions in this respect, and we often reason much in this way, which is a very scholastic one. "Such a system of institution would be intolerable with us, at the point we have reached; it must be so, therefore, everywhere, and it must have been so always." The simple minded carry this to a most delightfully ludicrous point, and a few months ago they wanted to make all Europe republican willy nilly. We want to establish everywhere the government which suits us and to which we are entitled. We think that we should be doing wonders if we established the constitutional *régime* among the savages of Oceania, and we shall soon be sending diplomatic notes to the Grand Turk to advise him to call together a Parliament.* We reason in the same way with regard to the emancipation of the blacks. Assuredly, if there is a reform which is urgent and ripe this is the one. But we conclude that we are without transition to apply to the negro the *régime* of individual liberty which suits civilized people like ourselves, not reflecting that what is above all things necessary is to educate these unhappy creatures, and that this *régime* is not suitable for doing so. The best system to be followed for the education of the negro races is that which Providence has followed in the education of humanity, for it is not, seemingly, by chance that it has made its selection. Look by how many stages the peoples have passed. It is certain that civilization cannot be improvised, that it requires a long course of discipline, and that it is doing a disservice to uncultivated races to emancipate them all at once. I imagine that they need to traverse a

* What M. Renan suggests as a joke in 1850 was actually done if not by France, at all events by other Western Powers in 1876. —Transl.

state analogous to the ancient theocracies. Slavery does not elevate the negro, nor does liberty. True, he will sleep all day, or will run, like a child, about the woods. Abolitionism carried to an extreme betrays a profound ignorance of the psychology of humanity. I imagine, moreover, that the scientific and experimental study of the education of barbarous races will become one of the most striking problems offered to the mind of Europe, when the attention of the continent can for a moment be taken off from itself.

The history of humanity is not only the history of its enfranchisement, it is above all the history of its education. What would humanity be if it had not traversed the ancient theocracies and the severe codes of a Lycurgus? The whip has been a necessity in the education of humanity. We have ceased to envisage these forms except as obstacles which humanity has been compelled to break. She has broken them, no doubt, but only after having turned them to her profit. And was it not she, after all, who had created them for herself? The effort which is made to destroy them renders us blind to their anterior use. The revolutionary histories make the mistake of presenting the destruction of ancient forms as the great resultant of the progress of humanity. To destroy is not an end. Humanity has lived in the ancient moulds until they have become too narrow, and then has caused them to burst, but does any one suppose that this was out of anger against these moulds? Do you suppose that when the bird breaks the shell of the egg, his object is to break? No; his aim is to pass to a new life. The most we can say is that if the egg resisted, he might show rather more temper. In the same way, the moulds of humanity having grown hard and as it were petrified, a great effort was required to break them; humanity was compelled to gather together its forces and to set itself to the work of destruction for destruction's sake. It is in the order of things that the moulds of humanity should acquire a certain solidity, that all thought should aspire to stereotype itself and to pose for being eternal (160). That becomes in the long run an obstacle, when the need comes for breaking; and in the same way one might say that only mud huts or tents which can be taken down in an hour and leave no ruins behind them should be erected because, if you build palaces, they will be very troublesome to demolish.

Alas, we are only too given to these ephemeral constructions. Humanity, in our day, is encamped beneath the tent. We have lost the long-sustained hope and the vast thoughts. The idea of demolition preoccupies and blinds us. Christianity, for instance, is no longer anything but a dam, a pyramid built across the road, a mountain of stones which stands in the way of the new buildings. But does it follow that those who built the pyramid were wrong? The mould, as it gains hardness, becomes a prison. No matter, for it is essential that, in order to impress its shape, it should be hard. It only becomes a prison when the object which has been moulded attempts to emerge. Then we have struggles and recriminations, for it is regarded as nothing but an obstacle. It is always the way; the matter is looked at from only one side, and becomes a partial one because of the practical aim in view. He who destroys cannot be just towards what he destroys, for he regards it only as a stumbling-block, a stupidity, an absurdity. But, if you reflect a little, you must see that it is humanity which has fashioned it. Take the most odious of institutions, the Inquisition. Spain made it and put up with it, and would apparently have got rid of it if she had been so inclined. No doubt if we looked at it from the Spanish point of view, we should understand it. The speculative man alone can be critical; liberals are not so; they are superficial. Humanity is responsible for everything. We only declaim against force because we fancy the chain has been imposed by a force foreign to humanity. Yet it is humanity alone which has forged fetters for itself.

There are in humanity elements which seem solely destined to arrest or moderate its progress. We must not assume them on that account to be useless. Reaction has its place in the plan of Providence; it works unwittingly for the general good. There are declivities down which the *rôle* of the traction engine consists solely in holding back. Those who seek to arrest a movement render it a double service: they accelerate and they regulate it. The aim of humanity is to get to the bottom of all the modes of life, to hatch them, to digest them, so to speak, in order to assimilate whatever they may contain of what is true and to cast out what is bad and useless. It is essential, therefore, that it should keep them for a certain time, in order to effect this analysis at leisure; otherwise too hasty a digestion of them would only result in weakening it; the accumulation of a number of really nutritive elements would be prevented. If the men who play this part were disinterested, that is to say if they had solely in view the highest progress of humanity, they would be heroes; for the part of reactionist is an ungrateful one and is not highly prized. The essential thing for humanity is to do well what it embarks upon, so that there may be no need to go back upon it. It is not by swaying hither and thither, by swallowing and rejecting all kinds of ideas with ungoverned voracity, without either masticating or digesting them, that so serious a work can be accomplished.

I repeat that if one only regarded in civilization the personal good which results from it for those who are civilized, one would perhaps hesitate to sacrifice for the good of civilization one portion of humanity to the other. But the object in view is to realize a more or less beautiful form of humanity, and for that the sacrifice of the individual is lawful. How many generations was it not necessary to sacrifice in order to raise the gigantic terraces of Nineveh and Babylon?

Positive minds regard that as quite absurd. No doubt, if the object in view had been to gratify the pride of some stupid tyrant this would have been true. But the object was to outline in stone one of the stages of humanity. Let there be no mistake about it; the generations buried beneath these masses lived a fuller life than if they had vegetated happy beneath their vine and fig-tree (161).

I see before me, as I write these lines, the marvel of Royal France, Versailles. I repeople in fancy its deserted corridors with all the century which passed away. The king in the centre; here Condé and the princes; there, in the pathway of the gardens, Bossuet and the bishops; here, in the theatre, Racine, Lulli, Molière and a few libertines; on the terrace of the Orange-House Madame de Sévigné and the great ladies, and, in the distance, within the gloomy walls of St. Cyr, Madame de Maintenon and her melancholy. There you have a civilization open to plenty of criticism, no doubt; but perfectly one and complete; it is one form of humanity among many. It would have been a great pity if it had not been represented, and yet it could not have been except at the price of great sacrifices. The degradation of the people, arbitrary and capricious conduct, court intrigues and *lettres de cachet*, the Bastille, the gibbet and the *Grands-Jours* are the essential parts of this edifice, so that if you repudiate the abuses you must also repudiate the edifice, for they are integral parts of its construction. I should prefer, for my own part, the age of Louis XIV. although it is very much against my individual taste and although I regard as rather silly the craze there was for this epoch during the last years of the *ancien régime*, to a perfectly regular state of things, with each person living at his ease, creating nothing, founding nothing, producing nothing. For the aim of humanity is not that individuals should live at ease, but that the beautiful and well-defined forms should be represented, and that perfection should be made flesh.

From the point of view of the individual, liberty and equality seem to be inherent rights. From the point of view of the human species, government and inequality are easily understood. It is better to have some brilliant personification of humanity, such as the king and his court, than a general mediocrity. It is desirable that the noble life should be led by a few as it cannot be by all. This privilege would be odious if only the self-indulgence of the privileged individual were regarded; it ceases to be so if we see in it the realization of a humanitarian form. Our petty system of bourgeois government, aspiring above all else to guarantee the rights and to secure the ease and comfort of each one, is conceived from the point of view of the individual, and has failed to produce anything grand. Would Louis XIV. have built Versailles if he had carping deputies to cut down his budgets? Only the accession of the people can revive these lofty aspirations of the ancient aristocratic world. It would be better no doubt that all should be great and noble. But so long as that remains impossible, it is important that the tradition of stately human life should be maintained among the *élite*. Would the lowly be greater because the great were brought down to their level? Equality will only be of inherent right when all men are able to be perfect in their measure. I say "in their measure;" for absolute equality is as impossible in humanity as it would be in the animal reign. Humanity, in short, would not exist as a unity, if it was formed of perfectly equal unities, without any relation of subordination between them. Unity only exists upon the condition that diverse functions work towards the same end; it presupposes the hierarchy of the parts. But each part is perfect when it is all that it can be, and that it does well all that it ought to do. Each individual will never be perfect, but humanity will be, and all will share in its perfection.

Nothing is capable of explanation in the moral world from the individual's point of view. All is confusion, chaos, revolting iniquity, if we do not envisage the transcendent resultant in which all things harmonize and justify themselves (162). Nature shows us upon an immense scale how the inferior species is sacrificed to the realization of a higher plan. It is the same in humanity. Perhaps, even, we ought to look beyond this too narrow horizon, and only look for justice, the perfect peace, the definite solution, the complete harmony in a vaster whole, to which humanity itself would be subordinated, in that mysterious *τὸ πᾶν*, which will still endure when humanity shall have disappeared.

CHAPTER XIX.

There is a tendency to believe that modern civilization must have a destiny analogous to ancient civilization and undergo, like it, an invasion of barbarians. It is forgotten that humanity never repeats itself, and does not twice employ the same methods. Upon the contrary, all goes to show that this fact of a civilization being nipped in the bud by barbarism will be unique in history, and that modern civilization is destined to propagate itself *ad infinitum.* It would probably have been so with the Greco-Roman civilization but for the great cataclysm which swept it away. The fourth and fifth centuries are only so meagre and superstitious in the Latin world because of the calamities of the times. If the barbarians had not come, it is probable that the fifth and sixth centuries would have presented to us a great civilization, analogous to that of Louis XIV., a grave and severe Christianity, tempered by philosophy. Certain persons find a malicious pleasure in putting their finger upon points in our literature and philosophy which recall the Greek and Roman decadence, and they draw from it this conclusion that the modern spirit, after having (as they say) had its brilliant epoch in the seventeenth century, is losing caste and is gradually dying out. Our poets remind them of Statius and Silius Italicus; our philosophers of Porphyrius and Proclus; eclecticism upon both sides closes the series. Our publishers, compilators, abbreviators, philologists and critics would answer to the rhetoricians, grammarians and scholiasts of Alexandria, Rhodes and Pergamus. Our lettered politicians would answer to the sophist statesmen, such as Dion Chrysostomus, Themistius and Libanius. Our pretty imitations of the classical style, and our pasticcios of exotic colour are just like Lucianus. But the true critics use with extreme caution this deceptive word "decadence." The rhetoricians who would have us believe that Tacitus, compared to Livy, is an author of the decadence, will doubtless also urge that Thierry and Michelet are of the decadence, by comparison with Rollin and d'Anquetil. The human mind does not proceed by so simple a path as this. How are you to explain by a decadence that prodigious development of German literature, which, at the close of the eighteenth century, opened up a fresh life to Europe? Say that St. Augustin, St. John Chrysostom and St. Basile are geniuses of the iron age. The human mind is only obscured in one of its aspects for another to glow with all the more brilliant light. Decadence only exists in the view of those narrow minds which hold obstinately to a single point of view in literature, in art, in philosophy, and in science. No doubt the man of letters finds St. Augustine and St. Ambrose inferior to Cicero and Seneca, the learned rationalist regards the legend tellers of the Middle Ages credulous and superstitious by comparison with Lucretius and Euhemerus. But he who invisages the totality of the human mind does not know what decadence means. The eighteenth century has neither Racine nor Bossuet, and yet it is very superior to the seventeenth; its science, its critique, its preface to the Encyclopædia, its luminous essays by Voltaire are its literature. There was but one form of life for these ancient States. To overthrow the ancient institutions of Sparta is to overthrow Sparta itself. In those times, in order to be a good patriot, it was necessary to be a thorough-paced conservative; the wise man of old is obstinately attached to the national customs. It is not the same with us, for the day upon which France destroyed her ancient institution was the day upon which her epic history began. For my part, I expect that five hundred years hence the history of France will begin with the Jeu de Paume, and that what preceded it will be treated as a background, as an interesting preface, much like the notions on ancient Gaul which are placed at the head of our own histories of France.

It is an easy commonplace to speak at large of social palingenesis, of renovation. It is not a question of being born again, but of continuing to live; the modern spirit and civilization are founded for ever, and the most terrible revolutions will serve solely to signal the infinitely varied phases of this development.

In accepting as a necessity the great fact of the invasion of the barbarians and passing an *à priori* criticism upon it, we find that it might have occurred in two ways. In the first (that in which it actually did occur), the barbarians, stronger than Rome, destroyed the Roman edifice, then, for long centuries to come, they endeavoured to reconstruct something upon the model of this edifice and with Roman materials. But another way would have been equally practicable. Rome had succeeded in perfectly assimilating the provinces and causing them to live after her way of civilization; but she had been unable to act in the same way upon the barbarians who poured in during the fourth and fifth centuries. It is impossible to believe that she could not, if she had desired, do so, when we observe the eagerness with which the barbarians, upon their entry into the Empire, embrace Roman forms, and drape themselves in Roman tinsel, titles of Consul and Patrician, in Roman insignia and costumes. Our Merovingians, among others, embraced the Roman mode of life with charming candour, and as to the two civilizations of the Ostrogoths and Visigoths, they are so thoroughly the direct prolongation of the

Roman civilization that they added an important chapter, though one containing little that was original, to the history of classic literature. The barbarians did not at first make any change in the order of things which they found established. Indifferent to learned culture, they regarded it without much attention and consequently without displeasure. Some of them, even, such as Theodoric and Chilperic, took it with a promptitude and ease which surprise one. I believe that if the Empire had had as great men in the fourth century as in the second, and especially if Christianity had been as strongly centralized in Rome as it was in the following centuries, it would have been possible to render the barbarians Romans before or immediately upon their entry, and thus to have maintained the continuity of things. The world was within a hair's breadth of having no Middle Ages, and of the Roman civilization continuing without interruption. If the Gallo-Roman schools had been strong enough to effect in a century the education of the Franks, humanity would have economized ten centuries. If that did not occur, the fault was with the schools and institutions, not with the Franks; the Roman mind was too much weakened to effect immediately this immense work. The question, in fine, was whether this ancient edifice, into which so many fresh materials were waiting to enter, should be renewed by a slow substitution of parts which would not break its identity, or whether it should undergo a thorough demolition, to be built anew with a combination of the old and new materials, but again upon the old plan.

As Rome was too weak at once to assimilate these fresh and violent elements, the change was effected in the other way. The barbarians overthrew the Empire, but in reality, when they attempted to reconstruct they reverted to the plan of the Roman society, which had struck them from the very first by its beauty, and which was the only one, moreover, they knew. Their conversion to Christianity was in plain truth their affiliation to Rome by the Bishops, who were the direct continuers of the Roman dress, language and habits. The Empire of which they caught up the idea for themselves was but a way of attaching themselves to Rome, the only source of all legitimate authority. And the Papacy what is its origin if it be not the selfsame idea, that all comes from Rome, that Rome is the capital of the world? The Roman Empire should not be regarded so much as a State which has been overthrown to make way for others as the first effort of universal civilization, being continued with a momentary extinction of reflection (which is the Middle Ages) in modern civilization. The invasion and the Middle Ages are really no more than the crisis provoked by the violent intrusion of fresh elements which vivified and enlarged the ancient circle of life: they are but accidents in the great voyage, accidents which may have caused unfortunate delays, well compensated for by the inestimable advantages which humanity has derived from them. All this may be applied, word for word, to the future of modern civilization. In the very improbable hypothesis of the barbarians (who are only to be looked for among ourselves) suddenly overturning it, and without its having time to assimilate them, it is unquestionable that after having overthrown it, they would come back to its ruins in order to extricate from them the materials of the future edifice, that we should become, in relation to them, classics and educators, that it would be the rhetoricians of the ancient society who would initiate them into the intellectual life and would be the occasion of another Renaissance, that there would again be a Martianus Capella, a Boethius, a Cassiodorus, an Isidore of Seville, putting into a portable and easily handled compass the civilizing data of the ancient culture, in order to form the intellectual aliment of the new society. But it is infinitely more probable that modern civilization will be sufficiently full of life to assimilate these new barbarians who are desirous of entering, and to continue its march in their company. For look how the barbarians appreciate this civilization, how they press around it, how they seek to understand it with their simple yet keen perception, how they study it with curiosity, how pleased they are when they have guessed its meaning. Who could fail to be touched at seeing the interest which our uninstructed classes take in this civilization which is there in the midst of them, but not for them. They remind me of the simple wonder of the barbarians, in presence of the Bishops speaking Latin and when brought face to face with the vast machine of Roman organization. It would, no doubt, have been difficult for Sidonius Apollinaris and the keen wits of the Gauls to have shouted: "Vive les barbares!" And yet they should have done so if they had the sentiment of the future (163). We who see things clearly, after fourteen centuries, take sides with the barbarians. What was it they asked for? Land to till, sunshine, civilization. Ah! blessed is he who asks only to increase the family of the children of light. The barbarians are those who receive these new-comers with blows, for fear that their own share may be diminished.

But, it will be said, your hopes rest upon a contradiction. You admit that intellectual culture, in order to become civilizing, demands a whole life of application and study. The immense majority of the human race, condemned to manual labour, is never therefore to enjoy the fruits of it.

No doubt, if intellectual culture was always to remain what it is among us, a profession of its own, a speciality, we might despair of seeing it become universal. A state in which no one would have any other profession but that of poet, man of letters or philosopher, would be the strangest of caricatures. Intellectual culture is as if it did not exist for humanity, when people study only with a view to write. Serious literature is not that of the rhetorician, who embarks upon literature for literature, who is interested in things spoken and written, and not in the things themselves, who does not care about nature, but who likes to read a description, who, unmoved by a moral sentiment, only understands it when expressed in sonorous verse. Beauty is in things themselves; literature is imagery and parable. Strange is the man of letters who troubles himself about morality or philosophy, not because that is human nature, but because there are works upon this subject, just as the man of erudition only concerns himself about agriculture and war because there are poems upon these subjects. This is to assume that the thing which is related is more real and important than the thing which is. Art, literature, and eloquence are only so far true as they are not empty forms, but as they serve and express a human cause. If the poet was merely, as Malherbe understood him an "arranger of syllables," if literature were merely an exercise, an attempt to do artificially what the ancients did naturally, it would, I admit, be a very trifling misfortune if all men could not be initiated into it.

We must, therefore, come to conceive the possibility of an intellectual life for all, not in the sense that all participate in scientific work, but that all participate in its results. We must, consequently, conceive the possibility of associating philosophy and the cultivation of the mind with a mechanical art.

This was what Greek society, so true and so unartificial, thoroughly realized. Greece was ignorant of our aristocratic prejudices which brand with ignominy any one exercising a manual profession and exclude him from what may be called the *distingué* world. A man might reach the noblest and most elevated life, though poor and working with his hands; or rather the morality of the individual so effaced his profession that regard was had only to the individual, whereas now we regard only the profession. Ammonius was not a porter who was a philosopher, he was a philosopher who happened to be a porter. May we not hope that humanity will some day revert to this beautiful and true conception of life in which the mind is everything, in which no one is defined according to his calling, in which the manual profession would merely be an accessory to which little attention would be paid, much as the trade of glass polishing was for Spinoza, a mere trifle which is done by the insignificant part of the individual without his attaching any importance to it or others doing so either. Work of this kind would not then be more servile than that which I am now doing in moving my fingers to write these lines.

What makes a manual labour brutalizing now is that it absorbs the individual and becomes his self, his all. The definition of this unfortunate being is, in fact, shoemaker, or carpenter. This word tells his nature, his essence; he is merely a human machine which makes boots and furniture. You cannot define in this way Spinoza as a maker of telescope glasses, or Mendelssohn as a shopman (164). The professional individuality only effaces the moral and intellectual individuality when the latter is insignificant. Imagine a well-instructed and noble-hearted man exercising one of these trades which require only a few hours' labour; so far from the higher life being closed to him, he is in a situation a thousand times more favourable to philosophical development than three-fourths of those who occupy the so-called liberal professions. Most of the liberal professions, in fact, absorb every hour of a man's time, and, what is worse, of his thoughts; whereas the trade requiring no reflection or attention leaves him who exercises it free to live in the world of pure spirits. For my own part, I have often thought that if I was offered a manual profession which, by my working four or five hours a day, would ensure me a living, I would abandon for it my calling of graduate of philosophy; for as this calling would occupy only my hands, it would not divert my thoughts so much as the necessity for speaking for two hours from what is not the actual object of my reflections. I should have four or five hours of delightful promenade, and I should have the rest of the time for the mental exercises which exclude all manual occupation. I should acquire during these hours of leisure positive knowledge, I should ruminate during the rest of the time upon what I had acquired. These are certain callings which ought to be reserved for philosophers, such as tiller of the soil, stone cutter, weaver and other occupations which require only the action of the hand (165). Any complicated work, anything which required the least attention, would be a depredation upon one's thought. Manufacturing work would, in this respect, be much less advantageous.

Do you suppose that a man, in this position, would not be much more free to philosophize than a lawyer, a doctor, a banker or a government official? All official positions are moulds more or less close; to enter them, you must break and bend by force all originality. The teaching profession is now the almost unique resort left to those who, having a vocation for mental labour, are reduced by the necessities of fortune, to accept some external profession, and teaching is very prejudicial to the higher qualities of the mind, as it absorbs and wears out much more than manual labour would do. None of us have forgotten the Lollards of the Middle Ages, those mystic weavers who, as they worked, hummed (*lollarent*) in cadence, and mixed the rhythm of the heart with that of the spindle. The *béguards* of Flanders, the *humiliate* of Italy, also reached a high degree of mystical and poetical exaltation, beneath the vivid pressure of that mysterious bow which causes new and candid souls to vibrate so powerfully. If most of those who exercise the so-called servile functions are really brutalized, it is because they are empty-headed, because they are only put to these functions as being fit for nothing else, because this function, purely animal and significant as it is, absorbs them altogether and still further degrades them. But if they had their heads full of literature, history, philosophy and humanism, if, in a word, they could, while working, talk to each other of higher things, what a difference that would make! Many men devoted to things of the mind set apart a few hours of each day to hygienic pursuits, sometimes differing but little from those which working men perform as a matter of necessity, and these do not appear to brutalize them (166). In the state of things which I should like to see, manual labour would be the recreation of mental labour. If I am told that there is no calling which would be sufficiently remunerative with four or five hours' work a day, I will reply by saying that in a properly organized society, where useless waste of time and unproductive superfluities would be eliminated, where everybody would work to some purpose, and especially where machinery was employed not to do without the workman, but to help him in his task and abridge his hours of labour, I am persuaded (incompetent as I am in such matters) that very few hours of labour would be sufficient for the good of the society, and for the wants of the individual; the rest would be for the mind.

"If every instrument," says Aristotle, "could upon an order being given or even guessed, work of its own accord, like the statues of Dœdalus or the tripods of Vulcan, which, as the poet tells us, went of their own accord to the meetings of the gods, if the spindles worked of themselves, if the bow played the violin without being held, the contractors would do without workmen and the masters without slaves (167)."

This simultaneousness of two lives, having nothing in common with each other, on account of the infinity which separates them, is by no means without example. I have often felt that I never lived more energetically by imagination and sensibility than when I was applying myself the closest to all that is most technical in science and seemingly the most arid. When the scientific object has some æsthetic or moral interest in itself, it quite absorbs the person who applies himself to it; when, upon the contrary, it does not appeal at all to the imagination and the heart, it leaves these two faculties free to vagabond at their ease. I can conceive, in the erudite man, a very active life of the heart, and all the more active when the object of his erudition offers less aliment to the sensibility; you then have two mechanisms quite independent of each other. It is the division which is fatal. A philosopher may exist in a state of things which requires only manual co-operation, such as work in the fields. He cannot exist in a position in which he must exercise his mind and concern himself seriously with petty affairs, such as commerce or finance. And, as a matter of fact, these professions have not produced a single man of mark in the history of the human mind.

Far be it from me to believe that such a system of society is applicable for the moment, or even that, if applied, it would serve the cause of intelligence. We must not forget that the immense majority of humanity is still at school, and that to let them out too soon would be to encourage them in idleness. Necessity, says Herder, is the weight of the clock, which causes all the wheels to turn. Humanity is only what it is owing to the severe course of gymnastics it has gone through, and liberty would only involve decadence if it resulted in diminishing its activity. I was only anxious to explain the possibility of a state of things in which the highest intellectual and moral culture, that is to say true religion, would be accessible to the classes now regarded as the lowest in society. Do you imagine that if the working man possessed education, intelligence, morality, a mild and beneficent culture, that he would complain of his outward inferiority? No, for, apart from the fact that morality and intelligence would infallibly secure for him order and ease, this culture would cause him to be highly considered and liked, would bring him within the limits of that enviable circle where delicate perceptions prevail and from which he grieves to see himself excluded. The peasant does not feel his moral and intellectual objection, but the working man in large towns sees our higher classes, he feels that we are more perfect than he is, he finds himself condemned to live in a fetid atmosphere of intellectual depression and immorality, he who has smelt the pleasant odour of the civilized world is fated to seek his enjoyment (for man cannot live without enjoyment of some kind, the trappist even has his) in ignoble haunts which are repugnant to him, repelled as he is by his lack of culture rather than by opinion, from more delicate pleasures. How could he do otherwise than rebel against such a state of things?

Chimerical as it may appear from the point of view of our present habits, I maintain that this simultaneousness of the intellectual life and of professional labour is possible. Greece is an illustrious instance of this; I do not speak of the most primitive societies, such as those of the Hindoos and the Hebrews, from which all idea of outward decorum and human respect was completely absent. The Brahmin of the forest, clothed in a few rags, feeding on leaves often dry, reaches a degree of intellectual speculation, a height of conception, a nobility of life unknown to the immense majority of those among us who call ourselves civilized.

There are men very highly endowed by nature but scantily endowed by fortune, who become proud and almost intractable, and who would die rather than accept for their livelihood what public opinion regards as an external humiliation. Werther leaves his ambassador because he meets silly and impertinent people in his salon; Chatterton commits suicide because a lord mayor offers him a situation as valet. This extreme sensitiveness as to externals proves a certain humility of mind, and testifies that those who feel it have not yet reached the highest philosophical summits. They are, in fact, upon the borders of the highly ridiculous, for if they are not in reality geniuses (and who can assure them that they are? How many others have believed that they were without being so in reality?), they run the risk of resembling the most stupid, ridiculous and self-conceited of men, the unappreciated young men of genius, who look upon everything as being beneath them, who anathematize society because society does not make a suitable allowance to those who devote themselves to sublime thoughts. Genius is in no wise humiliated because it works with its own hands. Of course, one cannot demand of it that it should surrender itself with all its soul to its task, that it should become absorbed in its office or its workshop. But dreaming is not a profession, and it is an error to imagine that the great writers would have thought much more if they had not had anything else to do but think. Genius is patient and full of life; I would almost say robust and hardy as the peasant. "The force of living is essentially a part of genius." It is amidst the struggles of an external situation that the great geniuses have developed themselves, and if they had not had any other profession than that of thinkers, perhaps they would not have been so great. Beranger was a clerk in a government office. The man who is really elevated has all his pride within him. To take account of the external humiliation is to show that one takes some account of that which is not the soul. The degraded slave, who felt his inferiority to his master, endured the blows inflicted upon him as due to fatality, without dreaming of resenting them. The cultivated slave, who felt himself superior to his master, could not have felt himself in the least humiliated at having to serve him. To have been irritated with him would have been to put himself on a par with him; it was better to despise him inwardly and say nothing. To be sparing of showing him respect and submission would have been equivalent to attaching importance to this. One only feels the insults of one's equals; those of a blackguard affect people of his own kidney, but do not reach us. In the same way those whose inward excellence renders them susceptible, irritable and jealous of an outward dignity in proportion to their worth, have not yet passed a certain level, or understood the true royalty of men of high intelligence.

The ideal of human life would be a state in which man had so mastered nature that material requirements were no longer a motive, in which these requirements were satisfied as soon as felt, in which man, the king of the world, would scarcely have to make an effort to maintain it under his dependence, in which all human activity would, in short, be directed towards the things of the mind, and in which man would merely have to live the celestial life. Then would prevail the true reign of the spirit, the perfect religion, the worship of God as spirit and truth. Humanity still has need of a material stimulus, and at present such a state of things would be prejudicial, as it would only engender idleness. But this drawback is merely relative. For us, men of intelligence, the labour of existence and material necessities are merely an obstacle; it is a part of the time we give to ransom the rest. If we were delivered from the thought of material wants, like the religious orders or the Brahmin, who plunges naked into the forest, we should be navigating with all sails set, we should conquer the infinite. . . .

Patriarchal life realized this lofty independence of man, but it was at the sacrifice of not less essential elements; civilization, in fact, can only exist where there is a parallel development of the intelligence, of the moral condition and of comfort. In ancient times, the same result was reached by slavery; the free man was really in a very fine position, dispensed from terrestrial cares, and at liberty to cultivate things of the mind. The careful organization of humanity will bring back this state of things, but in a much more complicated form than in patriarchal times, and without needing slavery. The work of the nineteenth century will have been the conquest of this material comfort, which, at first sight, may appear profane but which becomes hallowed if we reflect that it is one of the conditions of the enfranchisement of the mind. No one is more strongly opposed than I am to those who urge the rehabilitation of the flesh, and I believe nevertheless that Christianity was wrong in preaching the struggle, the revolt of the senses, the mortification of bodily desires. That may have been all very well for the education of humanity, but there is something more perfect still. That is not to think any more of the flesh, to live so energetically the life of the spirit that these gross temptations may have no hold. Abstinence and mortifications are the virtues of barbarians and materialists, who, subject to coarse instincts, can conceive of nothing more heroic than to resist them; so that these virtues are held in special esteem in sensual countries. In the eyes of those with coarse appetites, a man who fasts, who flagellates himself, who is chaste, who passes his life upon a column, is the ideal of virtue. For he, the barbarian, is gluttonous, and he quite feels that it would be a great sacrifice for him to live in this way. But in our eyes, such a man is not virtuous, for these pleasures of the palate, of the senses, are nothing to us; we think that there is no merit in foregoing them; affected fasting proves that one sets great store by the things which one goes without. Plato was not so mortified as Dominic Loricat, and apparently was more of a spiritualist. The Catholics sometimes assert that the disuse into which the fasting of the Middle Ages has fallen shows our sensuality; but, upon the contrary, it is owing to the progress of the mind that these practices have become unmeaning and out of date. We must destroy the antagonism of the body and of the mind, not by equalling the two terms, but by prolonging one of them *ad infinitum*, so that the other may be annihilated and become as zero. That being done, let the body have its enjoyments, for to refuse them to it would be to imply that these trifles are of some importance. The motto of the Saint-Simonians: "Sanctify yourself by pleasure" is abominable; it is pure gnosticism. "Sanctify yourself by abstaining from pleasure" also leaves something wanting. "Sanctify yourself, and pleasure will become insignificant, and you will not think of pleasure" is what we spiritualists would say. Holiness consists in living for the spirit, not for the body. Gross minds may have imagined that in cutting themselves off from the bodily life, they rendered themselves more apt to the spiritual life.

It is very possible that some day a still higher conception will be reached. The reason why pleasure is quite a profane thing for us is that we take it always as a personal enjoyment, whereas personal enjoyment has absolutely no suprasensible value. But if we took pleasure with the mystical ideas which the ancients attached to it, when they associated it with their temples and their festivals, if we succeeded in eliminatiug all idea of enjoyment, to see in it only the perfecting process which results from it for our being, the mystical union with nature, the sympathy which it establishes between us and things, I am not sure that it might not be elevated to the rank of a hallowed object. In my bare and cold room, abstemious and poorly clad, I seem to understand beauty from a somewhat lofty point of view. But I ask myself whether I should not understand it still better with my brain stimulated by a generous fluid, richly clad, perfumed and *tête-à-tête* with the Beatrice whom I have seen only in my dreams. If the creation of my fancy were incarnate at my side, should I not love and adore it the more? Doubtless, if there is one thing more revolting than another, it is that people should seek delight in intoxication. But if a man merely tries to aid his ecstasy by a very noble material element, and one which has called forth such noble poetry, the matter is quite a different one. I have read somewhere that a poet or a philosopher (a German, I think) made a point of getting intoxicated regularly once a month, in order to attain that mystical state in which one gets closest to the infinite. I should not like to say whether all pleasures could stand this purifying process, and become exercises in piety, from which all thought of self-indulgence was excluded.

The imperfection of the present state is that external occupations absorb all one's time, so that people devote themselves mainly to some profession, only cultivating the mind if they have time and inclination. The accidental thus becomes life itself, while the truly human and religious part almost disappears. If we come to look closely at the spectacle of human activity, we see that the greatest part of it is quite wasted. Raise yourself in fancy above a city like Paris, and try to analyze the motives which guide the hurried footsteps of so many thousands of men. You will find that the desire for gain, business, or material cares are the main cause of nine-tenths of these movements; that pleasure is the motive of perhaps a twentieth part of this agitation, that scarcely one-hundredth part of the crowd is guided by affection, and that barely one in a thousand is impelled by scientific or religious motives.

It seems as if external affairs are the main object of life, that the aim of the great majority of the human race is to live beneath the pressing and absorbing thought of how they are to gain their daily bread, so that life has no other aim than to sustain itself. What a vicious circle this is! In a better state of human society, a man would first of all be himself, that is to say that the first care of each person would be the perfecting of his nature.

Then, in quite a secondary way, to which scarcely any heed would be paid, one would belong to such and such a profession. That would be the antique idyl, the pastoral life conceived in their dreams by all the bucolic poets, a life in which material occupation is of such small account that no thought is given to it, and that a man is quite free for poetry and the beautiful. Then it will be said: "Our fathers had to place their paradise in heaven. But we can do without the paradise of God, for the celestial life is brought down to us here below."

Such a state of perfection would not exclude intellectual variety; upon the contrary, the originalities would be much more marked in it, owing to the free development of the individualities. And even if the variety of minds was destined to disappear in presence of a more advanced culture, where would the harm be? But let us hasten to say that uniformity would, as things now are, be the extinction of humanity. The hive has never been a centre of progress. We are traversing the age of analysis, that is to say of partial views, an age during which the diversity of views is necessary. When Plato desired, in his ideal Republic, that all should see with the same eyes and hear with the same ears, he purposely left out of the calculation one of the most essential elements of humanity. Humanity, in fact, is only what it is through variety. When two birds respond to each other, in what do their accents differ from an elegy? Only from their variety. So far from preaching communism in the present state of the human intelligence, we ought to preach individualism and originality. No two men ought to resemble each other, for those who are alike count only as one.

In the primitive syncretism, all men of the same race resembled one another like fish of the same kind. There are no individual characters in the primitive epics; what the ancient critics told us about the characters in Homer is very exaggerated, and, moreover, the Greek world, so full of life, so varied, so multiform, attained from the very first a very nice sense of distinction in these matters. The old Hebrew literature has little more than two categories of men, the good and the bad, and in the Indian literature there is scarcely even this much. All the characters are presented as being very much the same. Our more finely drawn types do not appear until much later.

As it is, education and the variety of the objects of study which create the varieties of mind, everything which tends to put all minds into one registered mould is prejudicial to the progress of human intelligence. Men's minds, as a matter of fact, differ much more by what they have learnt, by the facts upon which they base their judgments, than by their actual nature (168). The habits of French society, so severe upon all originality, are, from this point of view, altogether to be deplored. "As," says Madame de Staël, "what constitutes individual existence is always a peculiarity of some kind, this peculiarity excites ridicule, so that as people dread this above everything else they endeavour to avoid whatever would cause them to be remarked one way or the other." The truly noble and gifted natures are not those in which the opposing elements neutralize one another, but those in which extremes meet, not simultaneously, but successively, and according to the surface which has to be delineated. The perfect man would be one as inflexible as a philosopher, as weak as a woman, as hardy as a Breton peasant, as ingenuous and gentle as a child. The colourless natures, formed of a kind of proportional medium between extremes, are of no value in an epoch of analysis.

Analysis, in fact, exists only through the diversity of the points of view, and upon condition of all the sides of a subject being completely elucidated; to each his task, to each his atom to explore, such is its motto. What is required in a given state of things is the greatest possible variety among individuals; for each originality is the outline of one way of looking at things; it is one way of taking the world. But it may be that some day or other humanity will reach such a state of individual perfection, so complete a synthesis, that all men will have reached an equally advanced stage, and that they will make their fresh starting-point from there towards the future. And this harmony will be realized not by theocracy, not by the suppression of the individual, not by the "Father king" of the Saint-Simonians who regulated belief as well as all else, but by a mutual and unfettered aspiration, as is the case with the ideal in heaven. It is easy to bind the sheaves which have been cut. But it is another matter to bind the living sheaves! At the present time, all are harnessed to the same car; but some are pulling forward, some backward, others in different directions, and so there is no progress. But then all will pull in the same direction; then, science, which is now only cultivated by a few obscure men lost in the crowd, will be pursued by millions of men, seeking in unison the solution of the problems set before them. Oh! for the day when there will no longer be any great men, because all will be equally great, and when humanity, coming back to unity, will march like one man to the conquest of the ideal and of the secret of things (169)! What will be able to resist science when humanity itself is scientific and marches with one accord to the assault of the truth?

Why, it will be said, take account of these chimeras? Let the future take care of itself, and have to do with the present. My answer is that nothing can be done without chimeras. Man needs, in order that he may bring into play all his activity, to place before himself an object capable of rousing his energies. What is the use of labouring for the future if the future is to be colourless and insignificant? Would it not be better to think of his ease and pleasure in this life than to sacrifice himself for a void? The first Mussulmans would not have marched from one end of the world to the other if Abu Bekr had not told them that they had Paradise before them. The Conquistadores would not have undertaken their perilous expeditions if they had not hoped to find the El Dorado, the fountain of Jouvence, the Cipangu with its golden roofs. Alexander went in pursuit of the Griffins and the Arimaspes. Columbus, on returning from the islands of St. Brandon and the earthly paradise, discovered America. With the idea that the paradise is in the remote distance, the world marches on and on, and discovers something better than paradise. "The heart," says Herder (170), "beats only for that which is distant." Hopes, moreover, which may be chimerical in form are not envisaged as such when regarded as the symbol of the future of humanity. The Jews had the Messiah because they fervently looked for Him. No idea is realized without the inward working of faith and hope. The early Christians daily expected to see the New Jerusalem coming down out of heaven and Christ descending to reign. You will say that they were mad. But hope never deceives, and I am convinced that all the hopes of the believer will be fulfilled and even more than fulfilled. Humanity realizes perfection by desiring it and hoping it, as the woman is said to impress upon the child in her womb the objects which strike her senses. These hopes are so far from being objects of indifference that they alone explain and render possible the life of sacrifice and devotion. What use, indeed, is it to devote oneself to relieving misfortunes which only come into existence at the moment they are felt? Why sacrifice one's comfort to that of others, if, after all, only a mere question of enjoyment is at stake? My happiness is as precious to me as that of others, and I should be very stupid to sacrifice it for them. If I did not believe that humanity is called to a divine end, to the realization of the perfect, I should become an Epicurean, if I were capable of doing so, and if not, I should commit suicide.

CHAPTER XX.

I SHOULD be much misunderstood if it were concluded from what precedes that I had any intention of suggesting that science should descend from its eminence to put itself on a level with the people. I have a profound antipathy for popular science, because it cannot be the true science. Over the portals of an ancient school were inscribed the words: "Let no man enter here unless he knows geometry." The modern philosophical school should have for motto: "Let no man enter here unless he knows the human mind, history, literature, etc." Science loses all its dignity when it lowers itself to these childish formulæ and to a language which is not its own. In order to render these lofty philosophic theories intelligible to the vulgar, we are obliged to strip them of their true form, to subject them to the narrow measure of common sense, to distort them. It would be very desirable that the mass of the human race should elevate itself to the comprehension of science, but science must not abase itself in order to be understood. It must remain upon its lofty eminence and raise humanity up to it. I am not opposed to the literature of the working classes. Upon the contrary. I believe, with M. Michelet, that the people possess a true fund of feeling, superior in one sense to that of nearly all our aristocratic poets. The most original poems written since Lamartine and Victor Hugo ceased composing are, perhaps, those of working men. This is all the more meritorious seeing that the instrument which we put into their hands is extremely aristocratic and inflexible and the very antithesis of popular thought.

With respect to the social and philosophical writings, in which the form is of less consequence than in literature, working men often display more intelligence than the bulk of well-read men. The man who has only had primary instruction is nearer to positivism and to the negation of the supernatural than the bourgeois who has had a classical education, for this latter often inclines people to be content with mere words. But the working men often commit a really unpardonable error; that of discarding the line in which they might excel to treat of subjects in regard to which they are incompetent and which require a culture very different from that afforded by little school books. M. Agricol Perdiguier was original as long as he was content to be only the workman. What people liked in him was that he told them what was the way of thinking of a certain class of society, and also the ingenuous effort of the half-educated man to find an instrument for his thoughts. But one fine day, he set himself to compile a Universal History, a work in which Bossuet had failed. It is all very well for M. Perdiguier to say that it was a history for the working man; that all his predecessors had treated history from the point of view of the classics and like college pedants. There are not, so far as I am aware, two histories, one for the learned and the other for the unlearned, and I only know one class of men capable of writing it, and that is the savants who are broken by a long course of intellectual culture to all the fine points of criticism. Science and philosophy must preserve their lofty independence, that is to say only pursue the truth in all its objectivity, without troubling themselves about any popular or worldly form. Drawing-room science is just as little true science as is the science of petty popular treatises. Science degrades itself when it ceases to correspond directly, like poetry, music, and religion, to a disinterested requirement of human nature. How rare, with us, is this pure cult of all the parts of the human soul. Collecting by themselves, and as it were in a useless bundle of religious cares, we make vulgar interests the essential of life. Knowledge, we are told, does not suffice to ensure salvation; knowledge does not serve to make one's fortune; consequently, knowledge is useless (171).

The great misfortune of contemporary society is that intellectual culture is not recognized in it as a religious concern; that poetry, science, and literature are regarded as an artistic luxury appealing only to the classes privileged by fortune. Greek art produced for the sake of the country, for the national thought; the art of the seventeenth century produced for the king, which was, in one way, to produce for the nation. Art, in our day, produces only at the express or understood command of individuals. The artist corresponds to the amateur, as the cook does to the gastronomist. This is a deplorable situation at an epoch when, with rare exceptions, the subdivision of property makes it impossible for private individuals to achieve great things. Greece owed her poems, her temples, and her statues to her inward spontaneousness, they were due to her own fecundity and to her craving to satisfy one of the needs of human nature. With us, art is granted a few sparing subsidies, which are made not from a desire to see the national thought finding expression in great works, not by the inward impulsion which inclines man to realize beauty, but from a calm and critical consideration that art should have its place and from a reluctance to be behind the past. But if people merely obeyed the pure and spontaneous love of what is beautiful, little would be done. One of the reasons recently advanced in favour of the scheme for the completion of the Louvre was that it would be a means of giving employment to artists. I should be curious to know whether Pericles advanced this argument to the Athenians when the question of building the Parthenon was discussed.

Reflect for a moment on the consequences of this deplorable regime which subjects art, and more or less literature and poetry, to the fancy of individuals. In the productions of the mind, as in all other kinds, the question of supply and demand prevails, and it must necessarily happen that it is wealth which makes the demand. So that the man who contemplates living by intellectual production must first of all anticipate the demand of the rich man in order to comply with it. Now what does the rich man demand in the way of intellectual productions? Is it serious literature? Is it high philosophy, or, in the way of art, pure and severe productions, high moral creations? Assuredly not. It is amusing literature; serial stories, romances, clever plays in which his opinions are flattered, and so on. Thus, with the rich man regulating the literary and artistic production by his tastes, which are pretty well known, and these tastes being as a rule (there are a few noble exceptions) for frivolous literature and for art unworthy of the name, it was bound to happen that such a state of things would lower literature, art, and science. For with the rich man's taste setting the value of things, a jockey or a danseuse who correspond to this taste are persons of more value than the savant or the philosopher, whose works he does not want. That is why a composer of novels for serials may make a brilliant fortune and attain what is called a position in the world, while a real savant, had he achieved the distinction of a Bopper or Lassen, could not make a living out of his works.

I mean by plutocracy a state of society in which wealth is the principal thing, in which one can do nothing unless one is rich, in which the chief object of ambition is to become rich, in which capacity and morality are generally valued (and more or less accurately) by a money standard, so that the best criterion of the *élite* of a nation is the cess-rate.

It will not, I imagine, be questioned that the society of the present day combines these different characteristics. That being granted, I maintain that all the faults of our intellectual development come from plutocracy, and that it is in this respect above all that our modern societies are inferior to Greek society. In fact, when wealth becomes the principal aim of human life, or at all events the necessary condition of all other ambitions, let us consider what will be the direction given to the mind. What is needed to become rich? Is it to be a savant, a wise man, or a philosopher? Not at all; these are, upon the contrary, more obstacles than anything else. He who devotes his life to science may rest assured that he will die in want, unless he has a patrimony, or unless he finds a means of utilizing his science, that is to say, unless he can make a livelihood out of pure science. For it will be observed that when a man makes a living by intellectual labour, it is not as a rule his true science that he brings into play, but his inferior qualities. M. Lebronne has made more by compiling second-rate elementary books than by the admirable researches which have rendered his name famous. Vico earned his living by composing prose and poetic pieces of the most contemptible rhetoric for princes and nobles, and could not find a publisher for his *Science Nouvelle*. So true is it that it is not the intrinsic value of things which constitutes their value, but the relation which they bear to those who hold the purse-strings. I may without vanity consider myself to possess as much capacity as any clerk or shopman. Yet the latter are able, in serving purely material interests, to gain an honest living, while I, who appeal to the soul, I, the priest of true religion, do not, in sober truth, know where I am to look for my daily bread next year.

The profound truth of the Greek intelligence is derived, as it seems to me, from the fact that riches constituted, in their highly organized civilization, merely a motive of itself, but not a necessary condition to any other ambition. Hence arose the perfection of spontaneousness in the development of individual characters. A man was a poet or a philosopher because it was part of his human nature, and because he was specially endowed in that way. With us, upon the contrary, there is a tendency imposed upon whomsoever seeks to make a situation for himself in external life. The faculties which he must cultivate are those which serve to make rich: the industrial spirit and practical intelligence. Now these faculties are of very small value: they do not make a man better, or more elevated, or more clear-sighted in divine things; quite the contrary. A man devoid of worth or morality, selfish and lazy, will be more likely to make his fortune upon the Stock Exchange than one who concerns himself with serious matters. That is not just, and therefore it will disappear. Plutocracy, then, is not very favourable to the legitimate development of intelligence. England, the country of wealth, is of all civilized countries the one in which you find the minimum of the philosophical development of the intelligence. The nobles in former times regarded it as beneath their dignity to concern themselves with literature. The rich generally have coarse tastes and attach the idea of good form (*bon ton*) to matters which are ridiculous or purely of convention. A gentleman rider, however insignificant he is, may pass for a model of fashion. But I call him in so many words a fool.

Plutocracy, in another order of ideas, is the source of all our wars, because of the evil feelings it inspires among those whom fate has made poor. The latter, in truth, seeing that they are nothing because they possess nothing, direct all their activity towards this one aim; and as, in many cases, this is slow, difficult, and even impossible, evil thoughts germ; jealousy and hatred of the rich, the idea of stripping him of what he possesses. The remedy for this is not to contrive so that the poor may become rich, nor to excite this desire in him, but so to act that riches may become a secondary and insignificant thing; that without it one may be very happy, very great, very noble, and very handsome; that without it one may become influential and highly esteemed in the State. The remedy, in other words, is not to excite among men appetites which all cannot satisfy, but to destroy this appetite or to change the object of it, seeing that this object does not belong to the essence of human nature; but that, on the contrary, it impedes the proper development of it.

CHAPTER XXI.

Science being one of the true elements of humanity, it is independent of all social and fixed form like human nature. No revolution will destroy it, for no revolution will change the deep-rooted instincts of man. Doubtless one can, while devoting oneself to it, find spare moments for other duties; but this must be only a suspending not an abrogation of the worship. A man must maintain the elevated and ideal value of science even when attending to other and more pressing duties. There are, I admit, sciences which may be termed *umbratiles*, which are the better for security and peace. Only a M. de Sacy could have published in the terrible year of 1793, at the Louvre printing-press, a work upon Persian antiquities and upon the coins of the Sassanides kings. But, taking human intelligence in the mass, estimating progress by the movement accomplished in ideas, we are led to say: God's will be done! and we recognize the fact that a three days' revolution does more for the progress of the human mind than a generation of the Académie des Inscriptions.

If there is one commonplace to which facts give the lie more than to another, it is that a revolutionary period is very unpropitious for mental labour, that literature, in order to produce masterpieces, has need of calm and leisure, and that the arts deserve in reality the classic epithet of "friends of peace." History demonstrates, upon the contrary, that action, war, and rumours of war are the true medium in which humanity develops itself, that genius only puts out its full vegetation amid the storm, and that all the great creations of thought have appeared in troublous times. Of all ages, the sixteenth century is beyond doubt that in which the human mind displayed the most energy and activity in all directions; it is the creative century *par excellence*. There was a want of regularity in it, no doubt; it was a thick and luxuriant growth amid which art had not, as yet, traced paths. But what fertility we have in this the century of Luther and of Raphael, of Michael Angelo and of Ariosto, of Montaigne and of Erasmus, of Galileo and of Copernic, of Cardan and of Vanini. All branches of learning and knowledge are embraced in it: philology, mathematics, astronomy, physical science and philosophy. And yet this wonderful century, in which the modern spirit was definitely constituted, was the century of universal struggles and contentions: religious, political, literary, scientific. Italy herself, which then took the lead of all Europe in the paths of investigation, was the theatre of barbarous wars such as the future, let us hope, will never see again. The sacking of Rome did not disturb the brush of Michael Angelo; left an orphan at the age of six, mutilated at Brescia, Tartaglia worked out mathematics for himself. Rhetoricians alone can prefer the calm and artificial work of the *writer* to the burning and genuine work which was an *act*, and stood out in its day as the spontaneous cry of an heroic or impassioned soul. Æschylus had been a soldier at Salamis before being the poet. It was in camp and amid the risks of a life of adventure that Descartes thought out his method. Would Dante have composed in an atmosphere of studious ease those cantos of his which are the most original in a period of ten centuries? Are not the sufferings of the poet, his wrath, his passions, and his exile half the poem? Can one not feel in Milton the man who has been wounded in political strife? Would

Chateaubriand have been what he is if the nineteenth century had been an unbroken continuation of the eighteenth?

The customary state of Athens was one of terror. Never were political habits more violent, or the security of the individual less. The enemy was always a few leagues off; not a year elapsed but what he appeared at the gates, but what it was necessary to go out and fight against him. And within, what an interminable series of revolutions! To-day an exile, to-morrow sold as a slave; then regretted, honoured as a god, liable each day to be dragged before the most pitiless revolutionary tribunal, the Athenian, who amid this life of rush and uncertainty, was never sure of the morrow, produced with a spontaneousness which overwhelms us with surprise. Let us not forget that the Parthenon and the Propylœa statues of Phidias, the *Dialogues* of Plato, the stinging satires of Aristophanes, were the work of an epoch very similar to 1793, of a political state of things which entailed, in proportion to the number of people concerned, more violent deaths than our first revolution at its paroxysm. Where in these masterpieces do we find a trace of the terror? Some strange timidity seems to have taken hold of our minds. As soon as the smallest cloud appears upon the horizon, everybody withdraws himself inside his shell, so to speak, and is overcome with fear? "What can be done in times like these? We want security. There is no inclination to produce anything, when there is no certainty as to the morrow." But we might remember that the morrow has never been certain since the beginning of the world, and that if the great men whose labours have made us what we are had reasoned in this way, the human mind would have remained for ever sterile. Montaigne ran the risk of assassination as he walked round his château, but this did not prevent him from writing his *Essais*. This fatal craving for repose has come from the long period of peace we have gone through, and which has had so great an influence upon the current of our ideas. The masculine generation which grew to manhood in 1815, had the good fortune to be cradled amid great achievements and great perils, and to have had a high-spirited struggle to carry on as a training for its youth. But we, who began to think in 1830, born beneath the influence of Mercury, first saw the world as a machine in regular working order; peace seemed to us as being the natural medium of the human mind; our only experience of struggle and contention was under the petty proportions of a purely personal opposition. The least disturbance surprises us. Our horizon is limited to the timid preservation of what our fathers have accomplished. Woe to the generation which has only had experience of a regular state of administration, whose sole conception of life is as a rest and of art as an enjoyment. Great achievements never are accomplished in these lukewarm periods. But it would not do to deny all value to the productions of a period of calm and regularity. They are highly finished, sensible, reasonable, full of a delicate criticism; they are agreeable to read in hours of leisure; but there is nothing decided or original about them, nothing which breathes militant humanity, nothing which approaches the hardy works of those wonderful ages in which all the elements of humanity while in a state of ebullition appear alternately upon the surface. The universe created only in its primitive periods and beneath the reign of chaos. Monsters could not come into being under the peaceful régime of equilibrium which has succeeded the tempests of the earliest ages.

It is not, therefore, either material ease or even liberty which contribute much to the originality and the energy of intellectual development; it is the medium of great things, it is universal activity, it is the spectacle of revolutions, it is the passion developed in combat. The work of the mind would only be seriously threatened when humanity came to be too much at its ease. Thank God! that day is still far distant.

A newspaper some time since called upon the National Assembly to decree *the right to rest;* an ingenious metaphor the meaning of which no one could mistake. No doubt, if life were to be regarded merely as a period of rest and pleasure, we might well regard the agitation of thought as a curse, and as perverse those, who in order to satisfy their feeling of unrest, disturbed this pleasant slumber. Revolutions can be only absurd and odious disturbances in the eyes of those who do not believe in progress. Without the idea of progress, all the ideas of humanity are incomprehensible. If human life had no other horizon than to vegetate somehow; if society were merely an aggregation of persons each one living for himself and invariably subject to the same vicissitudes; if every one was born, lived, and died in much the same manner, the best thing to do would be to put humanity to sleep and patiently submit to this commonplace monotony. There are some who think it a good thing that the time of religious controversies is over. For my own part, I regret it, and I regret the disappearance of that most beneficent Protestant controversy which, for more than two centuries, sharpened up and kept on the *qui-vive* the mind of all civilized Europe; I regret the time when Turenne and Lesdiguière were in controversy, when a book by Claude or Jurieu was an event, when Caton and Turretin, matched against each other, arrested the attention of all Europe. The wars of religion are, after all, the most reasonable of all wars, and henceforward they will be the only ones such. We must be fair, and admit that never was life more free than from 1830 to 1848, and we shall perhaps have to wait a long time before we have a régime affording so large a share of liberty. But can it be said that during this period humanity was enriched by many new ideas, that morality, intelligence, and true religion made any real progress? Just as the monastic life, where all is cut and dried in advance, destroys the picturesque part of life and effaces all originality, so a regular civilization, tracing too narrow a road for existence, and placing constant fetters upon individual liberty, is more detrimental to spontaneousness than the arbitrary régime (172). "This formalist liberty," says M. Villemain, "engenders more petty vexations than great struggles, more intrigues than great passions." The human mind was infinitely more active in the years of compression of the Restoration than in the years of reasonable liberty which followed 1830. Poetry became egotistical, and its sole value was as a delicate accompaniment of pleasure; originality was out of place. It was admired and sought out with curiosity in the past, but it was regarded with contumely in the present. Keen interest was felt in the well-defined figures to be found in history, and no mercy was shown to the contemporaries whom the future will regard with the same interest. Thus a régime which realized the ideal of eclecticism will get the reputation of being a somewhat barren period as regards the history of human intelligence.

Upon the other hand, an epoch may, so long as it emerges from the commonplace, give birth to the most original and contradictory apparitions. Have we not seen the selfsame revolution producing at one and the same time the true formula of the rights of man, and the new symbol of liberty, equality, and fraternity, and, upon the other hand, the massacres of September and the scaffold in permanence? Did not the same century give birth to the Talmud and the Gospel, the most fearful monument of intellectual depression and the loftiest creation of the moral sense—to Jesus upon the one hand and Hillel and Schammai on the other? Anything may be anticipated in these great crises of the human mind, whether in the way of what is sublime or absurd. It is only the colourless productions of periods of repose which are consistent with themselves. The appearance of Christ would be inexplicable in a logical and regular state of things; it is comprehensible during the singular crises which at that period characterized the reasoning sense of Judæa. Those solemn periods during which human nature, in a state of exaltation, half-beside itself, emits the most extravagant sounds, are the periods of great revelations. If the same circumstances occurred again, the same phenomena would reappear, and we should again see Christs, not probably represented by individuals, but by a new spirit, which would be of spontaneous growth, without perhaps being personified so exclusively in any one particular person.

It must not be imagined that human nature is so clearly marked off as to be unable to extend beyond a commonplace horizon. There are rents in this horizon, through which the eye pierces the infinite; there are vistas which go direct beyond the goal. There may spring up among the stronger races and at periods of crises monsters in the intellectual order, which, while participating in human nature, exaggerate it so much in one sense that they catch a glimpse of the unknown worlds. These beings were less rare than is generally believed in the primitive epochs. It may be that more of these strange natures, placed upon the limit of man and open to different combinations, will one day appear. But assuredly these ancestors will not be born in our jog-trot time. A narrow and fixed conception of life weakens the creative faculties. Civilization, owing to the extreme definition of rights which it introduces into society, and by the fetters which it imposes upon individual liberty, becomes in the long run a very irksome burden, and deprives man in a great measure of the keen sense of his independence. I can understand German writers regretting from this point of view the old Germanic life and deploring the Roman and Christian influence which changed its rough-and-ready sincerity. Compare the modern man, swathed about with a thousand clauses of law, unable to take a step without being confronted by a sergeant or a regulation, with Antar in his desert, knowing no other law than the fire of his race, dependent only upon himself, in a world where no idea of penal law or of coercia exercised in the name of society exists.

All is fruitful except common sense. The prophet, the apostle, the poet of the early ages would be regarded as madmen in the midst of the colourless mediocrity in which human life is imprisoned. A man who shed tears without any apparent cause, who wept over the universal sorrow, or laughed with a long, mysterious laugh, would be shut up in a madhouse because he did not make his thoughts fit to our accustomed moulds. And yet I would ask why this man is not nearer to God than a self-sufficient little tradesman, huddled up in the corner of his shop. How touching is the custom prevalent in Judea and Arabia, where the lunatic is honoured as a favourite of God, as a man who sees into the world beyond! The Soufi and the Corybant believed, in losing their senses, that they touched the divinity; the instinct of different peoples has sought for revelations from the hallowed state of slumber. The prophets and the inspired men of ancient times would have been classed by our physicians as victims of hallucination. So true is it that an undetermined line divides the legitimate and the exorbitant exercise of the human faculties, and that they cover a serial scale of which the centre alone is attainable. The same instinct, in the one case normal, in the other perverse, inspired Dante and the Marquis de Sade. The greatest of religions was marked at its cradle by incidents of the purest enthusiasm and by the extravagances of convulsionists equal to anything to be seen among the most exalted of sectaries.

We must, therefore, resign ourselves to the fact that the most beautiful of things are born amid tears; sorrow is not too high a price to pay for beauty.

The new faith will only be born amidst terrible disturbances, and when the human intelligence has been checkmated, thrown off the rails, so to speak, by events as yet unparalleled. We have not yet suffered sufficiently to see the kingdom of heaven. When a few millions of men have died of hunger, when thousands have devoured one another, when the brains of the others, carried off their balance of these darksome scenes, have plunged into extravagances of one kind and another, then life will begin anew. Suffering has been for man the mistress and the revealer of great things. Order is an end, not a beginning.

So true is it that institutions bear their best fruit before they have become too official. It would be very foolish to imagine that now there is a discussion forum on the Quai d'Orsay, there will be greater political speakers. The first École Normale was unquestionably regulated with less care than that of to-day, and had no teachers to compare with the present ones. And yet the generation it produced is beyond all comparison with what is produced now. An institution only possesses its full strength when it corresponds to the real and clearly felt want which led to its establishment. At first, it may appear to be imperfect, and people are too ready to imagine that when the period of tranquillity and of peaceful organization comes, it will do wonders. This is all a mistake; the petty improvements spoil the work; the motive force disappears, and the whole becomes petrified. Official regulations do not impart life, and I am convinced for my part that an education like ours will always have the defects so often imputed to it of being mechanical and artificial. Regulations are supposed to take the place of the spirit, to achieve with men devoid of self-devotion or morality what could be accomplished with men who were devoted and religious. This is attempting the impossible; life cannot be simulated; no machinery however well combined will produce more than an automaton.

This evil is not to be corrected by regulations, inasmuch as they are the very source of the mischief. Rule existed, it is true, at the outset, but it was vivified by the spirit, just as the Christian ceremonies, which have become a mere series of closely regulated movements, were in the beginning true and sincere. What a difference there is between chanting a tag of Latin called the Epistle and reading in company the *Correspondence* of colleagues, between a piece of consecrated bread which no longer has any meaning and the feast of early times! The primitive gathering, the *agape*, had no need to be regulated, for it was spontaneous. Painting produced masterpieces before there was such a thing as a yearly exhibition. Men of letters and artists did not enjoy, in the seventeenth and eighteenth centuries, the amount of dignity they should have done, and therefore, now that they have obtained the place which is due to them, they will do much better. These are erroneous conclusions, for they presuppose that the regulating of the external conditions of intellectual production is favourable to it, whereas this production depends solely upon the abundance of the inner and living sap of humanity.

Some one said, speaking of the beatific calm in which Austria was previous to 1848: "What can you expect; these people are foolish enough to be happy." That is not so; it is not a vulgar thing to be happy. Only the high-strung souls know how to attain happiness. But to be at one's ease is the most prosaic of aspirations. None but a simpleton can be so eager for the régime of a full stomach.

As soon as a country begins to show signs of agitation, we are inclined to regard its condition as unsatisfactory. If, upon the contrary, it is in a dead calm, we say, and in this case with more truth, "This country feels the want of action." Agitation appears to be a regrettable transition and repose seems to be the aim; but repose never comes, and if it did that would be the worst thing that could happen. No doubt, order is a desirable thing, and all efforts should tend towards it; but order is only to be desired in view of progress. When humanity has reached its rational state, but not till then, revolutions will appear detestable, and pity will be due to the age which stands in need of them.

The aim of humanity is not repose; it is intellectual and moral perfection. How can people talk of repose, I should like to know, when they have the infinite to traverse and the perfect to reach? Humanity will only repose when it has reached the perfect. It would be too strange if a few profane persons could, from motives of £ *s.* *d.* or personal interest, arrest the progress of the mind, the true religious progress. The most dangerous state for humanity would be that in which the majority, finding itself quite at ease and not wishing to be disturbed, should retain its repose at the cost of thought and of an oppressed minority. When that occurred, the only safety would be in the moral instincts of human nature, which, no doubt, would not be found wanting.

The traction force of humanity has hitherto resided in the minority. Those who find themselves well off in the world as it is cannot like movement, unless they raise themselves above self-interested views. Thus the greater become the number of the well-to-do, the harder is it to stir humanity; it has to be dragged along. The good of humanity being the supreme aim, the minority must not scruple to lead along against its will, if needs be, the stupid or selfish majority. But in order to do that it must have reason on its side, otherwise it is an abominable tyranny. The essential thing is not that the will of the great majority should be carried out, but that right should be done. What! there are people who, in order to make a few sous the more, would sacrifice humanity and country, who would have the right to say: "You shall go no further; do not teach this, as it might stir up men's minds and be injurious to our business?" The only portion of humanity which deserves to be taken into consideration is the active and living part, that is to say, the part which is not well to do. It will, therefore, be quite in vain that our fathers, having become reasonable, will beg us no longer to think and to keep ourselves quiet, for fear of throwing the delicate machine out of gear. We ask for ourselves the liberty which they have taken to themselves. We will let them be converted, and we will appeal from Voltaire out of sorts to Voltaire in good health.

Reflect, then, for a moment upon what you are attempting to do, and remember that it is a sheer impossibility, that which since the beginning of the world all intelligent conservatives have attempted without success; to check the human mind, to deaden intellectual activity, to persuade youth that all thought is dangerous and leads to evil. You have had your liberty of thought, so will we; we will admire, as you have taught us to do, the great men of the past and the illustrious promoters of thought whom now you repudiate. We will remind you of what you have taught, we will protect you against yourselves. You are old and sickly, and you can become converted; but we, your pupils in liberalism, we, young and full of life, we to whom the future belongs, why should we accept a share in your terrors? How can you expect that a rising generation should be content to shrivel up with ill-humour and fear? Hope is a thing of our age, and we prefer succumbing in the struggle to dying of cold or fear.

There is something really comical in this ill-humour which has suddenly been manifested against the free thinkers, as if, after all, they could help the result of their speculations, as if they could have done different from what they have, as if they were free to see things different from what they are. One might have supposed that it was out of pure caprice and fancy that they suddenly attacked the beliefs of the past, and as if it depended upon them whether the world retained its faith or not. A book only succeeds when it responds to the secret thought of all; an author does not destroy faiths; if they apparently fall beneath his blows, it was because they were already very much shaken. I have met people who, imagining that the mischief comes from Germany, regretted that there was no inquisition against Kant, Hegel, Strauss. Sheer fatality this! You admire Luther, Descartes, and Voltaire, and you anathematize those who, without intending to imitate them, carry on their work, and if there was in our day a Luther, a Descartes or a Voltaire, you would treat them as enemies of society, as dangerous innovators. You condemn the eighteenth century which you formerly liked so well; you should, to be logical, condemn the Renaissance, the whole modern spirit and fatality into the bargain. You may condemn and curse as much as you please, but I defy you to deaden the human mind beneath a perpetual charm, I defy you to persuade it to remain idle, to rest motionless for fear of risk, for that is death. We will not allow it, we will cry out to the people: "It is false, it is false, you are being led astray," rather than tolerate this irreverent way of treating truth as being of less importance than the tranquillity of a few cowards.

The whole secret of the intellectual situation of the moment is, therefore, in this fatal truth, viz. that intellectual labour has been degraded to the rank of a pleasure, and that in the day of serious things it has become insignificant like the pleasures themselves. It is not, therefore, the fault of events, for they should rather have had the effect of awaking men's minds and stirring them up to thought; the whole fault is in the general depression brought about by the exclusive attention given to repose; the shameful hedonism of which we gather the fruits and of which communistic follies are, after all, but the extreme consequence. There are times when to amuse oneself is a crime, or at all events an impossibility. The silly literature of coteries and salons, the science of the inquisitive and of amateurs is much depreciated by these terrible spectacles; the serial story loses much of its interest when it appears in a paper containing the recital of the real and passionate drama of the hour; the amateur may well fear to see his collections carried off or blown all over the place by the wind of the storm. In order to appreciate these peaceful tastes, a man need have nothing to do and nothing to fear; to seek out such innocent diversions, time must hang heavy on the hand. But nothing of that which contributes to the awakening of humanity is lost to the true progress of the mind; philosophical thought is never more free than in the great days of history. Intellectual exercise is more vigorous at those times, for it is less tainted with amusement. It is imperative that the world should school itself to maintain, amid no matter what cataclysms, the value of intellectual culture, of science, of art, and of philosophy. What is good is always good, and if we wait for tranquillity, we may have to wait a long time. If our fathers had reasoned in this way, they would have folded their arms and done nothing, and we should not enjoy their inheritance. And what matters it, after all, whether to-morrow be sure or not? What matters it whether the future belongs to us or does not? Is the sky less blue, Beatrice less beautiful, or God less great? If the world were to collapse, it would still be well to philosophize, and I am convinced that if ever our planet is the victim of a fresh cataclysm, which at this moment seems impending, it will find some few minds of men which, amid this general upheaval and chaos, will have disinterested and scientific thoughts, and which, forgetting their impending death, will discuss the phenomenon, and endeavour to draw from it some consequences bearing upon the general system of things (173).

CHAPTER XXII.

I ASK the reader to pardon numberless views to some extent exaggerated which he will have encountered in the preceding chapter, and I beg him to judge this book not by an isolated passage, but by its general spirit. A man can only give expression to his mind by the successive delineation of various points of view, each of which is only true when taken as a whole. A single page is, as a matter of necessity, false; for it expresses only one thing, and truth is only the compromise between an infinity of things (174). Now what I have sought above all to inculcate in this book is faith in reason, faith in human nature. "I would have it serve to react against the sort of moral enfeeblement which is the malady of the rising generation; I would have it guide back into the true road of life some of those enervated souls which complain of being lacking in faith, which know not which way to go and which search everywhere in vain for an object of worship and devotion. Why complain so bitterly that in the world, as it is constituted, there is not enough employment for all intelligences? Is not serious and calm study available? and does not it contain a hope and a career within the reach of each one of us. Armed with this, we can endure the evil days, without feeling the weight of them, we can work out our own destiny, we can make a noble use of life (175). That is what I have done," added the noble martyr to science from whom I have borrowed these lines, "and what I would do again if I had to recommence my journey. Blind and suffering without hope of recovery, almost unceasingly, I can bear this testimony which, from my lips, will not be open to cavil: there is in the world something better than fortune, better than health itself, and that is devotion to science."

I know that in the eyes of many people, this faith in science and in the human intelligence will seem a piece of absurdity, and that it will not please those who, too clever to believe in the truth, regard scepticism itself as too doctrinal, and who, without paying too much heed to these cumbrous categories of truth and error, treat the enjoyments of life and the calculations of intrigue as the serious business of life. There is nothing but raillery for those who concern themselves with the reality of things, and who, in order to form an opinion upon morality, religion, social and philosophical questions, are so simple minded as to reflect upon objective reasons, instead of addressing themselves to the much easier criterion of interests and of good taste (176). Only the form of expression is taken into account; the intrinsic consideration of things is regarded as useless and in bad taste; or else, if it is thought distingué to assume the air of a believer, people accept some ready-made system, the absurdities of which they see very clearly, just because they think it amusing to admit these absurdities in order to make sport of reason. In this way, a man becomes all the grosser in the object of his belief as he has been more sceptical and has shown more levity in his motive for accepting it. It would be bad "form" to ask oneself for a moment whether it is true; it is accepted as one accepts a certain shape of coat or hat; one becomes superstitious to a degree because one has been sceptical —not to say frivolous. Pronounced scepticism has never been very prevalent in France; our sceptics, to begin with Montaigne and Pascal, have been either men of wit or believers; two forms of scepticism very close to each other and mutually interdependent. Pascal sought to borrow from Montaigne his sceptical arguments and give them the first place in his Apologia: "One cannot," he says, "fail to see with pleasure how, in this author, the magnificent reasoning is so invincibly rumpled by his own arms . . . and one would love with one's whole heart the ministry of so great a vengeance, if . . ." (177)

When scepticism has become a fashion, it does not imply penetration of mind nor keenness of criticism, but much rather dulness and incapacity to understand the truth. "It is very convenient," says Fichte, "to cover with the high-sounding name of scepticism lack of intelligence. It is pleasant to pass off this lack of intelligence which prevents us from seizing the truth for marvellous penetration of mind, which reveals to us motives of doubt unknown and inaccessible to the rest of mankind" (178). It is easy, by placing oneself as outside all dogma, to play the part of one in advance of his age, and the stupid, who are more afraid of being dupes than of anything else, carry this even further. Just as in the eighteenth century it was the fashion not to believe in feminine honour, so in our day there is not a provincial person with pretensions who does not make a glory of having no political creed and of not being a believer in the probity of those who govern. This is one way of having one's revenge, and also of making the world believe that one is initiated into deep secrets.

It is the honour of philosophy to have always had as its enemies the frivolous and the immoral. Who, not having in themselves the instinct of what is beautiful, boldly declare that human nature is ugly and bad, and embrace with frenzy any doctrine which humiliates man and keeps him entirely under dependence. Therein lies the secret of the faith of the "golden youth" of the Catholic creed, thoroughly sceptical, hard and scornful, which finds amusement in calling itself Catholic, because by so doing it puts an insult the more upon modern ideas. That dispenses it from harbouring noble thoughts; by dint of saying to oneself that human nature is vile and corrupt, one in the end resigns oneself to this and takes it in good part (179). The Church will be indulgent for errors of the heart, and then it is so easy for aristocratic fatuity to believe that the mass of mankind is absurd and ill-intentioned, and to have under control a potent authority to cut short the reasonings of these impertinent philosophers, who dare to believe in truth and beauty. O sordid souls, how dark you are within, seeing how few are the things you love! And you will be called believers, while we are called impious! This is more than can be endured.

God forbid that I should speak slightingly of those who, devoid of the critical sense and impelled by very powerful religious motives, are attached to one or other of the great established systems of faith. I love the simple faith of the peasant, the serious conviction of the priest. I am convinced, for the honour of human nature, that Christianity is, with the immense majority of those who profess it, a purely noble form of life. But I cannot help saying that, for a great part of aristocratic youth, Catholicism is merely a form of scepticism and frivolity. The first basis of that sort of Catholicism is scorn, malediction, and irony; malediction against anything which has caused the human mind to progress and which has broken the old chain. Compelled to hate whatever has aided the modern spirit to shake itself free of Catholicism, these frenzied partisans are full of hatred for everything and everybody: for Louis XIV. who, by constituting the central unity of France, worked so efficaciously for the triumph of the modern spirit, as for Luther, for science as for the industrial spirit, for humanity in short. They think that they are offering an apologia for Christianity when they make sport of all that is serious and philosophical.

It is impossible for me to express the physiological and psychological effect this sort of parody which has become so much the fashion in provincial circles of late years has upon me. It irritates and annoys me beyond endurance. It is so easy to elude in this way anything serious or original. Barbarians that you are, do you forget that we have had Voltaire, and that we might fling in your face Father Nicodemus, Abraham Chaumeix, Sabathier and Nonotte? We do not do so, because you have told us that it was unfair. But why, then, use against us a weapon which you have reproached us with employing? Do you suppose that if we wanted to make sport of theologians, we should not have as fine an opening as you, when, in order to raise a laugh from the triflers, you put silly arguments into the mouths of the philosophers? I happened to take up a pamphlet against eclecticism, in which Descartes is represented as a fool whose sole contribution to philosophy is the question "whether reason is not a thing which reasons falsely," Kant as an idol who does not know whether he exists and whether the world exists, Fichte as an impertinent fellow who asserts that "he, Fichte, is at once God, nature, and humanity," all the philosophers, in short, as greater lunatics than the magicians, the alchemists, and the astrologers. I can fancy the light laughter which the reading of these charming remarks would have excited at some country fireside. The author may be assured of making his fortune much better than dull-witted persons like us who are so stupid as to take things seriously. . . .

It is time that all parties who have truth at heart should abandon such unscientific ways. There is, I know, a philosophical form of merriment, which could not be suppressed without doing injury to human nature; and that is the merriment of the Greeks, who liked to weep and to laugh over the same subject, to see the comedy after the tragedy, and often the parody of the very piece they had just been witnessing. But pleasantry in scientific matters is always unreal, for it is the exclusion of elevated criticism. Nothing is ridiculous in the works of humanity; to give this aspect to serious things, they must be looked at from a narrow side, and what is majestic and true in them must be neglected. Voltaire makes mock of the Bible, because he has not the intuition of the primitive works of the human mind. He would, in the same way, have made mock of the Vedas, and no doubt of Homer. Banter compels one to regard things solely from their gross aspect; it is a bar against delicate distinctions. The first step in the philosophical career is to become proof against ridicule. If a man allows himself to become subject to vulgar jesters, and to take any account of their frivolities, he shuts himself out from all moral beauty, from all lofty aspirations, from all elevation of character, for all these can be turned into ridicule. The jester has the immense advantage of being dispensed from furnishing proofs; he is free to cast ridicule, at his will, upon what he pleases, and this without appeal, at all events in countries where, as in France, tyranny is accepted as legitimate authority. The only things which escape ridicule are those which are commonplace and vulgar, so that he who is weak enough to keep clear of anything which may lend itself to ridicule, *ipso facto* cuts himself off from all that is lofty. The ages of reflection risk seeing the noblest sentiments and the most sublime conditions of the soul distorted by stupid plagiarists, the ridicule of whom sometimes recoils upon the types which they aspire to imitate. A certain amount of courage is required to resist the reaction which these coxcombs excite among the right-minded. It is unreasonable to resign oneself to commonplace vulgarity, because in following up an elevated type, one runs the risk of resembling great men who have just missed their mark and the unsuccessful aspirants after genius. One may regret the time when a great man was fashioned without giving any thought to the matter, unconsciously so to speak; but the ridiculous attitude of a few weak-headed persons cannot suffice to condemn those who with well-resolved and deliberate will set themselves to accomplish something great and noble. The false Rénés and the false Werthers should not cause us to denounce the true ones. How many timid and modest souls have been kept back from attaining the beautiful for fear of resembling them! All praise to the Olympian thinker, who, pursuing in all things critical truth, has no need to become a dreamer in order to escape the platitude of the bourgeois life, or to become a bourgeois in order to avoid the ridicule of the dreamers.

I sometimes regret that Molière, when stigmatizing the absurdities of the Hotel de Rambouillet, laid himself open to the suspicion of setting up as models types inferior in one way to those which he ridiculed. The only defect in the pure affection of Armande and of Bélise in the *Femmes Savantes*, even that of Cathos and of Madelon in the *Précieuses Ridicules*, is that it is affected, and serves as a cover to vacuousness under the guise of a ridiculous pathos. If it were true, it would be preferable to the commonplace love of Clitandre and of Henriette. I prefer even the affectation of what is elevated to mere triteness. Boileau makes game of Clélie, "that admirable young woman, who so conducted herself that she had not a lover but was compelled to conceal himself under the guise of a friend; for otherwise they would have been turned out of the house." Doubtless, subtlety is not truth; but it is better to be ridiculous than vulgar, and taking refuge in commonplace is a very easy mode of escaping ridicule. It would be too much that superficial jesters should have the power of casting suspicion, whenever it pleased them, upon whatsoever is noble, pure and elevated, of treating enthusiasm as an extravagance, and morality as a dupery. One thing only does not lend itself to jest, and that is the atrocious. Let the mind travel over the scale of moral characters; jests may have been made of Socrates, Plato, Jesus Christ, God. Men may have made mock of savants, poets, philosophers, religious men, politicians, plebeians, nobles, rich traders. No one has ever made mock of Nero or Robespierre. Laughter cannot, therefore, be a criterion. Action appears to many a means of avoiding the dupery to which frivolous people suppose that men of thought and sentiment fall victims. It seems as if the warrior, the politician, the financier are not so open to attack as the philosopher or the poet. But this is a mistake. All things equally lend themselves to ridicule, and if there is one thing really serious, it is the critical thinker who looks at the objectivity of things; for things are serious. Who has not felt, in presence of a flower which blooms, of a rivulet which is murmuring, of a bird watching over its brood, of a rock in the midst of the sea, that these things are sincere and true? Who has not felt in certain moments of tranquillity, that the doubts which are raised as to human morality, are merely ways of searching beyond the limits of reason that which is within them, and of placing oneself in a false hypothesis for the pleasure of torturing oneself? Scepticism alone is entitled to jest, for it has no reprisals to fear. In what respect is it open to attack, inasmuch as it is the first to make light of all things? But how can a believer who makes mock of another believer fail to see that he lays himself open, by the very fact of his being a believer, to the same ridicule? Let us, therefore, leave to negation and frivolity the unenviable privilege of being invulnerable to attack, and let us glory in the fact that, by reason of our convictions and our seriousness, we lay ourselves open to the jests of the sceptical.

Extreme reflection thus inevitably leads to a species of insipidity and light scepticism, which would be the death of humanity if it became universal. Of all the intellectual states, this is the most dangerous and the most incurable. Those who are attacked by it are doomed to die. For how can they hope to survive it, the unhappy beings who do not believe in serious things, who, if they attempted to shake off this intellectual paralysis, would be stopped by the after thought that they, too, are going to be included in the number of people at whom they have been in the habit of laughing. But if there is no cure for over-refining, humanity has methods of rejuvenation and of oblivion which are impossible for individuals. Young and keen generations, and sometimes even new races, are constantly supplying it with sap, and moreover this, by its very nature, could not last more than a few years as a social evil. For, as its essence is to take a purely arbitrary view of things, those who come after do not feel themselves bound by the views of the former; upon the contrary, whatever is conventional provokes almost a reaction in the opposite direction; it is impossible that a fashion should be durable. Thus what is serious and frivolous follow one upon the top of the other in the records of fashion; frivolity in time becomes silly, and ridicule can be adapted to any subject. So that in time the world will make mock of these mockers and recover its liking for the serious life. Then will come an age made dogmatic by science; people will recommence believing in the certain, and of resting their feet firmly upon things, when they know that they are on solid ground.

Religion, philosophy, morality, and politics all have their sceptics; there are not any in the physical sciences (not at least as far as concerns that part of them which is definitely accepted or their method). The method of these sciences has thus become the criterion of practical certainty in the present day, and if the moral sciences seem to furnish results which are less positive, this is because they do not respond to that model of scientific certainty which they have formed for themselves. This is the sheet-anchor which will save our age from the shipwreck of scepticism; scientific certainty is accepted, the only complaint being that this certainty is only applicable to so few subjects. Effort should all be concentrated upon enlarging this circle, but the great thing is that the instrument is recognized, that people believe in the possibility of believing. My conviction is that results equally definite, though formulated in a different way and arrived at by different methods, will be attained in the moral sciences. There are some natures which find pleasure in torturing themselves and in setting themselves insoluble problems. The only proof of the morality and seriousness of life is in our nature. To carry one's investigations beyond that and to doubt of the bases of human nature is to vex one's spirit designedly, is like irritating the sensitive pulse for the doubtful pleasure of scratching oneself.

Those who make a jest of things will never be supreme. The day is not distant when these would-be fastidious people will be discovered so insignificant in presence of the immensity of actual facts, so incapable of producing, that they will collapse like an empty bag. Only the eternal is of any value; whereas these frivolous persons cling to successive offshoots, knowing that both will fade away. Like the worn-out stomachs which soon turn against nourishment and for which fresh culinary combinations have constantly to be found, they concentrate all their interest upon the various modes which succeed one another every ten years. Theirs is a literature of epicureans, well calculated to please a class which is rich and has no ideal, but which will never be the literature of the people, for the people are frank, strong and sincere. This literature has no care for truth, but as a mere matter of good taste and *bon ton*. The thing is not to say what is, but what ought to be said. "He who believes nothing is worth nothing," said M. de Maistre. The ancient faith is impossible, but there remains faith by science, the critical faith.

Criticism is not scepticism, much less levity. Criticism is keen and fine drawn, subtle and winged, without being frivolous. Germany was for a century the country of criticism, and yet who would say that Lessing, Kant, and Hegel were frivolous men? In France, it is not easy to conceive of a medium between the heavy erudition of the seventeenth century and the sharp-witted and sceptical manner of modern critics (180). When people talk of seriousness, they are thinking of the simple wit of Rollin, which is certainly not what we want. What we want is not the *bonhomie* which excites suspicion, because it infers shortness of view. What we want is complete criticism, at once learned and elevated, indulgent and pitiless. Narrowness of mind is very dangerous in France because of the suspicion which it awakens, and because this suspicion is extended to whatever is dogmatic and moral. What people most dread in France is being duped. A man prefers having the character of being easy of conscience and of scruples to that of an honest fool, and if any idea of dulness of perception is associated with morality that is sufficient for it to be regarded with suspicion. This is why the term *bon esprit* has fallen into disrepute. A title which should be the highest possible compliment has become almost synonymous with weak-mindedness, and is bestowed with surprising freedom. For, as a matter of fact, people are always ready to accord to others the qualities which they do not care about for themselves, and it is thought that, in applying to others the term *bon esprit*, one will pass for having a great or brilliant intellect oneself. We are so afraid of being duped that we are perpetually on the look out for attempts to deceive us, and we are ready to think that if our forefathers had been sharper, they would not have been so serious or so honest. And yet, if morality were only an illusion, it would be so fine a thing to allow oneself to be duped by it! *Domine, si error est, a te decepti sumus.* Oh thou who hast made sport of my simplicity, I shall thank thee for having stolen my virtue!

We reject in the same way frivolous scepticism and scholastic dogmatism; we are dogmatic critics. We believe in truth, although we do not claim to possess absolute truth. We do not desire to pen up humanity for ever within our formulæ; but we are religious, in the sense that we are firmly attached to the belief of the present, and that we are ready to suffer for this belief in view of the future. Enthusiasm and criticism are far from being incompatible. We do not force ourselves upon the future any more than we accept without verification the inheritance of the past. We aspire to the high philosophical impartiality which does not attach itself to any party, not because it is indifferent to them, but because it sees in each of them a mixture of truth with error; which has no feeling of exclusiveness or hatred for any one, because it sees the necessity of all these various groupings and the right which each one of them has, by virtue of the truth which it possesses, to make its appearance upon the world's stage. Error is not welcome to man; a dangerous error is a contradiction like a dangerous truth. The reasoning of Gamaliel (181) is unanswerable. If a doctrine is true, it is not to be feared; if it is false still less so, for it will explode of itself. Those who talk of doctrines as dangerous should add "for me." Cabet, I am sure, never provoked any man's anger. Pure error would only provoke in human nature, which after all is kindly constituted, a sentiment of ridicule or disgust.

What makes proselytism, what influences the world, are incomplete truths. The complete truth would be such a quintessence, would be so nicely poised, that it would not excite the passions enough, and would resemble scepticism. The breadth of mind which eliminated in the declaration of its views all limit or exclusion would seem sheer folly. The brain reels when one goes too close to identity; the human mind can only have play when within the limits of a finite frame and of antithetical negation. Passion, while adoring its object, must needs hate its counterpart. Would France be so thoroughly herself if she had not the antithesis of England to keep her personality at high pressure? Passion implies exclusion, antagonism and partiality. Every doctrine, like every institution, bears within itself the germ of life and the germ of death. Called to life by its truth, the doctrine develops with it a principle of death which becomes in time intolerable and kills it. The fruit bears in it from the very first hour the principle of its decay; hidden at first during the period of growth by the organizing forces, this principle declares itself at maturity, and then gains the mastery, until complete decomposition has set in. What any one system affirms is its share of truth, what it denies is its share of error. It only errs because it excludes whatever is not itself, because it participates in human weakness which cannot embrace all things at once and creates science by an analytical and gradual process. The critic is he who examines all affirmations, and perceives the reason of all things. The critic runs his rule over all systems, not like the sceptic, in order to find them false, but to find them true in certain respects. And that is why the critic is little adapted for proselytism. For that which is partial is strongest; men only become impassioned for that which is incomplete, or, to speak more accurately, passion, attaching them exclusively to one object, blinds them to all the rest. It is like the everlasting dupery of the lover who sees only his or her object. Exclusive love is parallel with hatred and anathema. The critic sees too clearly the distinctions to be energetic in action. Even when he takes a side, he knows that his adversaries are not altogether wrong. Now to act with vigour, one needs be more or less pitiless, believe that one is entirely in the right, and that those whom one is opposing are blind or bad. If M. Cavaignac or M. Changarnier had been as critical as I am, they would not have done us the service of saving us in June, for I confess that since February (the revolution of 1848) the question has never appeared clearly enough defined to my judgment for me to have ventured taking sides one way or the other. For, I said to myself, perhaps my brother is on that side; perhaps I shall be killed by one whose will is as mine.

Scepticism is thus graduated along the diverse degrees of human intelligence, alternating with dogmatism according to the more or less advanced development of the intellectual faculties. At the bottom of the ladder is the absolute dogmatism of the ignorant and the simple, who affirm and believe naturally, and have not seen any motives for doubt. —When the mind, which has for a long time been cradled in this simple faith, begins to discover that it may have been the sport of its belief, it becomes suspicious and imagines that the surest means of not being deceived is to reject everything; a primitive scepticism which is not devoid of *naïveté* (Sophists, Montaigne, etc.).—Then a more extensive learning, taking human nature in the mass, without concerning itself as to radical problems, endeavours to build upon common sense a reasonable but shallow dogmatism (Socrates, Th. Reid). More vigorous views soon show what little ground there is for this fresh attempt; the instrument itself is attached, and thence arises a great, a terrible, a sublime scepticism (Kant, Jouffroy, Pascal).—At last, the complete view of the human mind, the consideration of humanity aspiring after the true and enriching itself by the elimination of error, brings about the dogmatic criticism, which has no further dread of scepticism, for it has been through it, knows what its actual value is, and, very different from the dogmatism of the early ages, which had not even suspected the existence of any motives for doubt, is strong enough to live face to face with its enemy. Like all the men of this century, I have had my periods of scepticism; I have loved Lelia as much as Sténio did; but I have found a foothold in criticism, and even when such and such a creed does not appear to be as scientific as might be desired, I still say without hesitation: There is some truth in it, although I do not happen to possess the formula for extracting it. In the eyes of the schoolmen, Goethe is a sceptic, but he who is full of enthusiasm for all the flowers which he meets on his way and takes them to be good and genuine of their kind, is not to be confounded with one who passes disdainfully by without looking down at them. Goethe embraces the universe in the vast affirmation of love; the sceptic has nothing but narrow negation for everything.

While making the most liberal allowance to moral scepticism, while admitting that life and the universe are merely a series of phenomena of the same order, and of which all one can say is that they exist, while allowing that thought, feeling, passion, beauty and virtue are merely facts, exciting different sentiments in us, just like the various flowers of a garden or the trees of a forest (whence it would result, as Byron and Goethe thought, that everything is poetical), while admitting that, having reached the final atom, one is free, at will, to jest or to adore, so that the option would depend upon the individual character of each person, even from this point of view, from which morality ceases to have any meaning, science still would have. For what is certain is that these phenomena are curious, that this world of manifold movements interests and fascinates us. Morality is also lacking in the world of insects which is all alive in a sheet of water, and yet how delightful it is to watch some flitting upwards to the sun, others, like the salamanders, making for the bottom, and the little worms forcing their way into the mud to seek their prey. This is life, and ever life (182). This explains how science formed an essential part of the intellectual system of Goethe. To investigate, to discuss, to regard, to speculate in a word, will always have been the most attractive thing, whatever may be the case with the reality (183). However much one may be like Werther, there is so much pleasure in describing all this that life becomes as it were coloured with it. Goethe, I am sure, was never tempted to shoot himself. It is not impossible that humanity will some day come to an end, and that we shall have been working for the benefit of the sea or volcanoes, of ice or flame. But certain it is that the knowledge and the realization of the beautiful will have had their use, and that science, like virtue, imports into the world facts of unquestionable value. The mystical Christians have developed in all its various forms the favourite theory that Mary, the symbol of contemplation, has, even in this world, the better part, and that he who has embraced the perfect life finds here below a sufficient reward. This is true to the letter of science. One of the noblest minds of modern times, Fichte, assures us that he had reached perfect happiness, and that at times he tasted such delight as almost to tremble (184). And yet, poor man, he was at the same time almost dying of starvation. How often, in my modest room, surrounded by my books, I have tasted the plenitude of happiness, and have defied the whole world to procure for any one purer joys than those which I derived from the calm and disinterested exercise of my thoughts. How often, letting my pen drop, and surrendering my soul to the thousand sentiments which, as they come and go, bring about an instantaneous relief to all our being, have I not said to heaven: Give me only life, and I can see to the rest for myself.

Would to God that all living and pure souls were convinced that the question of the future of humanity is entirely one of doctrine and belief, and that philosophy alone, that is to say rational research, is capable of solving it. The really efficacious revolution, that which will give its shape to the future, will not be a political, it will be a religious and moral revolution. Politics have supplied all that they are capable of yielding, and they are now only a barren and exhausted field, a battle-ground of passions and intrigues to which humanity is quite indifferent and which interest only those who take a share in them. There are epochs when the whole question of the day is in politics; thus, for instance, at the dividing line of the middle ages and of modern times, at the epoch of Philippe-le-Bel, of Louis XI., the doctors and the thinkers were of little account, or had only a real value in so far as they served politics. It was the same at the beginning of this century. Politics then governed the world, the men of intelligence whose ambition went beyond amusing their contemporaries, had to become Statesmen, in order to exercise their legitimate share of influence upon their epoch. A thinker in the time of the Empire was obliged to keep his thoughts to himself. It was not a blameworthy ambition which carried off into the *melée* all the intellectual notabilities of the first half of this century; these eminent men did what it was their duty to do in order to serve the society of their time. But this age is reaching its close; the leading part, as it seems to me, is more and more devolving upon the thinkers. Beside the centuries in which politics have occupied the centre of the movement of humanity, there are others in which they have been left to the petty world of intrigue, and in which the main interest has been concentrated upon things of the mind. Take, for instance, the eighteenth century; and consider who had the upper hand of humanity during this grand epoch! What are the names which strike one at a first glance over the history of this period? Not Choiseul, or Richelieu, or Maupeou, or Fleury, but Voltaire, Rousseau, Montesquieu, and a whole school of thinkers who have their hold upon the century, mould it according to their will, and fashion the future. What are the War of the Austrian succession, the Seven Years' War, the Family Compact, compared with such events as the *Contrat Social* or the *Esprit des Lois?* The affairs of State were in the hands of an incapable king, of insignificant courtiers, and of great noblemen without views or grasp. The true historical personages of the period are writers, philosophers, men of intellect or of genius. But these latter did not take any active part in the direction of public affairs, their influence being indirect, and I imagine that, in the same way, those who will restore us the great originality which has been lost will not be politicians but thinkers. They will grow up and magnify outside the official world, not even taking the trouble to offer it any opposition, leaving it to expire within its worn-out circle (185).

Upon the poor pasturages of the Brittany islands, each ewe of the flock, tied to a stake in the centre, could only nibble the scanty grass within a certain narrow radius of the cord by which it is fastened. Such appears to me to be the present condition of politics; they have exhausted their resources for solving the problem of humanity. Morality, philosophy and true religion are not within their reach; they are powerless to get beyond a certain circle. Can it be honestly hoped that, if the salvation of the present century was to be due to cleverness, or ability, we shall find men more able than M. Guizot or M. Thiers? Who would not shrug his shoulders at the idea of inexperienced beginners having the presumption to think that they could, at the first attempt, do better than such men as these? No, they will be outrivalled not by doing like them, but by doing different from them. If such men have been rendered incapable, is this their fault, or was it not rather because no amount of cleverness is on a par with the situation?

Let us take again the first three centuries of the Christian era. Where do we find that the greatest events were happening, where was the future being founded, what were the names being marked out for the respect of future generations? Not, surely, those of Tiberius or Sejanus, of Galba, Otho or Vitellius. It was not they who occupied the centre of humanity, as was doubtless believed in their time; the centre of the world was the most despised corner of land in the East. The great men marked out for apotheosis were enthusiastic believers entirely strangers to the secrets of high politics. Five centuries later, the only men referred to in history as having been famous were Peter, Paul, John and Matthew, simple persons who, assuredly, made a very modest figure in the world. What would Tacitus have said if he had been told that all those personages whom he brings so skilfully on to the front of the stage would be completely effaced by the leaders of the Christians whom he treats with so much contempt; that the name of Augustus would only be saved from oblivion because at the head of the records of the Christian year would be read: *Imperante Cæsare Augusto, Christus natus est in Bethlehem Juda:* that Nero would only be remembered because, during his reign, Peter and Paul, the future masters of Rome, are said to have suffered martyrdom; that the name of Trajan would still be found in a few narratives, not for his victory over the Dacians and for having put back to the Tigris the limits of the Empire, but because a credulous Bishop of Rome in the sixth century took it into his head to pray for him? Here then we have a vast development quietly in preparation for three centuries, growing in magnitude parallel with the official society, persecuted by this latter, but which, all at once, puts an extinguisher upon the politics of the day, or, it may rather be said, remains full of life and strength, while the official world is dying of exhaustion. If St. Ambrose had remained governor of Liguria, supposing even that he had obtained promotion, and had become, like his father, prefect of the Gauls, he would by this time have been entirely forgotten. He did much better to become a bishop. How can it be said after this that there is no way of serving humanity except by joining in the *melée?* I say, upon the contrary, that he who embraces with his whole heart this humiliating labour proves by that very fact that he has not a call for the great work. What are politics in our day? Agitation without a principle or a law, a struggle of rival ambitions, a vast stage of cabals and personal competitions. What are the qualities requisite for success, for "becoming possible," as the saying goes? Is it great originality, an ardent and powerful train of thought, impetuous conviction? These are insurmountable obstacles in the way of success; to succeed, one must not think, or at all events not give expression to one's thoughts; one must make so much use of one's personality that one ceases to exist; one must always be careful to say not that which is, but that which it is expedient to say; in short one must shut oneself up within a lifeless circle of conventional phrases and official falsehoods. And you would argue that it is from this that can proceed what we so stand in need of: an original sap, a new method of feeling, a dogma capable of stirring anew the pulse of humanity? It would be as reasonable to hope that scepticism will engender faith, and that a new religion will be born out of the offices of a ministry or the lobbies of an assembly.

The most important question in politics is "Who is to be Minister?" But will humanity, let me ask, be any the better off if it is Mr. A instead of Mr. B who holds office? I assure you that Mr. A knows as little as Mr. B the true secret of things, that the problem will not be any nearer solution than it was before, that all this is as of little importance as when at Rome people speculated whether it would be Didius Julianus or Flavius Sulpicianus who would bid the higher, and that the 750 intelligent persons who are grouped around this arena, following eagerly all the different phases of the combat, waste their time and their trouble. Not there is to be sought the field for great achievements. What humanity needs is a moral law and a creed; and it is from the depths of human nature that they will emerge, and not from the well-trodden and sterile pathways of the official world.

Think for a moment how humiliating, in epochs like our own, is the *rôle* of the politician. Banished from the high regions of thought, disinherited of the ideal, he passes his life in fruitless and ungrateful labour, administrative cares, office complications, mines and countermines of intrigue. Is that the place for a philosopher? The politician is the off-scouring of humanity, not its inspired teacher. Who is there with any pride in his perfection who would let himself be inveigled into such an atmosphere of suffocation?

M. de Chateaubriand has, I think, maintained, in one of his writings, that the intrusion of men of letters into active politics denotes a decline of political intelligence in a nation. This is an error; it proves the weakening of philosophical intelligence, of speculation, of literature; it proves that the value and dignity of the intelligence are no longer understood, inasmuch as it no longer suffices to occupy the thoughts of distinguished men; it proves, in short, that supremacy has passed away from the mind and from doctrine to intrigue and petty activity. But this activity will in due course declare itself to be incompetent, and then it will be felt that the great revolution can only come not from men of action, but from men of thought and sentiment. This vulgar kind of labour will be left to the uneasy spirits, and all noble and elevated minds, leaving the earth to those who have a liking for it, regarding the form of government as a matter of indifference, will take refuge upon the lofty summits of human nature, and, burning with enthusiasm for the beautiful and the true, will create new force which, rapidly descending upon earth, will upset the frail erections of politics, and will become in its turn the law of humanity. It does not do to expect too much from governments. It is not for them to reveal to humanity the law of which it is in search. All that can be expected of them in epochs such as ours is to sustain as well as they can the conditions of outward life so as to render it endurable. It is more to be desired than anticipated, too, that they will not be too severe upon the efforts which are made in the new direction. Humanity will accomplish the remainder, without asking any one for permission. No one can say from what part of the sky will appear the star of this new redemption. The one thing certain is that the shepherds and the magi will be once more the first to perceive it, that the germ of it is already formed, and that if we were able to see the present with the eyes of the future, we should be able to distinguish in the complication of the hour the imperceptible fibre which will bear life for the future. It is amid putrefaction that the germ of future life is developed, and no one has the right to say: This is a rejected stone, for that may be the corner-stone of the future edifice. Could a sage of the early ages have ever imagined that the future belonged to that despised and unsociable sect, the ban of the human race, which. was associated in the imagination with darksome mysteries and odious orgies. The wits of the present day would have shown all the antipathy for this doctrine that they do for innovators of the modern age. These Christians would have seemed to them to be a vile, ignorant and superstitious set of people.

It is certain that several Christian sects justified the calumnies of the pagans. The distinction which has since been established between the orthodox Church and the gnostic sects was at that time very vague; they all formed one body, and there was a certain solidarity between them all. In the orthodox sect itself, there are a great many defects which we can see. The faculty has a name for those who believe that they possess the gift of tongues, of preaching and of prophecy. What are we to say of those who are daily expecting the end of the world: and the coming of a human body which will descend upon earth to reign? The extravagant ideas of our maniacs of the phalanstery are nothing by comparison with those of the early enthusiasts. Jean Journet has recently been sent to the asylum at Bicêtre, but Jean Journet does not believe he can perform miracles, or speak a language he has never learnt. The *Journal des Débats* would have made fine game of these people, and yet they succeeded, while four centuries later the sharpest wits were found to be their disciples, and even in the nineteenth century many gifted minds regard them as being inspired. The bad complexion of a movement is never a decisive argument. Even if I had before me a popular movement of the most odious kind, a regular Jacquerie, egotism saying to egotism "Your money or your life," I should still exclaim: "Long live humanity! here is the promise of great things in the future." Great apparitions are always accompanied by extravagances, they only reach a high degree of power when philosophical minds have given form and shape to them. Who can say that phalanstery will not have been the gnosis, the wild aberration of the new movement? It is at any rate beyond question that the region is clearly enough designated, that, in order to know whence will come the religion of the future, we must always look in the direction of Liberty, Equality and Fraternity.

It is, therefore, to the mind and to the thought that we must revert. Now thought will in future only find true scope in the form of rational science. It may seem, at first sight, as if science has had little influence hitherto upon the development of things. Reckon up the men of intelligence who have put their shoulders to the wheel, and you will find among them thinkers and writers like Luther, Voltaire, Rousseau, Chateaubriand and Lamartine, but very few savants or technical philosophers. The four words from Locke which Voltaire knew have had more influence upon the course of the human mind than the whole of Locke's book. The few fragments of German philosophy which have crossed the Rhine, put together in a clear and superficial way, have effected more than the doctrines themselves. That is the French method; you take three or four words of a system, just enough to indicate a tendency;

the rest is all guesswork, and the thing is done. Humanity, it must be confessed, has not hitherto marched with much method, and many things have (if I may be permitted the expression) been jumbled together, in the progress of the human intelligence. But the one thing certain is that if the human race were as much in earnest as it ought to be, enlightened and competent reason in each order of things would govern the world. But what is enlightened and specially competent reason if it is not science? Supposing even that the erudite man were never destined to have a place in the great history of humanity, his work and its results, assimilated by others and raised to their full height, will find their place in history by means of that secret influence and that inward infiltration which leads to no part of humanity being closed for the rest.

Contemporary Germany offers one of the rare instances of the direct effects of science upon the march of political events. The idea of German unity came through science and literature. That nation seemed resigned to death, it had lost all consciousness, and no longer counted as an individuality in the world, when an incomparable group of geniuses, Goethe, Schiller, Kant, and Beethoven came and revealed it to itself. These are the true founders of German unity; no sooner had the different parts of that great country found each other once more in the language, tongue, the glory and the genius of these great men, than they felt the tie which bound them, and were prompted to realize it in a political sense. This has given rise to a characteristic incident, viz. the learned, poetical and literary colour given to this movement from the time of Arndt, Kleist and Sand down to that gathering of doctors whose clumsiness and lack of adroitness may have made Europe smile and have compromised for a time but not ruined an idea which has been definitely set in motion.

CHAPTER XXIII.

I ONE day visited this palace transformed into a museum on the frontispiece of which was written in a spirit of broad electicism "*To all the glories of France.*" I had passed through the Gallery of Battles, the Hall of Marshals, those of various campaigns; I had seen the coronation of kings and emperors, royal ceremonies, the capture of towns, princes, great lords, faces foolish or insolent, when all of a sudden I asked myself: Where is the place of talent? Here are men born to greatness, coxcombs, men without ideas, without morality, who never did anything for humanity. But where is the gallery of saints, the gallery of philosophers, the gallery of poets, the gallery of savants, the gallery of thinkers? I see Louis XIV. founding I know not what order of nobility, and I do not see Vincent de Paul founding modern charity; I see court episodes of more or less insignificance, and I do not see Abelard in the midst of his disciples discussing the problems of the day on Mount Sainte Geneviève; I see the oath of the Tennis Court and I do not see Descartes shut up in his room and swearing not to relinquish his search until he had discovered the true philosophy. I see brutal and vulgar physiognomies with nothing ideal, and I cannot see Gerson, Calvin, Molière, Rousseau, Voltaire, Montesquieu, Condorcet, Lavoisier, Laplace, Chénier. Bossuet and Fénelon are there rather as courtiers than as men of talent. Can it be that Rousseau and Montesquieu did less for the glory of France than such or such obscure general or long-forgotten courtier? It is all over, said I to myself, talent has been disinherited. . . . But no. Over the uniform terraces of the palace-museum see where rises that majestic edifice crowned by the sign of Christ. Enter and tell me if any glory equals the glory of Him who is sitting there. Napoleon, whose name worked miracles, is not enthroned on an altar. Thank God! the chief place is reserved for talent. The others have the palace, He has the temple.

In the eyes of the philosopher the glory of talent is the only true glory and it is permitted to hope that the philosophers and the savants will inherit the glory, which during its period of brutality and antagonism, humanity was obliged to award to military exploits. I am unable to approve of the commonplace objection urged against conquerors; one must have a very superficial mind to see in Alexander a madman who laid Asia in ashes. War and conquest may have been in ages gone by instruments of progress; a manner, in default of any other, of bringing people into contact and of realizing the unity of humanity. Where would humanity be without the conquests of Alexander, without the Roman conquests? But when the world becomes rationalist, the greatest man will be the one who has done the most for ideas, who has searched the most, who has discovered the most. The battle will not be gastrosophical, as Fourier wished; it will be philosophical. From the beginning it is talent which has taken the lead in all things (Christianity, Crusade, Reform, Revolution) and yet talent has remained humble, misunderstood, persecuted. Napoleon did not trouble the world as deeply as Luther, and yet what was Luther all his life? A poor unfrocked monk who only escaped his enemies because it pleased some little princes to take him under their protection. If anything proves the intimate force of speculation which exists in the human mind it is that in spite of the hard lot endured up to the present by thinkers, there have been men capable of devoting their lives notwithstanding insult, persecution and poverty, to the disinterested pursuit of truth. When one reflects that the whole intellectual movement accomplished up to our day has been realized by men unfortunate, suffering, harassed by internal and external afflictions and that we ourselves preserve the tradition, with agitated heart, in the midst of fear and anguish, one conceives a greater esteem for that human nature, capable of pursuing an ideal object with so much energy.

It is time to return definitively to the simplicity of life and to renounce all that artifice of convention, remnant of our aristocratical distinctions and the artificial society of the seventeenth century; it is time to return to the simplicity of antique customs. Take Plato, Socrates, Alcibiades, Aspasia; fancy them alive, acting according to the ravishing description handed down to us by antiquity, especially Plato. Do they possess that cold and insignificant pride which constitutes the tone of our aristocratic *salons?* Have they that silly air, that vulgar laugh, that dull and prosaic appearance, that manner of treating life as an affair of business, like the middle class? Have they that coarseness, that heavy look, that degraded expression which, I say it with sadness and without any idea of a reproach, marks our people? No. They are simple, they are men.

Those honest minds of refined ages, Rousseau, for example, Tacitus perhaps, through reaction against what was artificial and false in their time, often looked back with complacency on barbarous times which they called the age of nature. Innocent illusion which converted no one and which only inspires refined persons with an easy sort of resignation. One reads with pleasure those eloquent declamations: one accepts them as given themes, but no matter what Voltaire says, no one after reading Rousseau experiences the desire of walking on all fours. It is puerile to compare a barbarous state to refined civilization; it should be compared to the true civilization of which Greece offers us an incomparable example. What we require, in the way of civilization, is Greece without the slave system. Where can one find more freedom given to the individual, more personal originality, more spontaneousness, more dignity? We understand only royal or aristocratical majesty. The majesty of the ideal is blended by us with that of religion which we place beneath humanity, and as for the majesty of the people we do not understand it, because it does not exist. Athens, on the contrary, is pure humanity. M. de Maistre declared *majesty* to be entirely Roman. Certainly not. The Olympian Jupiter and the Grecian Pallas, Salamis and the Pireus the Pnyx and the Acropolis have their majesty; but that majesty is real and popular; whereas the Roman majesty is made up and worked by machinery. There were not two fashions at Athens; on the contrary, the fine manners of the time of Augustus were much the same as those of our aristocracy, and alongside of all this was to be found a ridiculous population.

Majesty is only to be found in true humanity, poetry, religion, morality. All other prestige at a certain moment becomes ridiculous. It is in the natural order of things that what has been imposed by force excites laughter as soon as its prestige is destroyed. One likes to revenge oneself for one's past respect as soon as the scaffolding has been stripped of its hangings. One requires, for the vulgar illusions of external respect, a simplicity which we no longer possess; we are too cunning not to lift up the veil. We have demolished the old idol of respect: an idol is not to be restored. How, I pray, can one give oneself respect? How can one recall to life by means of reflection that which only existed owing to the absence of reflection? The child may be afraid of the face he has besmeared; but, once he has laughed at it, will he not always remember that it was to frighten himself that he besmeared it?

The essential condition for a show of performing dolls is not to see the wire. The simple minded look on the thing as serious, as if the dolls were real persons; the clever people are amused even when they get a glimpse of the wire for they know that there is one. But if the demi-clever people have the misfortune to perceive it they laugh at the performance to show that they are not dupes. It is the same with respect: respect is natural with the simple, superficial people resist it with comical self-conceitedness; it flourishes among the wise in consequence of their perspicuity. The sages know of the existence of the wire but do not think it worth while to make a fuss over so simple a discovery. The superficial minded, on the contrary shout and storm that humanity must at any price be delivered from these prejudices. "One must have a hidden thought," says Pascal, "and judge of everything by that, at the same time, however, speaking like the people." But when the number of artful people is too large cheating is impossible, for it then becomes the fashion to pretend to be knowing and to say to the simple, "Ah! how foolish you are to allow yourself to be caught." Therefore one must go to work with simplicity and demand respect only for things which are respectable.

The advent of the middle classes has operated, it must be admitted a great simplification in our manners. Our costume is very narrow and very artificial when compared with the simple and noble fulness of the antique costume however it is no longer false like that of the old aristocracy. There is still much to be done; it is necessary to simplify and ennoble. The middle class has at times committed the error of wishing to return to the old airs of the nobility; it in no way succeeded and merely rendered itself ridiculous. For nothing is so ridiculous as a false imitation of majesty. What we require is real politeness, real gentleness, simplicity of life, virtue as shown by amenity and grace of manners. The Republicans who pretend to be austere delude themselves strangely in believing that the idea of majesty can be banished from humanity. Better the ancient idolatry, surrounding some individuals with splendour, than that colourless life in which the majesty of humanity is not represented. But it were better still to return to the truth and to recognize no other majesty than that of the nation and the ideal.

These manners I would willingly call *democratic manners*, in this sense that they do not repose upon any artificial distinction (186) but simply upon the natural and moral relations of men among themselves. People often imagine that these democratic manners are the manners of the wine shop and this is a little the fault of those who have confiscated this expression for their profit. But true democratic manners would be the most charming, the most gentle, the most amiable. They would be morality itself, more or less attractive, more or less harmonious, according as the individuals were more or less happily endowed. They would be the manners of poems and of ideal romances, where human sentiments would appear in their primitive simplicity, without a *bourgeois* or refined air. The real democratic manner would necessitate upon one hand the abolition of the aristocratic *salon* and the *café* and on the other hand the extension of family relations and of public gatherings. It is true that with regard to the latter our society offers a hiatus difficult to fill up. We have nothing analogous to the antique *school*. Our school is exclusively destined for children and therefore condemned to be demi-ridiculous like everything which is pedagogical; our club is entirely political and yet man requires intellectual assemblies. The ancient school was for persons of all ages the gymnasium of the mind. The sage, like Socrates, Stilpo, Antisthenes not writing but speaking to their disciples or frequenters (οι συνοντες) is now impossible. The philosophical conversation such as Plato has given us in his dialogues (187), the antique *Sympasie* is not conceivable in our days (188). The Church and the press have killed the school. Now that the Church is no longer anything for the people, what will replace her?

What is called society is far from being favourable to good manners and noble characters. I would not dare to say it, if M. Michelet had not said it before me: "After the conversation of men of genius and savants that of the people is certainly the most instructive. If one cannot talk with Béranger, Lamennais or Lamartine, one must go into the fields and converse with the peasants. What has one to learn from the middle class? As for the *salons* I never left them without finding that my heart had been diminished and had grown colder." The impression which I carry away with me on quitting a *salon* is despair of civilization. If civilization were destined to terminate in this abortion, if the people in their turn were to exhaust themselves in this way, and, at the expiration of a few centuries to grow insipid in the bosom of vanity and pleasure, Cato would be right, it would be necessary to regard as instruments of feebleness and wisely to break everything which in our eyes is an instrument of culture and progress, but which, in this hypothesis, would only serve to create generations greedy of servitude in order to live at their ease. Nothing can equal, especially in the provinces, the life of the middle class, and I never see without sadness and a sort of terror the physical and moral deterioration of the rising generation; and yet these are the grandsons of the heroes of a great epopee! I get on better with the simple, with a peasant, with a working man, with an old soldier. We speak to some extent the same language, I can at a pinch converse with them: this is radically impossible with a vulgar citizen: we are not of the same clay.

Herman lived only for himself, his family and a few friends. With them he is simple, natural, and full of life; he reaches to heaven. In society he is insupportably stupid and condemned to mutism in the round of conversation which does not allow him to insert a single word. If he tries the strange sound of his voice causes every head to be raised; it is discordant. It does not know how to give change; does he wish to indulge in repartee, he takes from his pocket gold instead of coppers. In the Academy or the Portico, he would have held his own, he would have had favourite disciples; he would have figured in a dialogue of Plato like Lysis or Charmides. If he had seen Dorothea lovely courageous and proud standing by the fountain he would have dared to say—Allow me to drink. If, like Dante, he had seen Beatrice with downcast eyes coming out of the church at Florence, perhaps a ray of light would have traversed his life and perhaps the daughter of Falco Portinari would have smiled at his trouble. Well! in presence of a young lady he only experiences and causes awkwardness.—Your Hermann, you will say, is a countryman let him go to his village—not in the least. At his village he would find vulgarity, ignorance and the impossibility of comprehending delicate and beautiful things. Now, Hermann is polished and cultivated, more refined even than the gentlemen who frequent the *salons*, but not of an artificial and fictitious refinement. There is in him a world of thought and sentiment which would be unable to understand coarse stupidity or frivolous scepticism. He is a true and sincere man taking a serious view of his nature and adoring the inspirations of God in those of his heart.

Intellectual work therefore only possesses all its value when it is purely human, that is to say when it corresponds to this fact in human nature: man does not live by bread alone. The grand scientific and religious feeling will only revive when people return to a conception of life as true and as little mixed with what is fictitious as if it were formed alone in the midst of the forests of America, or by some Brahmin, when, finding that he has lived long enough, he takes off his drawers, ascends the Ganges and goes to die on the summits of the Himalayas. Who has not experienced these moments of inward solitude when the mind descending from stratum to stratum pierces one after the other all the superposed surfaces until it arrives at the real bottom, where all convention expires and where one faces one's self without fiction or artifice? These moments are rare and fugitive: we habitually live in presence of a third person who hinders the fearful contact between me and himself. Life is only sincere on condition of piercing this intermediary veil and of constantly reposing on the true depth of our nature, in order to listen to the disinterested instincts which lead us to learn to adore and to love.

This is why the sincere man so greatly admires and tires himself out in adoration before *simple life*, before the infant who believes in and smiles at everything, before the young girl who does not know that she is beautiful, before the bird which sings on the branch merely for the sake of singing, before the hen which struts out proudly in the midst of her chickens. It is because God is seen there in simplicity. The refined man considers as foolish the things in which the people and men of genius take the most interest, animals and children. Genius is to possess at the same time the critical faculty and the gifts of the simple. Genius is infant; genius is people; genius is *simple*.

The Brahminic life offers the most powerful model of life exclusively devoted to religion, or rather the serious conception of existence. I do not know if the picture of the life of the first Christian recluses of the Thebaid, so admirably traced by Fleury, offers such a halo of idealism. Besides the Brahminical life has this superiority over the cenobitical and hermitical life that it is at the same time human life, that is to say family life and that it is allied to positive life, without lending it a value which it does not possess: the Christian ascetic received his food from a celestial raven; the Brahmin goes into the forest and cuts his own wood; he must have his hatchet and his basket to collect his wild fruits. During the sojourn of the sons of Pandou in the forest their wife Draupadi offers to the strangers whom she has received in her hermitage the game which her husbands have killed.* The Lives of the Fathers of the desert offer nothing which can be compared to the following sketch extracted from the Mahabharata: "The king advanced toward the sacred grove, image of the celestial regions: the river was filled with bands of pilgrims while the air resounded with the voices of pious men who each one repeated fragments of the sacred books. The King followed by his minister and his high priest, advanced towards the hermitage, animated with the desire of seeing the holy man, inexhaustible treasure of religious science; he looked at the solitary asylum, similar to the region of Brahma; he heard mysterious sentences, taken from the Vedas, pronounced in rhythmic harmony. . . . This spot sparkled with glory owing to the presence of a certain number of Brahmins. . . . some of whom sang the Samaveda while another band sang the Bharoundasama. . . . All were men of cultivated mind and imposing appearance. . . . These places resembled the dwelling of Brahma. The king heard on all sides the voices of these men instructed by long experience in the rites of sacrifice, of those who possessed the principles of morality and the science of the faculties of the soul, of those who were skilled in conciliating texts which do not harmonize, or who knew all the private duties of religion; mortals whose minds tended to free their souls from the necessity of regeneration in this world. He heard also the voices of those who, by indubitable proof, had acquired a knowledge of the supreme being; of those who knew grammar, poetry and logic and who were versed in

* The sons of Pandou are supposed to have had but one wife—Draupadi.—TRANSL.

chronology; who had penetrated the essence of matter, of movement and of quality; who knew causes and effects, who had studied the language of birds and that of bees (the good and the evil omens) who believed in the works of Vyasa, who offered models for the study of books of sacred origin and the principal personages who search out the trials and troubles of the world (189)." India in fact represents to me the truest and most objective form of human life, that where man, struck with the beauty of things, pursues them without any personal feeling and simply owing to the fascination which they exercise over his nature.

Religion is the word under which has been resumed, up to the present, the life of the mind. Take the Christian of the first ages; religion is his whole spiritual life. Not a thought, not a feeling which is not attached to it: material life is almost entirely absorbed in this great movement of idealism. *Sive manducatis, sive bibitis*, says St. Paul. What a superb system of life, all ideal, all divine and really worthy of the children of God. There is no exclusion there, the chain is not felt; for, although the limit is narrow our wants do not go beyond it. The law, severe as it is, is entirely the expression of man. In the Middle Ages this great equation still existed. The fairs, the meeting for business or pleasure are religious fêtes; the scenic representations are mysteries; voyages are pilgrimages; wars are crusades. Take, on the contrary, a Christian, even the most severe, in the time of Louis XIV., Montausier, Arnauld, Beauvilliers, you will find two divisions in his life: the religious portion which, although the principal, is not sufficiently strong to assimilate itself with the rest; the profane portion to which one must accord some value. Then, but not before, the ascetics began to preach renunciation. The first Christian had no need to renounce anything for his life was complete; his law was adequate to his wants. Afterwards, religion, not being able to provide for everything, cursed what escaped it. I am sure that Beauvilliers took a very delicate pleasure in the tragedies of Racine, and perhaps even in the comedies of Molière; and yet it is certain that in going to see them he did not consider that he was performing a religious act, perhaps he even thought that he was sinning. This separation was a matter of necessity. At that epoch religion was received as a letter closed and sealed, which was not to be opened, but which one was bound to receive and transmit, and yet, human life always opening up, it was necessary that new wants should overcome all scruples, and that, not being able to find a place in religion, they should take up a position opposite to it. Hence a system of life both colourless and indifferent. Religion is respected but people guard against its *invasions;* they give it its share, to it which is only something on condition of being everything. Hence these petty theories concerning the separation of the two powers, of the respective rights of reason and faith.

The result must be that religion, being isolated, cut off from the heart of humanity, no longer receiving anything from the general circulation, like a limb which is bound, must wither and become an appendage of secondary importance, while on the contrary profane life, in which all the actual and living feelings, all discoveries, all new ideas, are concentrated, must become the master portion. Without doubt the great men of the eighteenth century were more religious than they thought; what they banished under the name of religion was clerical despotism, superstition, narrow forms. The reaction however carried them too far; the religious colour was almost entirely wanting in that century. The philosophers placed themselves without knowing it in the position of their adversaries, and, under the empire of the association of obstinate ideas, appeared to suppose that the secularization of existence would bring about the elimination of all religious habits. I think, like the Catholics, that our society founded upon a supposititious pact, and our atheistical law, are temporary anomalies, and that until we can speak of "*our holy constitution*" stability will not be conquered. Now, the return to religion can be nothing else than the return to the great unity of life, to the religion of the intellect without exclusion and without limit. The sage has no need of praying at certain hours for his whole life is a prayer. If religion is to have a distinct place in life it must absorb life altogether. The most rigorous asceticism is alone consistent. Only superficial-minded or weak-hearted people, once Christianity admitted, can take any interest in life in science in poetry, in the things of this world. The mystics look with pity upon this weakness, and they are right. The true philosophical religion would not reduce this great tree, which has its roots in the soul of man, to a few branches, it would only be a manner of spending one's whole life in seeing beneath everything the ideal and divine sense, and in sanctifying one's whole life by the purity of the soul and the elevation of the heart.

Religion, as I understand it, is far removed from what the philosophers call *natural religion*, kind of petty theology without poetry, without action upon humanity. All the attempts made in this direction have been and will remain fruitless. Theodicy has no meaning regarded as an individual science. Is there any man of sense who can hope to make discoveries in such an order of speculations? True theodicy, is the science of things physics, physiology, history looked upon in a religious light. Religion is to know and to love the truth of things. A proposition is only of value in so far as it is understood and felt. What signifies this sealed formula, this unknown tongue, this $a \times b$ theology, which you present to humanity saying "This will preserve your soul for life eternal: eat and you shall be healed,"—a pill which you must not bite on pain of feeling a cruel bitterness? Well! what matters it to me if I do not taste it? Give me a leaden bullet to swallow, that will have the same effect. What to me are stereotyped phrases devoid of sense, like the formulas of the alchemist and the magicians which operate of themselves *ex opere operato*, as the theologians say? Black and scholastic doctors occupied only with your Incarnation and your real Presence, the time has come when you shall worship the father, not on this mountain nor in Jerusalem, but in spirit and in truth (190).

M. Proudhon is certainly a distinguished philosopher of great intelligence. But I cannot pardon him his airs of atheism and irreligion. It is to commit suicide to write such phrases as this—"Man is destined to live without religion. A number of symptoms show that society, by an internal work constantly tends to shake off this envelope henceforward useless." That if you practice the worship of what is noble and true, if the sanctity of morality speaks to your heart, if all beauty, all truth, all goodness leads you to the threshold of a holy life, to intelligence; that if, arrived there, you refuse to speak, you wrap yourself up, you purposely mix up your thought and your language in order to say nothing limited in presence of the infinite, how do you dare to speak of atheism? That if your faculties, resounding simultaneously, have never uttered that grand and unique rite, which we call God, I have nothing more to say, you are devoid of the essential and characteristic element of our nature.

Humanity is only converted when it falls in love with the divine charm of beauty. Now beauty in the moral order is religion. This is why a religion dead and outstripped is still more efficacious than all the institutions which are purely profane; this is why Christianity is still more creative, comforts more suffering, acts more vigorously upon humanity than all the principles acquired in modern times. The men of the future will not be mean disputatious reasoners, insulters, men of party, intriguers, without ideal. They will be noble, they will be amiable, they will be poetic. I, inflexible critic, I shall not be suspected of flattery for a man who searches the

Trinity in all things and who believes, God pardon me, in the efficacy of the name of Jehovah; well! I prefer Pierre Leroux, mistaken as he is, to these pretended philosophers who would recast humanity in the narrow mould of their scholastic ideas and triumph with politics over the divine instincts of the human heart.

The word God having taken possession of the respect of humanity, this word having a long prescription in its favour and having been employed in beautiful poems, it would perplex humanity to suppress it. Although it is not very univocal, as the scholastics say, it corresponds to an idea sufficiently clear: the *summum* and the *ultimum*, the limit where the mind stops in the scale of the infinite. Suppose even that we philosophers should prefer another word, *reason* for example, in addition that these words are too abstract and do not sufficiently express real existence, there would be a great inconvenience to deprive us thus of all the poetic sources of the past and to separate us by our language from the simple who adore so well after their fashion. Tell the simple to live by aspiration after truth and beauty; these words would have no meaning for them. Tell them to love God, not to offend God; they will understand you perfectly. God, providence, soul, good old words, rather heavy, but expressive and respectable, which science will explain but will never replace with advantage. What is God for humanity if not the transcendent epitome of its suprasensitive wants, *the category of the ideal*, that is to say the form under which we conceive the ideal, as time and space are categories, that is to say forms under which we conceive bodies (191)? Everything reduces itself to this fact of human nature; man in presence of the divine is no longer himself, he clings to a celestial charm, he lays aside his paltry personality is carried away and absorbed. What is that if it is not to adore?

If one views the matter as regards substance and asks oneself: This God does He or does He not exist?

—Oh, God would I reply, it is He who is and all the rest which appears to be. If the word *to be* has any meaning, it is assuredly applied to the ideal. What, you would admit that matter exists, because your hands and your eyes say so, and you would doubt of the divine being which all your nature proclaims from the first? What is the meaning of this phrase: "Matter is"? what would remain of it in the hands of a strict analysis? I do not know and to tell the truth I consider the question senseless; for one must confine oneself to simple notions. Beyond is the gulf. Reason only attains a certain mean region; above and below it loses itself as a sound which by dint of becoming sharp or flat ceases to be a sound or at least to be perceived. I like, for my own part to compare the object of reason to those foaming or frothy substances where the substance is hardly anything and which only exist thanks to their effervescence. If one pursues too closely the substantial foundation nothing remains but the bare unity; as mathematical formulas too closely pressed render all identity fundamental and only mean something on the condition of not being too simplified. Every intellectual act, like every equation, reduces itself at bottom to $A = A$. Now, with this limit, there is no more knowledge there is no more intellectual work. Science commences only with details. In order that there should be any effort of the mind a superfices is necessary, something variable, diverse, otherwise one loses oneself in the infinite One. The One only exists and is perceptible when developing itself in diversity, that is to say in phenomena. Beyond, it is repose, it is death. Knowledge is the infinite poured into a finite mould. The knot alone has any value. The faces of the unity are alone an object of science.

There is not a word in the philosophical language which may not give rise to great errors if one takes it in its substantial and vulgar sense, instead of using it to design classes of phenomena. Realism and abstraction touch each other; Christianity may have been turn about with good reason accused of realism and of abstraction. Phenomenalism alone is genuine. I hope that no one will ever accuse me of being materialist and yet I regard the hypothesis of two substances joined together to form man as one of the most clumsy inventions made by philosophy. The words body and soul remain perfectly distinct in so far as they represent the orders of irreducible phenomena, but to make this diversity, entirely phenomenal, synonymous to an ontological distinction, is to fall into a ponderous realism and to imitate the ancient hypotheses of physical sciences, which suppose as many causes as different effects, and explain by those real and substantial fluids facts in which a more advanced science sees nothing but various orders of phenomena. Of a truth it is much more absurd to say in a spirit of exclusion: man is a body: the truth is that there is a unique substance, which is neither body nor soul, but which reveals itself by two orders of phenomena, which are the body and the soul, that these two words have no meaning except in their opposition, and that this opposition exists only in acts. The spiritualist is not him who believes in two substances coarsely united; it is he who is persuaded that the acts of the mind alone have a transcendental value. Man is; he is matter, that is to say expanded, tangible, endowed with physical properties; he is mind, that is to say thinking, feeling, adoring. The mind is the goal as the goal of the plant is the flower; without roots, without leaves, there are no flowers.

The most simple act of intelligence comprehends the perception of God; for it comprehends the perception of being and the perception of the infinite. The infinite exists in all our faculties and constitutes, it is true, the distinctive feature of humanity, the unique category of pure reason which distinguishes man from the animal. This element may become effaced in the vulgar acts of intelligence; but as it is to be found indubitably in the acts of the mind, this is a reason to conclude that it is to be found in all those acts; for that which exists in one degree exists in all the others; and besides the infinite shows itself much more energetically in the acts of primitive humanity, in that life vague and without conscience, in that spontaneous state, in that native enthusiasm, in those temples and pyramids than in our age of polished reflection and analytical view. This is the God of whom we have an innate idea and who does not require demonstration. Against this God atheism is impossible; because people affirm Him while denying Him. Everywhere man has outpaced nature; everywhere, beyond the visible, he has supposed the invisible. This is the only feature which is truly universal, the identical foundation upon which divers instincts have embroidered infinite varieties, from the multiple forces of savages to Jehovah, from Jehovah to the Indian Oum. To look for a universal consent on the part of humanity to anything else but this psychological fact is to misuse terms. Humanity has always believed in something beyond the finite, this something, it is suitable to call God. Therefore all humanity has believed in God. Very well. But do not, misusing a definition of words, pretend that humanity has believed in such and such a God, in a moral and personal God formed by anthropomorphic analogy. That God is so little innate that the half of humanity has not believed in Him and that it has required ages to formulate this system in a complete manner, in ordering man to love God. It is not that I entirely blame the method of anthropomorphic psychology. God being the ideal of every one, it is right that every one should fashion Him after his manner and on his own model. One must not therefore fear to employ all the goodness and beauty that can be imagined. But it is contrary to all criticism to pretend to erect such a method into a scientific method and to raise, out of an ideal construction, a discussion on the qualities of a being. Let us say that the supreme being is eminently possessed of all that is perfection; let us say that he has in him something analogous, to intelligence, to liberty; but do not let us say that he is intelligent that he is free: this would be trying to limit the infinite, to give a name to the ineffable (192).

One is accustomed to consider monotheism as a definitive and absolute conquest, beyond which there can be no ulterior progress. In my eyes monotheism is, like polytheism, only an age in the religion of humanity. This word, besides, is far from designating a doctrine absolutely identical. Our monotheism is only a system like another, inferring it is true very advanced notions. It is the Jewish system, it is Jehovah. Neither the ancient polytheism, which also contained a great portion of truth; nor India, so learned in its conception of God, understood things in this manner. The deva of India is a superior being to man, by no means our God. Although the Jewish system has entered into our intellectual habits it should not make us forget all that was profound and poetic in other systems. No doubt, if the ancients had understood by God what we ourselves understand, polytheism would have been a contradiction of terms. But their terminology in this matter reposed upon notions quite different from ours respecting the government of the world. They had not yet arrived at the conception of unity of government in the universe. The Greek worship representing at bottom the worship of human nature and the beauty of things, and that without any orthodox pretension, without any dogmatic organization, is only a poetical form of universal religion perhaps not far removed from that to which philosophy will return (193). This is so true that where the moderns have wished to make some trials of natural worship they have been obliged to approach it. The great moral superiority of Christianity makes us too easily forget the breadth, the toleration, the respect for all that was natural which existed in Grecian mythologism. The origin of the severe manner in which we have judged it lies in the ridiculous way in which mythology has been presented to us. One imagines a religious body obliged to enter our conceptions by force. A religion which has a God for thieves, another for drunkards, appears to us the height of absurdity. Now, as humanity has never lost its common sense, we must admit that until we can conceive these fables naturally we do not possess the key to the enigma. Polytheism only appears absurd to us because we do not understand it. Humanity is never absurd. The religions which do not pretend to repose upon a revelation, so inferior as machines of action to religions organized dogmatically, are, in one sense, more philosophical, or rather they only differ from truly philosophical religion by a more or less symbolical expression. These religions are, at bottom, only the State, the family, art and morality elevated to a high and poetic expression. They do not divide life in two; they have no sacred and profane. They know nothing of mystery, renouncement and sacrifice since they accept and sanctify nature at first sight. These were bonds, but bonds of flowers. There lies the secret of their feebleness in the work of humanity; they are not strong but also less dangerous. They do not possess that prodigious psychological subtilty, that spirit of limit, of intolerance, of *particularism*, if I dare say so, that force of abstraction, veritable vampire which has gone on absorbing all that is gentle and mild in humanity ever since it was given to the wan image of the crucified One to fascinate the human conscience. It sucked everything even to the last drop out of poor humanity: juice and force, blood and life, nature and art, family, people, country; everything went down, and on the ruins of an exhausted world there remained but the phantom of the Me, tottering and distrustful.

Up to the present, men as far as religion is concerned have been divided into two categories: religious men believing in a positive dogma, and irreligious men holding themselves aloof from all revealed belief. This is insupportable. Henceforth they must be classed thus: religious persons taking a serious view of life and the sanctity of things; frivolous men, without faith, without seriousness, without morality. All those who adore something are brothers, or certainly not such great enemies as those who adore simply pleasure and interest. It is indubitable that I resemble more a Catholic or a Buddhist than a sceptical laugher and my intimate sympathies are a proof of this. I love one, I detest the other. I can even call myself a Christian, in this sense that I admit being indebted to Christianity for most of the elements of my faith, just as M. Cousin might have called himself Platonician or Cartesian without accepting all the inheritance of Plato or Descartes, and above all without feeling himself obliged to regard them as prophets. Do not say that I am twisting words when I thus arrogate to myself a name the acceptation of which I have greatly altered. No doubt if one understands by religion a number of imposed dogmas and external practices, then I admit that I am not religious; but I also maintain that humanity is not essentially so and will not always be so in that sense. What is a part of humanity, and will consequently be as eternal as itself, is the religious want, the religious faculty, to which up to the present the great ensembles of doctrine and ceremonies have corresponded but which will be sufficiently satisfied with the pure worship of good and beautiful things. We have therefore the right to speak of religion, since we have the analogy, if not the thing itself; since the want which was formerly satisfied by positive religions is now satisfied by something equivalent which has the right of being called by the same name. If people absolutely persist in taking this moral in a more restricted sense, we will not dispute over a free definition, we will merely remark that religion thus understood is not essential and that it will disappear from humanity, leaving vacant a place which will be filled up by something analogous.

A great deal has been said within the last few years of a religious revival, and I willingly admit that this revival has generally shown itself in the form of a return to Catholicism. This is as it should be. Humanity feeling the imperious want of a religion will always cling to that which it finds already made. It is not to Catholicism, as Catholicism, that this century has returned but to Catholicism as a religion. It must also be admitted that Catholicism with its harsh and absolute forms, its rigorous rules and its perfect centralization, must please a nation which saw in it the most perfect model of its own government. France which finds it quite natural that a law emanating from Paris should become at once applicable to the Breton peasant, to the Alsatian workman, to the nomad shepherd of the Landes, must also find it quite natural that there should be at Rome an *infallible being* who regulates the belief of the world. This is very convenient. Delivered from the care of making one's creed and even of understanding it one can, after that, attend in full security to one's affairs saying—that does not concern me; tell me what I must believe; I believe it. Strange contradiction, for, formulas having no value except the *sense* they contain it is of no advantage to say—"I trust to the Pope; he knows what to believe and I believe as he does." People believe that faith is like a talisman which saves by its own virtue; that they will be saved if they believe some unintelligible proposition, without taking the trouble to understand it; they do not feel that these things are only of value according to the good which they do the soul, by their personal application to the believer.

If a return towards Catholicism has taken place it is therefore in no way because progress in the way of criticism has brought it back, it is because the want of a religion has been more sharply felt and because Catholicism alone was ready at hand. Catholicism for the immense majority of those who profess it, is no longer Catholicism; it is religion. It is repugnant to pass one's life like the brute, to be born, to contract marriage, to die without any religious ceremony consecrating these holy acts. Catholicism is there to satisfy this want; then let us have Catholicism. People do not examine matters more closely; they do not enter into details of dogmas, they pity those who undertake so sterile a task; they are heretics a hundred times over without being aware of it. What has made the fortune of Catholicism in our days, is that it is little known. It is only seen through certain imposing externals, one only takes into consideration what is elevated and moral in its dogmas; one does not enter into the brush-wood. What is more, one bravely rejects or complacently explains those dogmas which are too openly opposed to modern ideas. If one were obliged to accept as an article of faith every text of Scripture and every decree of the Council of Trent, it would be a different matter; one would be surprised to find oneself incredulous. Those who have been led by peculiar circumstances to wage a death-struggle on this ground have reasons for not being accommodating.

This then is the explanation of the return to Catholicism which appears to be so strongly opposed to philosophy. The eighteenth century, having had for its mission to destroy, found in it that pleasure which every being experiences in accomplishing its object. Scepticism and impiety were pleasing in themselves. But we who are not intoxicated with this first burst of joy, we who, having returned to the soul, have found in it the external want of religion, which is at the bottom of human nature, we have looked round us, and, rather than remain in this penury which has become intolerable, we have returned to the past, and we have accepted, as it stands, the doctrine handed down to us. When one no longer knows how to create cathedrals one imitates them. For one can do without religious originality; but one cannot do without religion.

Individuals pass through analogous phases in their inward life. In the age of force, when the critical spirit is in all its vigour, when life appears like an appetizing prey, and when the sun of youth sheds its golden rays on everything, the religious instincts are easily satisfied; one enjoys life without any positive doctrine; the charm of intellectual labour tones down everything, even doubt. But when the horizon comes closer; when the old man endeavours to chase the cold terrors which assail him; when the sick man has exhausted the generous force which allowed him to think boldly, then there is no rationalist however firm who does not turn towards the God of women and children and ask the priest to comfort him and to deliver him from the phantoms which beset him under this pallid sun. Thus may be explained the *weaknesses* of so many philosophers in their last days. The death-bed requires a religion. Which? no matter; but one is necessary. It seems to me at this moment that I should die contented in the communion of humanity and the religion of the future. Alas! I would not swear to this were I to fall ill. Each time that I feel myself enfeebled I experience a nervous excitability and a kind of return to piety.

Mole sua stat: such in our days is the reason why Christianity exists. Who has not stopped, while passing through our ancient towns, become modern, at the foot of those gigantic monuments of the faith of ages past? Everything has been renewed around them; there is no longer a vestige of the dwellings and customs of former times. The cathedral has remained, somewhat damaged perhaps as far as the hand of man can reach, but deeply rooted in the soil. It has resisted the deluge which has swept everything away around it, and the family of ravens which have built their nests in the steeple have not been disturbed. Its magnitude is its right. Strange prescription! Those converted barbarians, those builders of churches, Clovis, Rollo, William the Conqueror tower over us still. We are Christians because it pleased them to be so. We have reformed their political institutions, become superannuated; we have not dared to touch their religious establishment. It is considered wrong that we who are civilized should meddle with the dogma created by barbarians. And what right have they which we do not possess? Peter, Paul, Augustin lay down our law, much as if we were still subject to the Salic law and the Gombette law. So true is it that as far as religious creation is concerned centuries are given to calumniate themselves, and to refuse to themselves the privileges which they freely accord to distant ages!

Hence the immense disproportion which can, at certain epochs, exist between religion and the moral social and political state. Religions are petrified and customs are continually modified. Like those granite rocks consolidated in swallowing up in their still liquid mass foreign substances, which will form a portion of their body for ever, Catholicism has solidified itself for ever and henceforward no purification is possible. I know that there is a milder Catholicism which has known how to compound with the necessities of the times and to throw a veil over truths too unpalatable. But of all the systems that is the most inconsistent. I can conceive orthodox and incredulous people, but not the neo-catholics. The profound ignorance which exists in France, outside of the clergy, of biblical exegesis and theology, has alone given rise to that superficial school so full of contradictions. It is in the Fathers and in the Councils that true Christianity must be sought for and not among those weak and light-minded spirits who have perverted it in toning it down, without rendering it more acceptable.

For the great majority of men the established

religion is only the ideal portion of human life and looking at it in this light it is supremely respectable. How charming to see in the cottages or in vulgar houses, where everything appears to be buried under the weight of useful preoccupations, pictures representing nothing real, saints and angels! What consolation amid the tears of our state of suffering, to see unfortunate people, bowed down under the weight of six days' labour, come on the seventh day to repose themselves on their knees, to look at lofty columns, vaults, an altar, to hear and enjoy the singing, to listen to a moral and consolatory sermon. Oh! barbarians are those who call this lost time and speculate on the gain of suppressed Sundays and fête-days! We who have art, science and philosophy, we have no need of the church. But the people, the temple is their literature, their science, their art. The people do not see what is dangerous and fatal in Christianity. The mind which aspires to a high and reflective culture must first of all shake off Catholicism; for there are in Catholicism dogmas and tendencies inconsistent with modern culture. But what is this to the simple minded? They only pluck the flower; what does it matter to them that the roots are bitter? I feel indignant on seeing a man however little initiated in the culture of the nineteenth century still preserve the belief and the practices of the past. On the contrary when I travel through the country and I see at the angle of each road and in every cottage signs of Catholic superstition my heart is touched and I would rather hold my tongue all my life than scandalize one of those children. A Holy Virgin in the dwelling of a reflective man and in that of a peasant, what a difference! In that of a reflective man it appears to me to be a revolting absurdity, the symbol of an exhausted art, the amulet of a degrading devotion; in that of the peasant it appears to me as a ray of the ideal which has penetrated beneath the cottage roof. I love this simple faith as I love the faith of the Middle-Ages as I love the Indian prostrating himself before Kali or Kristna, or placing his head under the wheels of the car of Juggernaut. I adore the ancient sacrifice; I have no distaste for the foolish taurobolium of Julian. The peasant without religion is the ugliest of brutes, without any distinctive sign of humanity (*animal religiosum*). Alas! a day will come when they will undergo the common law and pass through the hateful period of impiety. It will be for the good of humanity; but, God, for nothing in the world would I labour at such a work. Let the improper undertake it! These good people not belonging to the nineteenth century must not be blamed for belonging to the religion of the past. This is my manner of acting: in the village I go to mass; in the town I laugh at those who go there.

I am sometimes tempted to shed tears when I think that, by the superiority of my religion I isolate myself in appearance from the great religious family in which are all those I love, when I think that the purest minds in the world must consider me impious, wicked, damned; must do so, be it remarked, owing to the very necessity of their faith. Fatal orthodoxy, thou which formerly caused the peace of the world, thou art only good now to work separation. The man of a ripe age can no longer believe what the child believes; the man can no longer believe what the woman believes; and what is terrible is that the woman and the child join their hands to say: In the name of heaven, believe as we do or you will be damned. Ah! not to believe them one must be very savant or very hardhearted!

A souvenir is recalled to my mind, it fills me with sadness, without making me blush. One day at the foot of the altar and under the hand of the bishop, I said to the God of the Christians: *Dominus pars hæreditatis meæ et calicis mei; tu es qui restitues hæreditatem meam mihi.* I was very young then and yet I had reflected a great deal. At every step that I took towards the altar doubt followed me; it was science and, child that I was I called it the demon. Assailed by contradictory ideas; tottering at twenty years on the bases of my life, a luminous idea entered my mind and for the moment re-established calm and comfort: Whoever thou art, I exclaimed in my heart, O God of noble minds I receive thee as the portion of my lot. Up to the present I have called thee by the name of a man; I believed on his word him who said: I am the truth and the life. I will be faithful to him in following the truth wherever it may lead me. I will be the true Nazarene, while, renouncing the pomps and vanities of the world, I shall love only what is good and shall exert my activity for nothing else. Well! I do not repent of these words to-day and I willingly repeat them: *Dominus pars hæreditatis meæ*, and I am pleased to think that I pronounced them during a religious ceremony. The hair has grown again on my head; but I still belong to that holy militia, the disinherited of the earth. I shall not look upon myself as an apostate until material interests usurp in my mind the place of what is holy, the day when, in thinking of the Christ of the Gospel I no longer feel myself his friend, the day when I prostitute my life to inferior matters and when I become the companion of the jovial of the earth.

Funes ceciderunt mihi in præclaris! My lot will always be with the disinherited: I shall belong to the league of the poor of spirit. Let all those who still adore something unite together in the object they adore. The day for little men and little things has passed; the time for saints has arrived. The atheist is the man who is frivolous; the impious and the pagans are the profane; the egotists those who understand nothing concerning the things of God; branded souls who affect to be clever and who laugh at those who believe; base and terrestrial souls destined to grow yellow from egotism and to perish from nullity. How, O disciples of Christ, can you enter into an alliance with those men? Oh! would it not be better for us to sit down side by side with poor humanity, seated gloomy and silent on the side of the dusty road, to raise its eyes to the mild heaven which it no longer regards? For us the die is cast: and even should superstition and frivolity, henceforth inseparable auxiliaries, succeed in deadening human conscience for a time, it will be said in the nineteenth century, the century of fear, that there were still men, who, in spite of common contempt, liked to be called men of the other world; men who believed in the truth, who were ardent in its search, in the midst of an age, frivolous because it was without faith and superstitious because it was frivolous.

I was formed by the Church, I owe all to her and I shall never forget her. The Church separated me from profane men, and I thank her. He whom God has touched will always be a being apart; he is, no matter what he does, out of place among men, he is known by a sign. For him young men have no joys to offer, and young girls have no smile. Since he has seen God his tongue is embarrassed, he no longer knows how to speak of terrestrial things. O God of my youth I have long hoped to return to Thee with colours flying and in the pride of reason and perhaps I shall return humble and vanquished like a feeble woman. Formerly Thou listened to me; I hoped some day to see Thy face for I heard Thee answer my voice. And I have seen Thy temple crumble away stone by stone; the sanctuary has no longer an echo, and, instead of an altar ornamented with lights and flowers, I have seen rise before me an altar of brass against which prayer, severe, unadorned, without images, without tabernacle, blood-stained by fatality shatters itself. Is it my fault? is it Thine? Ah! how willingly I would beat my breast, if I could hope to hear that beloved voice which formerly made me tremble. But no, there is only inflexible nature; when I search Thy fatherly eye I find only the orbit of the infinite empty and baseless, when I search Thy celestial brow I dash myself against a vault of brass which coldly sends back my love. Farewell then, O God of my youth! Perhaps wilt Thou be the God of my death-bed. Farewell; although Thou hast deceived me, I love Thee still!

NOTES.

1. This tendency to place the ideal in the past is peculiar to ages that repose on unassailed and traditional dogma. Ages of upheaval like our own, on the other hand, in which the continuity of doctrinal teaching has been broken, must of necessity appeal to the future, seeing that to them the past is merely a mistake. All ancient peoples placed the ideal of their nation at its origin ; the ancestors were more than men (heroes, demi-gods). On the other hand, during the Augustan period, when the disintegration of the ancient world begins to manifest itself, observe the aspirations towards the future, so eloquently expressed by the incomparable poet, in whose soul the two worlds were locked in close embrace. Oppressed nations do the same ; "Arthur is not dead ; Arthur will come again ;" they exclaim. The most puissant cry towards the future, ever uttered by any nation is the belief of the Jewish nation in the Messiah. That belief had its birth and grew during the iron grip of alien persecution. The embryo is formed at Babylon, it gathers strength and assumes a distinct character under the persecution of the Syrian monarchs, it finds its climax under Roman oppression.

2. I have seen men of the people transported with genuine ecstasy at the sight of the graceful movements of swans on a piece of water. It is impossible to say at what depth the feelings of these two simple lives were interpenetrated. But it is evident that the people, face to face with the animal, regards it as his brother, as leading a life analogous to his own. Lofty intellects, whose sympathies reunite them with the masses, experience the same feeling.

3. How modest and amiable, for instance, is that declaration of savants—often eminent—at the beginning of their works that they have no intention of encroaching on the domain of religion, that they are not theologians, and that the theologians can have no objection to their attempts at unambitious natural philosophy. There are in France, men who vastly admire the religious "establishment" of England, because it is the most conservative of all. In my opinion this system is the most illogical and the most irreverent with regard to things divine.

4. This seems to me to be the true definition of the accidental in history rather than *Et quia sæpe latent causæ, fortuna vocatur.*

Gustavus Adolphus is struck down by a cannon ball at Lutzen, and his death changes the face of things in Europe. Here we have a fact, the cause of which is by no means unknown, but which nevertheless may be termed chance or the irrational part of history, because the direction of a cannon ball a few centimètres one way or the other is not a fact proportionate to the immense consequences resulting from it.

5. Life is nothing else but this ; the aspiration of the being to be all it can be ; the tendency to pass from potentiality to act. Dante, who in his book "De Monarchia" expresses ideas on humanity almost as advanced as those of the boldest humanitarians, had an enlightened perception of this fact. *Proprium opus humani generis totaliter accepti est actuare semper totam potentiam intellectus possibilis.* (De Monarchia, I.) Herder says in the same way, "The perfection of a thing consists in its being all that it should and can be. Hence, the perfection of the individual is that he should be himself in the whole of the successive phases of his existence" (*Ueber den Charakter der Menschheit*).

6. The year 1789 will be in the history of humanity a holy year as having been the first to trace the outline of this previously unknown fact—with marvellous ingenuity and incomparable energy. The place in which humanity proclaimed itself, the Tennis Court, will be one day a temple. It will be visited by the pilgrim, like Jerusalem, when distance shall have sanctified and characterised particular facts in symbols of general facts. Golgotha only became hallowed ground two or three centuries after Jesus.

7. See as eminently characteristic the Declaration of Rights in the Constitution of '91. It is the whole of the Eighteenth Century with its claims to the control of nature and of things established, its analysis, its craving for clearness and logical evidence.

8. What, for instance, shall we say of our university education, reduced to pure outward discipline without the smallest regard for the soul and moral culture ? As for other matters, is it surprising that Napoleon should have conceived a college as a barracks or a regiment ? Our system of education is still, without our being aware of it, feature for feature, that of the Jesuits, based upon the idea that man can be "licked" into moral shape by bringing outward influences to bear upon him, totally forgetful of the soul that imparts life, treating him, in short as a piece of intellectual mechanism.

9. Languages afford a curious instance in point. Languages, manipulated, twisted, remodelled by the hands of man, like French show the indelible stamp of their treatment in their want of flexibility, in their laboured construction, in their lack of harmony. The French language, made by logicians is a thousand times less logical than Hebrew or Sanskrit created by the instincts of primitive man. I have developed this point in an "Essay on the Origin of Language," published in the philosophical review, "*La Liberté de Penser*" of the 15th September and the 15th December 1848.

10. See for instance, "les Considerations sur la France" of M. de

Maistre. The ingenious writer has plainly perceived the defects of the reformers, the artificiality, the formalism, the rage for writing and publishing that which is much stronger when left unwritten. But he has failed to perceive that these defects were a necessary condition of ulterior progress.

11. Voltaire never professed to say anything else in his numerous attacks on optimism ; they are just satires on the absurdities of his century.

12. "De la Démocratic en France" ; p. 76. A little further on, the principle is laid down that landed property is superior to any other, because the proceeds of it depend less upon the exertions of men, and more on blind causes.

13. The greater or lesser extent of a people's belief in fate is the test of her civilization. The Cossack blames no one for being whipped, it is his fate ; the Turkish rayah bears no one a grudge, on account of the burdens imposed upon him ; it is his fate. The poverty-stricken Englishman nurses no grievance ; if he starves to death, it is his fate. The Frenchman revolts if he suspects that his misery is the consequence of a social organisation capable of reformation.

14. By reason I do not solely mean human reason, but the reflection of every thinking being, extant or to come. If I could believe in the endless perpetuation of humanity, I would unhesitatingly infer that it must attain perfection. But it is physically possible that humanity may be fated to perish or become exhausted, and that the human species itself may gradually perish of atrophy when the fountains of living force and new races shall have dried up. (Lucretius has some weighty arguments on this point, V., 381, et seq.). In that case it will have only been a transient form of the divine progress of all things and of the evolution of the divine conscience. For even if humanity should not exercise a direct influence on the forms that will succeed it, it will have played its part in the graduated progress as a branch necessary to the growth of higher branches. For though these may not be offshoots of the first branch, they will spread outwards from the same trunk. Hegel has no foundation for attributing an exclusive rôle to humanity, which is doubtless not the only conscious form of the divine, though it may be the most advanced within our knowledge. To find the eternal and perfect we must go beyond humanity and plunge into the deep sea. Were I here to disclaim any tendency to pantheism, it would look as if I did so in deference to a suspicious timidity, and that I admitted the right of somebody to demand a profession of orthodoxy ; I will, therefore not do so. Sufficient be it for me to state my belief in a living reason for all things, and that I admit human freedom and personality as evident facts ; consequently every doctrine *logically* advanced in order to deny them would, in my opinion, be false. I should add that if pantheism appears so absurd to most people, it is because they do not understand it, and because they interpret the principle ; "*All is God*," in a *distributive* and not in a *collective* sense. In this instance *all* is not synonymous with

every any more than in the sentence; "*All* the departments of France constitute an area of so many square leagues." There would be few absurdities comparable to this: "Every object is God." Hegel has very well explained this. (*Cours d'esthétique*, t. ii., p. 108; Bénard's translation.)

15. What else in fact is the science of the Middle-Ages but disputation? Wrangling is so dear to the schoolmen that they preserve it like game, and provide one another with the opportunity for sport; they dispose their canons so as never to lack material for it. There are propositions, acknowledged to be false, but which are, nevertheless, not condemned in order to afford an opportunity for disputation. Read the treatise the theologians call; "*Les lieux Théologiques*," and you will get an idea of that strange method. Never mind the truth, the thing is to hit upon something lending itself to controversy; to know is nothing, to wrangle everything.

16. If you wish for a typical instance of this irreverent manner of treating science, of taking it as a *jeu d'esprit*, fit at most to beguile the tedium of an aimless life or to raise the inane laughter so dear to those who are debarred from laughing genuinely, read the *Journal de Trevoux* and in general the scientific works issued by the same brotherhood, which, be it said by the way, has not produced a single serious *savant* (except perhaps Kircher, who also drifts into sheer folly at times, though his folly is at any rate that of his time), but which on the other hand has produced some matchless types of scientific charlatanism, Bougeant, Hardouin, etc. All this belongs to the same order of things as the thoroughly innocent and twaddling minor poetry of the members of the society; Du Cerceau, Commire, Rapin, etc. And though the works of the Benedictines are of an altogether different order, they do not disprove my thesis. The need of beguiling the leisure of a tranquil and retired existence with useful work, a taste for study, the instinctive love for compiling and collecting may render immense service to scholarship, but they do not constitute a love of science.

17. Let us suppose that the considerations of Descartes for theology were not solely inspired by political motives, which I do not admit; the intellect of Descartes was of the absolute order, altogether devoid of the critical faculty; and it is quite possible that he may have fully believed in Christianity.

18. This is so true that semi-critical intellects only resign themselves to admitting miracle in antiquity. Tales that would raise a smile if they were related as contemporary, pass muster in virtue of the enchantment lent by distance. It seems to be tacitly admitted that primitive humanity lived under natural laws different from our own.

19. The way in which every nation naïvely reflects herself in the physiognomy of her miracles is truly marvellous. Compare the miracle of the Hebrews, grave, severe, without variety like Jehovah Himself; the miracle of the Gospel, beneficent and moral; the Talmudic miracle, disgustingly vulgar; the Byzantine miracle, dull and devoid of poesy; the miracle of the Middle-Ages, graceful and

sentimental; the Jesuit and Spanish miracle, materialistic, enervating, immoral. This is not surprising seeing that each people only puts upon the stage in its miracles the supernatural agents of the government of the Universe, as it understands them; and these agents are fashioned by each race after its own model.

20. The study of Greek science and philosophy had already produced an analogous result among the Mussulmans in the Middle-Ages. Averroes may be considered a rationalist pure and simple. But this splendid onward movement was checked by the rigid Mussulmans. The numbers and the influence of the philosophers were not sufficiently large to carry the day, as was the case in Europe.

21. See the admirable description of the pietist reaction in the beginning of the seventeenth century in Michelet's "Du Prêtre, de la Femme, de la Famille," Ch. I. and throughout the book; a thoroughly vivid and original description of the most delicate and indescribable facts. It contains a whole world of which people scarcely care to speak with bated breath. See also the delicate psychological analysis which M. Sainte-Beuve so unfortunately entitled "Volupté." Nor should we forget *Das ewig Weibliche* at the end of Faust, and Mephistopheles, though blaspheming, vanquished with roses, and the admirable episode of Dorothée and Agnès in the "Pucelle."

"The damsel o'ercome by a contrite emotion
Determined a father confessor to seek;
For there's only one step between love and devotion,
And dear are the weaknesses both—to the weak."

A rigorous psychological analysis would class the innate religious instinct of women in the same category with the sexual instinct. The first manifestation of all this occurs in a characteristic manner in the Middle-Ages in the case of the lollards, béguards, fraticelli, "poor men of Lyons," *humiliati*, flagellants, etc.

22. This opposition sometimes produces strange effects. Certain weaknesses of the fiercest rationalists can only be explained in this way. There are in life "melting moments" when everything thaws —becomes moist, limp, deliquescent. I have often thought that this type (of fierce intellectual pride combined with the most feminine weakness) might be taken as the subject for a psychological novel. Faust only corresponds to a part of my idea. The ancients distinguished between dry heat and moist heat by one of those distinctions banished by our physical system, because they are not based on sufficiently accurate facts, but which contained nevertheless a great deal of truth. This distinction is perfectly just, at any rate in psychology.

23. I have heard a very excellent person rejoice over the cholera; "for," said he; "those calamities are sure to bring in their wake a return to religious ideas." After all, this is perfectly consistent. What matter, so long as souls are saved?

24. In reality, the divergences between religious sects are not less

great. But they do not strike one so much, because they do not exist simultaneously in the same country, while philosophy is looked at synoptically, as inter-connected in all its parts. As a matter of course in countries where several sects confront one another, scepticism is never far behind.

25. Finding it impossible to define these ideas with accuracy, I refer to the hymn, in which in my earliest youth, I tried to express my religious ideas, at the end of the volume. (*It has been suppressed.*)

26. Such for instance are Descartes' proofs of the existence of God. No mind laying claim to any subtelty has taken them seriously, and I should deeply pity the man whose religious faith has no better basis than this scholastic scaffolding. Still these proofs are really true, all equally true, however narrow the spirit in which they are expressed.

27. It is in this that Germany excels. The views of her writers are thoroughly individual and absolutely untranslatable. Change the form in which they are expressed however slightly, they vanish, like the essence that evaporates in being transferred from one vase to another. Certain German works of the highest order are intolerably heavy in French; take away the fragrance of rosewater and it becomes worse than ordinary water. Take, for instance, the admirable introduction of Wilhelm von Humboldt to his essay on Kawi, in which the most subtle views of German writers on the science of language are brought together; well, this essay, translated into French, would have lost all meaning and would emerge simply as a monumental platitude. This constitutes its very claim to praise, because it proves the delicacy of the style.

28. Fichte, for instance, in his "Method to attain a happy life," is never tired of repeating; "Is not this as clear as daylight? Can any well-ordered mind fail to understand this?" When a sincere man speaks in this tone, I always believe him. For how can a straightforward mind, applying itself seriously to its object, fail to see right? It is, therefore, certain that Fichte's system was perfectly true to him, from his own standpoint.

29. Thus the hypotheses on electricity and magnetism, *afford an explanation* of the phenomena; they supply a convenient connection between the facts; but they are not to be taken as possessing an absolute value, and as correspondent to physical realities.

30. "I see the sea, rocks, islands;" says he who looks through the windows on the northern side of the castle. "I see fields and trees and meadows;" says he who looks through the windows on the south. They would make a mistake to dispute; both are right.

31. The typical representative of this kind of intellect is assuredly Joseph de Maistre, a *grand seigneur* who has no patience with the slow discussions of philosophy. "In God's name, give us a decision, and let there be an end of it. True or false it matters not, so long as I am at peace. An infallible pope, that is the shortest way and the best! What do I say, an infallible pope? That would be honouring those vile mortals too much! No, no; a pope from whom there is no appeal."

32. The most naïvely touching thing I know is the effort made by the faithful, forcibly carried away by the scientific current of the modern spirit, to reconcile their old doctrines with that formidable power which dominates them, do what they may. If one could lay bare this or that conscience, one would find in it hoards of pious subtelties, truly edifying and indicating an exceedingly amiable code of morality.

33. One of those who have most vigorously insulted human nature in the interests of revelation has said somewhere (See *L'Univers* of 26th March, 1849) that he greatly preferred Rabelais, Parny and Pigault-Lebrun to Lamartine. I can easily believe it. Voltaire also got along better with the Curé of Versailles who petted and fleeced his flock in turns than with St. Vincent de Paul or St. Francis de Sales.

34. A curious inquiry might be made into *the higher or lower price of human life* at different phases of the development of humanity. It would be found that this price has always been estimated according to the real value, that is, that human life was much more respected at the periods when it really was worth more. Human conscience is very gradually developed, and traverses sundry different stages. The value of a conscience therefore is in direct proportion to the advancement of its development. Civilized man who is so energetically conscious of himself is much more man, if I may be permitted to say so, than the savage who is scarcely conscious of his own existence, and whose life is only a small comparatively valueless phenomenon. This is why the savage sets so little store by life, relinquishes it with such strange unconcern and deprives others of it in mere sport. With him the feeling of individuality has scarcely commenced. The animal, and to a certain extent the child, looks upon the death of one of his fellows without alarm. The price one sets upon one's own life is generally the price one sets upon that of others. Several facts of our Revolution can only be explained on this theory. Human life had become dreadfully cheap.

35. Christianity by its universal and catholic tendencies has been instrumental in diminishing the antique love of country. The Christian forms a part of a much more extensive and holier society, which if needs be he must prefer to his country.

36. Heaven forbid that I should insult so distinguished an intellect as Franklin. But it is difficult to conceive how a man endowed with the least moral and philosophical feeling could have written chapters, entitled ; "Advice how to make a fortune."—"Necessary advice to those who wish to be rich."—"The means always to have money in your pocket." "Thanks to these means," he adds, "the sky will be more bright to you, and a feeling of pleasure will cause your heart to throb. Make haste to adopt these rules and to be happy." Truly a charming way of ennobling human nature.

37. Of all the usages of antiquity libation seems to me the most poetical and the most religious ; it is the sacrifice (sheer waste the positivists would say) of the first fruits to the invisible powers.

38. The same irrational, but withal energetic and beautiful appli-

cation of human nature may be noticed in the ideas of the religious on expiation. The need for expiation after an immoral or frivolous life is indeed very legitimate ; the error consists in having entertained the belief that it was a question of punishing one's self. The only rational penance is repentance and a more impassioned return to a beautiful and earnest existence.

39. Small minds which conceive perfection as a state of mediocrity resulting from the reciprocal neutralisation of extremes call this excess, but this is a narrow and paltry way of explaining such facts. The blame lies, not in the abundance of energy, but in the wrong direction given to powerful instincts.

40. These harmonious complaints have become one of the most fruitful themes of modern poesy. With the exception of that of Jouffroy, I know of none more sincere than those of Louis Feuerbach, one of the most advanced representatives of the ultra-Hegelian school ("Recollections of my religious life"—a continuation of "The Religion of the Future"). This regret is not noticeable among the first sceptics (such for instance, as the philosophers of the eighteenth century) who destroy with a marvellous joy without feeling the need of any belief, engrossed as they are with their work of destruction and the vivid consciousness of exerting their strength.

41. Heraclitus conceived the stars as meteors lighting themselves at certain times at receptacles prepared for the purpose, as a kind of cauldrons which by turning their dark side to us produced phases, eclipses, etc. Anaxagoras thinks that the vault of the sky is of stone and conceives the sun and stars as so many stones on fire. Cosmos Indicopleustes pictures to himself the world as an oblong chest, of which the earth constitutes the bottom, at the four sides rise strong walls and the sky forms the arched lid. The Hebrews conceived the sky as a molten looking glass (Job xxxvii. 18) supported by pillars (Job xxvi. 11) ; above which are the upper waters, dropping through it by means of channels or grated windows and thus making the rain (Psalms lxxviii. 23 ; Gen. vii. 11 ; viii. 2). Strepsiades concocts for himself a system of similar meteorology, only a little more burlesque (Aristoph., Clouds, line 372).

42. Shall I say that we are justified in already suspecting something of the kind ? The final term of progress being, in fact, a condition in which there will be only one being, a state in which all existing matter will beget a unique resultant, which will be God ; in which God the universe will be the *soul* of the universe, and the universe the *body* of God, and in which, the period of individuality having been traversed, the unity which is not the exclusion of the individuality, but the harmony and combination of individualities, shall reign alone ; we may conceive, I say, that in such a condition, which will be the result of the blind efforts of all that has lived, in which exact individuality down to the tiniest insect will have had its share, every individuality will be found again, as in the distant sound of an immense concert. This, at any rate, is the way in which I like to understand it. See some admirable pages of *Spiridion*, though they are presented in too substantial a form.

43. An admirable expression of Schiller.

44. I am specially alluding here to France. The successes of M. Ronge and of the *German Catholics* prove that a religious movement is not altogether impossible in Germany. The constant apparition of new sects, with which the Catholics twit the Protestants as a sign of weakness, proves on the contrary that the sentiment of religion is still alive among them, seeing that it still possesses the power of creating. There is no danger of such a thing happening in France, everything has been battened down for ever and aye. There is nothing more dead than that which no longer stirs. Several facts also attest that the religious power of production is not extinct in England. As for the East, the Arabs show that the list of prophets is not closed, and the successes of the Wahhabites proved that the advent of another Mahomet is not among the impossibilities. I have often thought that a clever European acquainted with Arabic and presenting a legend professing to have some connection with a branch of the Prophet's family, and in addition to this preaching the doctrines of fraternity and equality, so likely to be properly understood by the Arabs, might with eight or ten thousand men conquer the Mussulman East, and create a movement analogous to that of Islamism.

45. Fichte in the work in which he shows at its best his admirable moral sense, has forcibly expressed this priestly mission of science. ("Of the Destiny of the Savant and the Man of Letters," 4th Lesson. See also his "Method to attain a happy life," 4th Lesson.)

46. This is so true that entire peoples have been without such a religious system, for instance the Chinese who have never known anything but a code of natural morality, without the slightest mythical belief. The worship of Fo or of Buddha is, as is well known, utterly foreign to China.

47. How can one help regretting at the same time the deplorable nullity to which the provinces seem condemned, for want of a great literary movement and institutions? When we come to consider that every small town in the Italy of the sixteenth century had its grand master painter and master musician, and that every town of 3000 inhabitants in Germany is a literary centre with a printing press, devoted to works of science, a library and often a university; when we consider all this we feel grieved at the want of initiative of a great country, reduced to the servile imitation of her capital. The distinction between Parisian good taste and provincial bad taste is the consequence of the same intellectual organization; but it so happens that this distinction is as mischievous to the capital as to the provinces; it invests the question of *taste* with an exaggerated importance. All this is a proof of the somewhat melancholy proposition that art, science and literature do not flourish among us in consequence of an innate and spontaneous need, as in ancient Greece, as in fifteenth century Italy; for with us, in the absence of stimulation from without there is no production.

48. The Germans who have studied our system of public education maintain that there are only certain courses at the lyceums, such

as for instance those of philosophy, that remind them of German University teaching. See L. Hahn, "Das Unterrichtswesen in Frankreich," Breslau, 1848. 2te Theil.

49. The following is the programme of a University Commemoration at Koenigsberg. "*Conditi Prussiarum regni memoriam anniversariam die* XVIII *Jan.* MDCCCXL *in auditorio maximo celebrandam indicunt, prorector, director, cancellarius et Senatus Academiæ Albertinæ. Inest dissertatio de nominum tertiæ declinationis vicissitudine. . . .* G. B. Winer made up the programmes of a dozen academical solemnities with a series of dissertations *on the use of verbs compounded with a preposition in the New Testament.*

50. See the Transactions of the Annual Congresses of the German philologists. Verhandlungen der Versammlungen Deutscher Philologen und Schulmænner.

51. Malebranche in his admirable but too severe chapter on Montaigne, had already called him *un pedant à la cavalière* (a free and easy pedant). Pascal, the Logicians of Port-Poyal and Malebranche thoroughly appreciated this innocent pretension on the part of the author of "the Essays."

52. This is so true, that the same sentiment can furnish poesy, eloquence, philosophy, according to the manner in which it is made to vibrate, almost in the same way as the divers vibrations of the same fluid produce heat and light.

53. "Stobæus, Apophth.," 8. ii. p. 44, Edit Gaisford.

54. Quintilian was perfectly right when he said; *Grammatica plus habet in recessu quam fronte promittit.*

55. See the history of classical philology in antiquity ("Geschichte der Klassissichen Philologie im Altherthum"), by M. Graefenhan, Bonn, 1843–46. The following are the various subjects he includes in it: 1° GRAMMAR and its various branches; Rhetoric, Lexilogy (Etymology, Synonymy, Lexicography, Glossography, Onomatology, Dialectography).—2° EXEGESIS, allegorical, verbal, Commentaries of the Rhetoricians, of the Grammarians, of the Sophists, Scholia, Paraphrases, Translations, Imitations.—3° CRITICISM of Texts, literary criticism (authenticity, etc.) criticism, aesthetics.—4° ERUDITION, Theology, Mythology, Politics, Chronology, Geography, Literature (Compilers, Abridgers, Bibliography, Biography, Literary History), History and Theory of the Fine Arts. —M. Haase in the *Jena Journal* smartly criticises the use of so vast a syllabus. (*Neue Jenaische Literatur-Zeitung*, Febr. 1845, N° 35–37). The school of Heyne and of Wolf understood by philology the thorough knowledge of the antique world (Greek and Roman) in all its aspects, so far as it is necessary to the perfect understanding of these two literatures.

56. This is how antiquity understood it. Grammar was the encyclopedia not for positive science itself, but as a necessary means to the understanding of authors. Everything was reduced to this literary aim. The most complete enumeration of all that an antique grammarian had to know will be found in the elogium of Statius on his father ("Sylv.")

57. An epigram of Crates of Mallos ; "The grammarian is the mason, the critic is the architect." Wegener, "De aula Attalica ;" Collection of the fragments of Crates.

58. I am speaking only of the Scholastic Middle-Ages—from the eleventh to the fifteenth centuries. The rhetoricians of the Carlovingian period are truly the successors of the Roman grammarians and are, if anything, too *philological* in the narrow and literal sense. Roger Bacon in whom we find the first spark of the modern spirit and who almost alone, during the space of ten centuries, understood science as we understand it, already foresaw the benefits of philology. He devoted the third part of the "Opus Majus" to demonstrating the usefulness of the study of ancient languages (Greek, Arabic, Hebrew) and propounds perfectly just views on this delicate subject. The study of languages is to him no longer a means of exercising the *trade* of interpreter or translator, as it nearly always was in the Middle-Ages ; it is an instrument of scientific and literary criticism.

59. We can say as much with regard to the knowledge of Greek literature possessed by the Syrians, the Arabs and other Orientals, (except perhaps the Armenians). It was crude in the extreme, because it was not philological.

60. These are his words : "I have placed the prince of poets by the side of Plato, the prince of philosophers ; and I am obliged to content myself with looking at them, seeing that Sergius is absent and that death has deprived me of Barlaam, my old master. Sometimes I console myself by casting a glance at that masterpiece, at others I embrace it and exclaim with a sigh. "With what pleasure, oh, great man, would I listen to thee, if death had not closed one of my ears (Barlaam) and absence had not rendered the other useless (Sergius) !" (Epist. Var., xx. Opp. pp. 998, 999).

61. To get a clear understanding of the character of ancient criticism, see the excellent article of M. Egger on Aristarchus (Revue des deux Mondes, 1 Feb. 1846).

62. *Aristarchus Homeri versum negat quem non probat.* One could have wished that Porson, Brunck, and a good many other German critics had not chosen this strange means of becoming Aristarchi.

63. It is thus that European students of Arabic literature are quite justified in believing that they understand the Koran better than the Arabs. It is thus again that the modern Hebrew scholar corrects several explanations of ancient texts given in Hebrew books of more modern composition, such as for instance in the "Chronicles" (or Paraleipomena) and point out in the ancient books themselves etymologies more than doubtful. None of our philologists pretend to know Greek better than Plato, or Latin better than Varro, yet not one among them scruples to correct the etymologies of Plato and Varro.

64. The real "manuals" of antiquity are the compilations of the fifth and sixth centuries, those of Marcianus Capella, of Isidore of Seville, of Boethius, etc. The deluge of elementary books is also with us but a recent fact and decidedly not a sign of progress. In

a system of education pretending to any vitality the child has to perform for himself the labour spared to him by these artificial means—a labour of immense advantage to his originality. The seventeenth century acquired a better knowledge of Latin in the authors themselves, or even in Despautères than we did in Lhomond or than others are likely to do in even better grammars. In this as in many other things people have been beguiled by the sophism; "Our fathers did wonders with comparatively imperfect methods. What will not our children do when everything is regulated and perfected." In gymnastic exercises the perfection of the dumb-bell or Indian club is of no importance.

65. Polybius devotes a book of his history to the most elementary notions of geography and pauses to explain the four points of the compass, etc., as curiosities of great interest. Strabo ("Géogr.," Book VIII. init.) tells us that Ephorus and several others did the same. Let us suppose for a moment M. Thiers beginning his "History of the Revolution" with a short course of cosmography. The undergraduate of to-day smiles at the animated controversy of Cicero against Tiro on the knotty question whether all the cities of the Peloponnesus are seaports and whether there are any ports in Arcadia. ("Letter to Atticus," Book iv. 2.)

66. The ancients never definitely departed from the narrow point of view according to which aesthetics are supposed to supply the rules of literary composition, as if every work ought to be judged according to its conformity with a given type, and not by the amount of positive beauty it presents. One single rule may be given for the production of the beautiful; "Elevate your soul, feel nobly and say what you feel." The beauty of a work lies in the philosophy it contains.

67. The reformers of the sixteenth century are philologists. In the eighteenth century the work is accomplished under the banner of the positive sciences. D'Alembert and the "Encyclopédie" are characteristic of this new spirit.

68. What, then, would be the result if to scientific experiment one could add *practical experiment on life?* Saint-Simon as an introduction to philosophy led the most active life possible, trying all kind of conditions, all kind of enjoyments, all the ways of seeing and feeling, nay, creating for himself fictitious relations that do not exist or rarely present themselves in reality. There is no doubt that the habit of life *does teach* as much as books, and constitutes a *culture* to those who have no other. The only uncultivated man (*inhumanus*) is he who has not been able to partake of either practical or scientific culture.

69. To avoid a misunderstanding which would strangely distort my real views, I must repeat that in the whole of the foregoing I have taken the word "philology" in the sense of the ancients, as synonymous with polymathy; ὡς φιλόλογός ἐστι καὶ πολυλόγος (Plato, Legg. i., 641, E).—"*Quæ quidem erant φιλόλογα et dignitatis meæ,*" says Cicero speaking of certain demands he had addressed to Cleopatra. ("Ad Atticum," lib. xv., ep. xv.)

70. Thus (tom. v., pp. 47, 48) M. Comte prophesies *a priori* that the comparative study of language will lead to the recognition of their unity as a historical fact; "for," he says; "each kind of animal has only one cry." As a matter of fact, the result has been precisely the reverse.

71. The visions of the pseudo Daniel are in my opinion the most ancient essay on the philosophy of history, and on that account remain very interesting.

72. The trouble taken by M. Jouffroy to invest the word "philosophy" with a special meaning arises from his not having paid sufficient attention to the conventional sense attributed to the word in France. (See his memoir on "The Organisation of the Philosophical Sciences.")

73. "Cicero, Tuscul," v. 3, there attributed to Pythagoras.

74. M. Villemain, after having read the general part of his "Cours sur les Mammifères," wrote to Geoffroy Saint-Hilaire; "Natural history thus understood is the foremost of all philosophies." One might say the same of all the sciences, were they treated by Geoffroy Saint-Hilaires.

75. This must be admitted even in the ideas of ancient theism, since according to that conception of the system of things, God is considered as no longer creating in time, but as having created everything in the beginning.

76. The true psychology is the poem, the novel, the comedy. There are a great many things that can only be expressed in that way. What we call psychology, for instance that of the Scotch, is only a ponderous and abstract way, without any corresponding advantages, of expressing that which subtle intellects had felt long before the theorists reduced it to formulas.

77. Says M. Michelet; "Let us gather round, and listen to, this young master of olden times. He has no need, in order to instruct us, to go very deeply into what he says, but he is like a living eye-witness; he was there, he knows the whole story best." ("Du Peuple," p. 212.)

78. M. Ozanam shows clearly enough, without any special pleading, that Dante conceived the unity of humanity in a manner almost as advanced as that of the moderns. Christianity in virtue of its *catholicity* made a long stride in the direction of this idea. Nevertheless it is only towards the latter end of the eighteenth century that it appears to us in distinct outline. Old French *humanity* was a virtue or a moral quality but with a good many shades that explain the transition. "I give it you in the name of humanity;" says Molière's Don Juan. "I know no word written in the seventeenth century which conveys a more advanced idea."

79. M. de Maistre pushes the paradox so far as to deny the very existence of human nature and its unity. "I know Frenchmen, Englishmen, Germans," he says; "I do not know *men*." We outsiders are under the impression that the aim of nature is enlightened man, be he French, English, or German.

80. The psychological analysis of the faculties as given by the

Indian philosophers is utterly different from our own. The names of their faculties are untranslatable for us; at times their faculties comprise several of ours under a common name; at others they subdivide ours. I have heard M. Burnouf compare this divergence to that of pieces cut by a punch out of the same surface, or, better still, out of two maps of the same region, drawn at different periods, and placed above one another. Place a map of Europe according to the treaties of 1815 above a map of Europe in the sixth century; the rivers, the seas, the mountains will coincide, but not the ethnographical and political divisions, though even there certain groups remain unchanged.

81. "A Dissertation on the Theology of the Chinese, with a view to the Elucidation of the most appropriate term for expressing the Deity in the Chinese Language, by M. H. Medhurst, 1847, in 8°." See also the report of M. Mohl, *Asiatic Journal*, Aug. 1848, p. 160.

82. "Cours de Littérature Dramatique," vol. i. ch. xvii.

83. Καλός in the Greek sense.

84. The defect of most of our elementary grammars lies in their substituting rules and processes for a rational history of the mechanism of the language. This is especially disgusting in the case of ancient languages, which, properly speaking, had no *rules* but a living organization the actual consciousness of which still existed.

85. "When once one has found what is fitting and beautiful, one ought never to change," says Fleury. There are still people who regret that the world no longer writes in the style of Louis XIVth's time, as if that style were suitable to our mode of thought.

86. The same progress has occurred in mathematics. The ancients considered quantity in its actual being, the moderns take it in its generation, in its infinitesimal element. It is the immense revolution of the differential calculus.

87. India alone deserves in some respects serious consideration as capable of furnishing positive documents to science. We have still much to learn from Indian metaphysics. The most advanced propositions of modern philosophy which here are within the ken of only a very small minority, are there official doctrines. India ought to have nearly as good a right as Greece to furnish themes for art. I have not given up all hope that one day our painters will borrow subjects from Indian mythology as they do from Greek mythology. Narayana lying on his lotus bed, contemplating Brahma who slowly emerges from his navel, Lachmi reposing under his eyes, would not this afford a picture comparable to the most beautiful Greek conceptions? Mathematicians will also find in the Indian theory a number of highly original algorithms.

88. The modern East is a corpse. There has been no education for the East. It is as little ripe to-day for liberal institutions as in the first days of history. It has been the lot of Asia to have enjoyed a charming and poetic childhood, and to perish before arriving at manhood. It seems like a dream to think that Hebrew poesy the Moallacat and the admirable literature of India have sprung from a soil, in our own day so dead, so utterly burnt up. The sight of a

Levantine excites in me the most painful feelings when I reflect that this pitiful personification of stupidity or cunning hails from the country of Isaiah and of Antara, from the country of the mourners for Thammuz, of the worshippers of Jehovah, where Mosaism and Islamism first appeared, where Jesus preached!

89. Hence the aversion to, or the suspicion of, the literatures of the East, to profess which is considered good taste in France, an aversion no doubt due to a certain extent to the worthless criticism too often brought to bear upon these literatures, but still more to our national habit of thought, which is too exclusively literary and not sufficiently scientific. "Do what we will," says M. Saint-Beuve, "we in France do not care to lose sight of the Hellenic horizon without knowing the reason why." Be it so, but why this incurable distrust in presence of methods offering every guarantee? Dugald Stewart in his "Philosophy of the Human Understanding" (1827) is still under the impression that Sanskrit is a worthless jargon compounded haphazard of Greek and Latin.

90. After all Voltaire only followed the track of the apologists. The latter took the Bible in the light of an absolute work, irrespective of time and space; Voltaire criticizes it as he would have criticized a work of the eighteenth century, and from that point of view he finds in it, as a matter of course, not a few absurdities.

91. Hence the pedantry of all classical pretension. We must leave each century to create its own form and original expression. Literature goes on devouring its forms in proportion to their becoming exhausted by the wear and tear to which it subjects them, for literature is bound to be contemporary with the nation. M. Guizot justly points out that the true literature of the fifth and sixth centuries consists no longer of colourless essays by the belated rhetoricians of the Roman schools, but of popular works embodying the Christian legend.

92. "Lecture of M. Burnouf at the Meeting of the Five Académies," 25th October, 1848.

93. The great progress of literary history in our days consists in its having drawn attention to origins and declines. That which occupies us most was never so much as thought of by Laharpe.

94. "Verhandlungen der Versammlungen Deutscher Philologen and Schulmänner," Bonn, 1841.—See also a lecture of M. Creuzer on the same subject, at the Congress at Mannheim, 1839.

95. That which interests us most in ancient writings is the very thing to which their contemporaries never gave a thought; namely, peculiar manners and customs, historical traits, linguistic facts, etc.

96. It is not unusual in Brittany to enclose the head of the dead in a box of wood shaped like a small chapel, with a heartshaped opening in front, and it is through this that the head is supposed to look upon the outer world. Care is taken to dispose the head in such a fashion as only to show the eye from the outside. From time to time those relics are buried and the procession passes round the place of interment every Sunday.

97. It is on this account that the man of the people is far more

sensitive to patriotic glory than the more deeply thinking man, who possesses a pronounced individuality of his own. The latter may stand out from the masses by his personal qualities, by his talents, his titles, his wealth. The man of the people, on the contrary, who possesses none of these takes unto himself, as a patrimony, as it were, the national glory and identifies himself with the masses who have accomplished these great things. It is his own, his patent of nobility, and herein lies the secret of this almost universal adoption of Napoleon by the people. The glory of Napoleon is the glory of those who can lay claim to no other.

98. And, again, those who know how most of these reviews are written are of opinion that in many cases, the monographer cannot reckon on having a single reader. The great art of reviewing consists no longer, as in the time of Fréron, *in judging of the whole from the preface.* The modern reviewer merely takes the title as a peg on which to hang any amount of rigmarole on the same subject as the author.

99. The historians of the seventeenth century who professed to write and flattered themselves that they were read, Mézerai, Velly, Daniel, are nowadays completely shelved, while the works of Du Cange, Baluze, Duchesse and the Benedictines who never professed to do more than collect materials are still as fresh as on the day they were published.

100. The perfection of the Parthenon consists above all in the fact that the parts not intended to be seen are as carefully executed as those intended to be seen. So in science.

101. Eugène Burnouf. "Comment on the Yaçna," pref. p. v. See also in the *Journal des Savants*, Ap. 1848, some excellent reflections by M. Biot on the respect due to anterior works.

102. We are compelled to say as much with regard to the knowledge of Greek literature of the Arabs of the Middle-Ages.

103. Herewith an example which will prove interesting to others besides theologians. With reference to the celebrated passage; "*Regnum meum non est de hoc mundo.*" NUNC AUTEM *regnum meum non est hinc* (John xviii. 36) several schools with widely different intentions have insisted on the *νῦν δέ*, and, translating it by *now* have deduced different consequences from it. This inaccurate remark would not have been so frequently repeated if it had been known that this idiom *νῦν δέ* is the literal translation of a Hebrew locution (*ve-atta*) which serves as an adversative conjunction connoting no idea of time. The same locution in Greek and Latin is employed to denote; *and, moreover, but.* The passage, therefore, should simply be translated; "*But* my kingdom is not of this world." Another of the most important and liveliest discussions in the whole of Biblical exegesis ("Isaiah," ch. liii.) turns entirely on the use of a pronoun (*lamo*).

104. Translation of the *Bhagavata-Purana* by M. Burnouf, tom. i. pref. pp. iv., clxii., clxiii.

105. M. Auguste Comte has paid great attention to this difficult problem and proposes as a remedy for the dispersion caused by

specialities the creation of an additional speciality, namely that of *savants* who without being specialists in any particular department, should occupy themselves with the generalities of all the sciences. ("Cours de Philosophie Positive," tom. i., 1st Lesson, pp. 30, 31, etc.)

106. By the way, I can see only one means of saving this precious collection and keeping it available for use, that is, to close it to all further publications and to declare that no book posterior, say, to 1850 shall be admitted. A separate library should be established for more recent publications. There is evidently a limit beyond which the wealth of a library becomes an obstacle to research, when its abundance becomes truly poverty on account of the impossibility of finding one's way about it. This limit, I believe has already been reached.

107. The burdens imposed on the taxpayer to this intellectual end are in reality a service rendered to him. He profits by the outlay of his money in a way for which he was not sufficiently enlightened to wish deliberately. The state affords the taxpayer, often an inveterate materialist, the opportunity—a rare one in his life—of performing an idealistic work. The day he pays his taxes is really the best in his life, for on that day he expiates his egotism and sanctifies his property, often ill-gotten and as often ill-applied. As a rule rates and taxes are the best applied part of the fortune of the layman and sanctify the rest. It is analogous to libation among the customs of antiquity—an art of high idealism, a touching offering to the invisible, the useless, the unknown, which transforms a vulgar into an ideal act. Taxes, almost wholly applied to civilizing purposes, legitimize, as it were, by their supra-sensible significance the wealth of the farmer and the tradesman ; at any rate they are the best employed part of it. Wealth, under such conditions, from being profane, becomes to a certain extent sacred. In our days taxes are the equivalent of the part devoted under the old system by each man to the Church and to pious works, "for the sake of his soul." For the sake of the taxpayer himself we ought to make this part as large as possible, without, however, giving the taxpayer the true reason which he would never understand.

108. It must be admitted that in that case they would not have existed at all. The thinker never lives by the produce of his thought. Copernicus did not live by his discoveries, he lived by his strict attendance in the choir at Thorn, where he was a canon. The Benedictines of the seventeenth century lived upon ancient foundations designed solely for monastic practices. In our days the thinker and the *savant* live by their teaching, a social employment which has scarcely anything in common with science.

109. The typical sample of that science of the *grand seigneur* who flourishes a horsewhip is M. de Maistre. One could make a collection of the blunders he perpetrates with the infallibility of a perfect gentleman. *Oratio* he tells us, comes from *os* and *ratio* (the reasoning from the lips, which he fancies is admirably profound), *cæcutire*, *cæcus ut ire ; sortir*, *sehorstir ; maison* is a Celtic word ;

sopha comes from the Hebrew, from the root *saphan*, which, he says, means to elevate, whence comes the word *sofetim*, judge, the educators of the peoples (another profound meaning)! Unfortunately the root *saphan* is unknown to any student of Hebrew, and the root *schafat*, whence is derived the word "judges" does in no way mean to elevate or to educate. What matters? It is thus that genius coruscates.

110. See a magnificent page at the end of Laplace, "Système du Monde;" 1st Edition.

111. See in the work of an English missionary, Robert Moffat ("Twenty-three Years in South Africa," pp. 84, 157, 158), some curious instances of myth improvised on the spot. "I saw one day a child who after thinking for some time all at once seriously maintained, and with strange persistency that a few days before it had seen a human head in the sun. Now, it was very patent that this idea had just sprung up in his mind, in combination perhaps with some reminiscence of a passage in an almanac." This is the process that presides at the formation of the most ancient myths; *the dream affirmed.*

112. Where is life more simple than in the animal? Malebranche, one day kicks a bitch who is about to litter. Fontenelle is shocked. "What does it matter?" replies the hard Cartesian, "do not you know she does not feel it?" Father Poirson thus proves that animals have no soul; suffering is the penalty of sin; now, animals have not sinned, ergo, they cannot suffer; ergo, they are mere machines. Father Bougeant traverses the proposition by supposing that animals are demons, consequently, that they have sinned.

113. No one has demonstrated these laws better than M. Fauriel. See the analysis of his course of lectures in 1836 by M. Egger in a series of articles in the *Journal de l'Instruction Publique* of that year, and the excellent notice by M. Ozanam of his illustrious predecessor (*Correspondant* 10 May 1845).

114. "Antar" though it has become the centre of a very characteristic cycle, is not an epic. Everything in it is individual and though the national pride of Arabia is its primary texture, no national cause is sufficiently brought into play to justify us in ranking this beautiful composition in any higher category than that of the novel.

115. As a set-off, the Semites with remarkable facility have conceived in God other relations, such as paternal, filial, distinctions of power, of attributes ("Cabbala," etc.).

116. The efforts that have been made to trace back the laws which determine the succession of the Greek systems to Indian philosophy, are almost chimerical. It cannot be maintained that the law of the development of the Semitic languages is from synthesis to analysis, as is the case in the Indo-Germanic languages. In the same way modern Armenian appears to have much more syntax and synthetic construction than ancient Armenian in which the dissection of thought is pushed to a much more extended limit. Nor can we say that modern Chinese is more analytic than ancient Chinese, seeing

that on the contrary, the inflections in the latter are richer, and the expression of relations more exact. On these different sides the laws are analogous but not the same, albeit they are always perfectly rational, because of the individual element of each race which modifies the result. All formulas are partial because they are only moulded on certain particular cases.

117. M. Auguste Comte, for instance, claims to have found the definite law of the human understanding in the succession of three conditions, *theological*, *metaphysical* and *scientific*. This is, no doubt, a formula containing a great part of the truth; but how can we credit it with explaining everything? M. Comte commences by saying that he only treats of Western Europe ("Philosophie Positive," tom. v. pp. 4, 5). Everything beyond is a mere impertinence not worth considering. And in Europe he only concerns himself with the development of science. Poetry, religion, imagination, all these are ignored.

118. By taking the history of philosophy to mean the history of the human intellect, and not the history of a certain number of speculations.

119. Most of the popular judgments and proverbs are of this kind, and express a true fact, complicated by a fictitious cause. The simple statement of fact is one of the most difficult things to the people; they always mix some apparent explanation with it. When nursemaids aver that *an angel watches over children*, they express a true fact, viz.; that little children do not hurt themselves in the least under circumstances in which grown up people would hurt themselves severely, but not perceiving the cause of this, they consider it more easy to ascribe the cause to a *guardian angel*. The explanation of illnesses by attributing them to devils, which is so continually taken for granted in the Gospels has its own origin in the same intellectual process.

120. Islamism only began to gather strength one or two centuries after the death of the prophet, and since then it has always gone on consolidating itself by the power of established dogma. It has been proved that the immense majority of those who followed the hardy Koreishite had not the slightest religious faith in him. After his death it was seriously discussed whether they should not abandon his religious work and only continue his political work.

121. This, of course, does not impair the originality of that divine product. The learned Jews often try to prove that Jesus has *stolen* the whole of his doctrine from Moses and the prophets, and that what has been called *Christian morality* is in reality nothing but *Jewish morality*. This would be true if a religion consisted of a given number of dogmatic propositions, and morality of sundry aphorisms. Most of those aphorisms being very simple and of all ages, there are no new discoveries to be made as regards morality. Originality in morals lies merely in an indefinable touch and in a new way of feeling. In order to test this we have merely to place side by side the Gospel and the collection of moral apophthegms of the rabbis contemporary with Jesus, the *Pirké-Avoth*, and to compare *the moral impression* resulting from these two books.

122. See in Voltaire's "Dictionnaire Philosophique" the charming article "Gargantua" in which by arguments similar to those of the apologists it is proved that the marvellous exploits recorded in the history of Gargantua admit of no doubt. Rabelais bears witness to them; no historian has refuted them; the sceptical Lamotte Le Vayer was inspired with such respect for them, as not to have breathed a single word concerning them. These prodigies were performed in the sight of the whole world. Rabelais, who testifies to having seen them, was neither deceived nor a deceiver. If he had departed from the truth, the newspapers would have soon brought him to book. And, if that history were not true, who would have dared to imagine it? The great proof of its being worthy of belief is that it is incredible, etc. In fact, the defect of the critical system of the supernaturalists is to judge all the periods of the human understanding by the same test.

123. When the Arabs had adopted Aristotle as the grand master of science they wove around him a miraculous legend as if he had been a *prophet*. They pretended that he had been taken away from heaven on a column of fire, etc.

124. It is strange that Europe should have adopted as the basis of her spiritual life the books least adapted to her, the literature of the Hebrews, the work of another race and emanating from a spirit different from her own. As a matter of course she only accommodates herself to them by entirely misconceiving their meaning. The Vedas would by far have a better claim to be the sacred book of Europe than the Bible. They at least are truly the work of our forefathers.

125. In the East an *ancient* book is always *inspired*, whatever may be its contents. There is no other criterion with regard to the canonicity of a book. As for primitive epochs, every book, from the very fact of its being a written book, was *sacred*. For did it not treat of things divine? Was not its author a priest, in direct communication with the gods? The conception of the profane book, the individual work, good, bad or indifferent, of this man or that, belongs to a later period.

126. A few months ago, I heard a much-admired preacher in the pulpit of Notre-Dame classify the religions in the following manner; "There are three religions: Christianity, Mahometanism, Paganism." This is exactly as if someone were to classify the animal kingdom by saying; "There are three kind of animals: men, horses, and plants."

127. I am not referring to China. That strange nation is perhaps the least religious and the least supernaturalistic of all. Her sacred books are nothing more than *classics*, much the same as "the ancients" are to us, or at least as they were to our classical scholars. In this lies perhaps the secret of Chinese mediocrity. It is a beautiful thing to dream, not like India, for ever, but to have dreamt during one's childhood; there remains a beautiful perfume of it during our waking hours and a whole tradition of poesy on which to fall back when age has chilled the imagination.

128. The religion of the Nomadic Semites is exceedingly simple. It is the patriarchal worship of the only God, pure, chaste, without creed or symbols, without mysteries, without orgies. All those grand systems of Assyrian, Persian, Egyptian symbolism are not of Semitic origin, and disclose a different kind of spirit, much deeper, bolder, more inquiring. It is only in the sixth century before the Christian era that ideas of this kind were introduced among the Semites. There is a world-wide difference between the all-ruling and solitary God of Job, of Abraham, of the Arabs and those grand pantheistic poems, disclosed to us by the Egyptian and Assyrian monuments. It appears, moreover, that the primitive worship of Egypt came very near Semitic simplicity and that the polytheistic symbolism was a foreign importation.

128*a*. The Arabs, to use the accepted phrase, have certainly shown a philosophic and scientific development, but their science is wholly borrowed from Greece. We should, moreover, point out that Greco-Arab science by no means flourished in Arabia ; it flourished in the non-Semitic countries under Islamite sway which adopted Arabic as a learned language, in Persia, in the provinces of the Oxus, in Morocco, in Spain. The Arabian peninsula has remained almost free from Hellenism and has never understood aught but the Koran and the ancient poems.

129. The real mythology of the moderns ought to be Christianity, the monuments of which are still alive among us. But the age of Louis XIVth, which dogmatically took this mythology as a theology, could not make a poetic machinery of it. Boileau is right. *To invest sacred truths with the semblance of fable*, is a sin. I paid a visit one day to M. Michelet, he led me round his drawing-room and pointed out for my admiration the most beautiful Christian subjects of the great masters, the Saint-Paul of Albert Dürer ; the Prophets and Sibyls of Michael-Angelo, the "Disputa del Sacramento," etc., and then he began to comment upon them. I am certain that Racine, who was a believer, had Pagan images in his drawing-room. If he had had Christian engravings he would have treated them as *devotional* images. Syracuse did not consider it an act of bigotry to stamp her medals with the beautiful head of Arethusa, nor Athens with that of Minerva. Why then should there be an outcry about encroachment if we were to put Saint Martin or Saint Remi on our moneys? Until people ceased to look upon Christianity as a Theology, they were unable to begin looking upon it as a Poesy, and I have often asked myself whether Chateaubriand aimed at anything more than at a *literary* revolution.

130. The Mosaic prescriptions, for instance, on the abstention from the flesh of certain animals killed in a certain fashion, respectable enough when looked at as a means of educating humanity, and all of which had a highly moral and highly political justification with an ancient tribe of the East, to what do they amount when transferred to our modern States? Merely to a good deal of inconvenience which obliges people of a certain religion to have their particular butchers who are obliged to purvey cattle according to

certain rules; simply a question of the slaughter-house and the kitchen.

131. The Latin authors of the Decadence, the tragedies of Seneca, for instance, often sound better when translated into French, than the masterpieces of the grand epoch.

132. As a typical instance of this imbecile admiration consult the Preface to the Translation of the Psalms by La Harpe. M. de Maistre has very naïvely remarked, "To get a conception of the beauties of the Vulgate, select a friend *who has no knowledge of Hebrew* and you will find how a simple syllable, a simple word, and an *undefinable lightness of touch imparted to the phrase*, will cause to spring forth under your very eyes beauties of the highest order." ("Soirées de Saint-Petersbourg," 7th Conversation.) With such a system, and especially with the aid of a friend who knows nothing of Greek, I will undertake to find beauties of the highest order in the most worthless translation of Homer or Pindar, independent of those that are there. This reminds one of Madame Dacier going into ecstasies about a certain passage of Homer, because it is susceptible of five or six different interpretations, all equally beautiful.

133. I will only point out one trait among many. We shall not disparage the glory of the illustrious author of "Le Génie du Christianisme" by refusing him the title of Hellenist. He admires ("Génie du Christ.," Book v., ch. i. or ii.) the simplicity of Homer in describing the grotto of Calypso by the simple epithet, "carpeted with lilac." And now let us look at the passage; ἐν σπέσσι γλαφυροῖσι, λιλαιομένη πόσιν εἶναι (Od. i. 15). I believe, Heaven forgive me, that he saw lilacs in λιλαιομένη.

134. Unless one has read the exegetical works of this great man one can conceive no idea of his radical want of the critical faculty. He is exactly on a par with Saint Augustin, his master. To quote only one instance in point; did not he write a book to justify the policy of Louis XIV. by the Bible? The annoyance shown by Bossuet at the works of Ellies Dupin, Richard Simon, Doctor Lannoy in which they sounded the first notes of higher criticism, and the persecution he aroused against those intelligent pioneers are, with the exception of the revocation of the Edict of Nantes, the most deplorable episode in the history of the Gallican church in the seventeenth century.

135. Says M. Michelet; "The simple-minded are fond of connecting and linking; they rarely divide and analyse. Not only is all kind of division repugnant to their mind, but it pains them, they look upon it as a dismemberment. They do not like to divide life and to them everything seems endowed with life. Not only do they not divide, but the moment they find a divided or partial thing, they either neglect it or else join it mentally to the whole from which it is divided. In this lies their existence as the *simple of the earth*." See the whole of this admirable passage ("Du Peuple," pp. 242, 243). One consequence of this simple mode of taking life is the perception of the *physiognomy* of things, which is never vouchsafed to the

learned analysts, who only see the inanimate element. Most of the categories of ancient science which the moderns have excluded corresponded to outward characteristics of nature which are no longer considered, though indeed they contain their share of the truth.

136. Poetry itself shows an analogous onward movement. In primitive poesy, all the styles were confounded ; the lyrical, elegiac, didactic, epic elements co-existed in it in a confused harmony. Then came the epoch of distinguishing styles, during which objections would have been raised to the introduction of lyricism into the drama, or of the elegy into the epic. This was succeeded by the higher form in the grand poesy of Goethe, of Byron, of Lamartine, admitting the simultaneous introduction of all the styles. "Faust," "Don Juan," "Jocelyn" do not fall into any particular literary category.

137. This turn, peculiar to the German genius, explains the strange progress of ideas in that country for something like the last quarter of a century and how after the lofty and ideal speculations of the grand school, Germany is now enacting her eighteenth century after the French manner ; hard, bitter, negative, scoffing, swayed by the instinct of the finite. For Germany, Voltaire comes after Herder, Kant, Fichte, Hegel. The writings of the young school are definite, destructive, realistic, materialistic. They boldly deny "*the beyond*" (*das Jenseits*), that is, the supra-sensible, the religions under all its forms, they declare that it is fooling mankind to make him live in that fantastic world. Such is the sequel of the most idealistic literary movement presented by the history of the human intellect, a sequel not arrived at by logical deduction or as a necessary consequence, but by a deliberate contradiction, and in virtue of this foregone conclusion ; the great school has been idealistic, we are going to produce a reaction towards the realistic.

138. Languages present an analogous development. Let us take a group of languages, comprising several dialects, such as for instance, the Semitic group. Certain linguists suppose that, at the origin, there was only one Semitic language, from which all the dialects are derived by alteration ; others suppose all the dialects equally primitive. The truth is, it seems, that at the origin the various characteristics, which, by forming themselves into groups, became the Syriac, the Hebrew, etc., existed syncretically, though without, as yet, constituting independent dialects. For instance ; 1° Confused and simultaneous existence of the dialectal varieties ; 2° Isolated existence of the dialects ; 3° Fusion of the dialects into a more extensive unity.

139. The divine Spherus of Empedocles, in which everything exists at first in the syncretic condition under the domination of the φιλία, previous to passing under that of Discord, νεῖκος (analysis) presents a beautiful picture of this grand law of divine evolution.

140. "Le Peuple," p. 251.

141. The most curious instance of this is M. de Talleyrand becoming converted at the close of his life. He had been sharp enough to outwit all the diplomatists of Europe, and bold enough to celebrate the mass of liberty and to constitute himself a schismatic ;

but when it came to a theoretical question, he becomes weak-minded and credulous, seeing nothing strange in Nebuchadnezzar being changed into an animal, Balaam's ass conversing with his master, or the diplomatists of the Council of Trent receiving the aid of the Holy Ghost. Talleyrand, it will be said, did not admit all this. No ; but he would have had to admit it, if he had been consistent with himself.

142. Fichte, who (in France, I of course mean) would have been set down as impious, always had family prayer in the evening ; after which a few verses of a hymn were sung to the accompaniment of the piano. The philosopher then delivered a short homily to his family upon some few pages of the Gospel of St. John, adding, as the opportunity suggested itself, some words of consolation or pious exhortation.

143. Can a rag-picker, as he passes in front of the Tuileries, exclaim : "This is my work ?" Can we realize the sentiment of the artisans and the land tillers of Attica in presence of these monuments which belonged to them, which they appreciated, and which were in reality the expression of their thought ?

144. One of the benefits of the Empire was that it gave the people heroic souvenirs and a name easy to understand and to idolize. Napoleon, so frankly adopted by the popular imagination, providing it, as he did, with a grand subject for national enthusiasm, will have powerfully contributed to the intellectual elevation of the ignorant classes, and has become for them what Homer was for Greece : the initiator of great deeds, causing the pulse to vibrate and the eye to sparkle.

145. It is needless to say that this excuse, if it be one, never applies to the silly plagiarists, who imitate in cold blood the furies of another age. I am glad to say once for all that those who credit me with sympathies for any political party, especially with that one, would quite misinterpret my ideas. I am for France and the right ; for them and for nothing else.

146. How can a member of the Academy of Moral and Political Sciences have come to write, now in the middle of the nineteenth century, such axioms as these : "Society is not human kind ; it is only their union. Men live for themselves and not for this chimera, this vain abstraction styled humanity. The destiny of a free state could not possibly be subordinated to any other destiny." (See *L'Homme et la Société*, pp. 53, 81. And yet fifty years after Herder had written : "Man could not, even if he so willed it, live for himself alone. The beneficent influence of man upon his fellow man is the aim of all human society. In addition to the individual fund, which each one brings into play, there is the mass of capital, which, ever accumulating, forms the common fund of the species, etc." See *Ueber den Charakter der Menscheit.*) The bee's cell could not exist without the hive, so the hive has a claim to make upon the bee.

147. What folly to take any interest in such degraded creatures, says the master when speaking of the negroes, though it is he who keeps them in a state of degradation.

148. (*Polit.*, book i. ch. ii. p. 8 and foll.). Aristotle goes so far as to say that if beauty were a mark of individual worth, the least handsome ought to be the slaves of those who are the handsomest.

149. If it were only for political or external reasons that such important engines should be kept a close watch over. No doubt, but there is another question. Let me add that it is somewhat strange to find modern and unconcerned politicians giving salaries to their mortal enemies, to those who have fought them to the death, to those who embrace them only to stifle them or to get a profit out of them.

150. The Inquisition is the logical consequence of the whole orthodox system. The Church will be bound, when she has the chance, to bring back the Inquisition, and if she does not, it is because she cannot. For why is this kind of repression less necessary to-day than it was in former times? Is our opposition less dangerous? Assuredly not. The reason must, therefore, be that the Church is weaker. We are tolerated because we cannot be strangled. If the Church became once more what she was in the Middle Ages, an absolute Sovereign, she would be bound to resume her maxims of the Middle Ages, inasmuch as it is declared that these maxims were good and beneficent. Power has always been the measure of the Church's toleration. In reality, this is not a reproach; this is as it should be. It is a mistake to rate the orthodox with regard to tolerance. Ask them to renounce orthodoxy, if you like, but do not ask them, while remaining orthodox, to put up with heterodoxy. With them, it is a question of life or death.

151. See Bossuet's admirable sermon upon the profession made by Mlle. de la Vallière and for the festival of Presentation.

152. The first impression which Christianity produced upon barbarous peoples, subject to aristocratic and gross prejudices, was one of repulsion for the spiritualistic and democratic part of its precepts. In the Irish legends, Ossian, singing of wars, heroes, grand hunts, etc., is often brought forward in comparison with St. Patrick and his psalm-singing flock. Mihir-Nerseh, in a proclamation addressed to the Armenians, in order to dissuade them from becoming Christians, asks them how they can put any faith in ill-clad vagabonds, who prefer men of humble estate, to persons of quality, and who are foolish enough to attach little importance to wealth.

153. This change generally occurs in this wise. A day comes when the retrograde party is obliged to pose as being persecuted and to lay claim to the principles which it had fought against; for instance, the principles of the sovereignty of the people and of liberty. Even those who had so strongly repudiated them when they were contrary to them have found themselves brought by the force of things to invoke them and to insist that the heresies which they had dethroned should be carried to their extreme consequences. The new ideas can only be vanquished by themselves, or rather it is they which vanquish their opponents by compelling the latter to have recourse to them in order to triumph. Children that you are,

do you not see that, when you think you are drawing the chariot of humanity backward, that it is the chariot which is dragging you along with it?

154. Cadit et sic aperiuntur oculi ejus (Num. ch. xxiv. v. 4).

155. How singular! a month after the constitution has begun to work, it stands in need of being interpreted. It is violated, says the one side.—Not at all, says the other. Who is to decide? M. de Maistre is right; to cut down disputes at the root, you must have infallibility. The worst of the business is that infallibility does not exist. Principles only apply to a certain extent. So we must give up the idea of discovering the definite ulterior, and maintain well-considered reason as the final authority. Yet it is so easy to find repose in the infinite, to embrace with one's whole soul a narrow and finite formula. The immensity of humanity excites awe; the brain reels before this deep abyss.

156. The result would be a very poetical situation and one as yet unknown; a system of slavery felt and endured with delicacy and resignation. The slave of old was not poetical, because he was not regarded as a moral being. The slave of ancient comedies is an infamous and vile character; he has only his baseness to console him; he is not susceptible to virtuous feelings. The slave as here conceived would be the superior of his master, because he would have a better perception of what is divine and would find in love an escape from the hideous reality.

157. One is sometimes tempted to ask oneself whether humanity was not emancipated too soon. Strong and intellectual consciences like our own are with much more difficulty brought to set themselves to a great work, being too much attached to their own will and to life. How is humanity, with an individual liberty as highly developed as ours, to conquer the desert places? Will it be said that humanity has become incapable of subduing the whole universe because it has been prematurely set free? Any great enterprise of this kind demands a first supply of men. Think of what the English colonies cost, those of the Presbyterians and Methodists in the United States, for instance. Such sacrifices have become impossible now, for the price of human life has gone up; the world has got into the way of counting the cost too closely. If a score or so of settlers fall ill when a colony is founded, there is at once a great outcry. Yet one must remember that the first generations of colonists have nearly everywhere been scarificed. The Icaria of M. Cabel might have succeeded two hundred years ago; but in our day, and especially with Frenchmen, it was an absurdity. The greatest things cannot be done without sacrifice, and religion, which prompts sacrifice, no longer exists. I sometimes delude myself with the hope that machinery and the progress of applied science will one day compensate for what humanity has lost in the way of aptitude for sacrifice by the progress of reflection. Man is always ready to run a risk, but he is less ready to face certain death.

158. Let me assume, for instance, that chemistry were now to discover a means for rendering the acquisition of good food so easy

that one would only have to stretch out the hand to take it; it is certain that three-fourths of the human race would give itself over to idleness, that is to barbarism. One might then use the whip to compel them to build great social monuments, pyramids, etc.; tyranny would be legitimate to secure the triumph of the mind.

159. We are indignant at the way in which man is treated in the East and in barbarian States, and at the small value set on human life. This is not so revolting when it is remembered that the barbarian has but little command over himself and possesses, as a matter of fact, much less value than civilized man. The death of a Frenchman is an event in the moral world; that of a Cossack is little more than a physiological fact; a machine was in motion which is in motion no longer. And as to the death of a savage, it is scarcely of more importance in the march of events than when a watch-spring breaks, and even this latter occurrence may have much graver consequences, owing to the very fact that the watch in question arrests the thought and excites the activity of civilized men. The deplorable thing is that a portion of humanity should be so degraded that it scarcely counts for more than the animals; for all men are called to have a moral value.

160. For instance; it was essential for humanity that the Jewish nation should exist, should be hard, indestructible, made of bronze as it has been. By the second or third century, they had answered their purpose; humanity had no further need for the Jews. The Jews nevertheless continue to exist like a dead branch, while, if the matter is looked at rather more closely, it will be seen that this dead branch has not been so useless as may be imagined.

161. The picturesque is not taken sufficiently into account in the guidance of humanity. And yet this is of at least as much importance as happiness. I have heard of an engineer who, in the tracing of roads, endeavoured to secure for those who travelled over them pretty views, at the cost of convenience and of time. That is the sort of man I should have liked.

162. I do not allow that it is an unanswerable proof of immortality to say that it is a necessity for divine justice to repair, in another world, the acts of injustice which the general order of the universe entails in this. Our forefathers suffered and we inherit the fruits of their suffering. The future will gain by this. Who knows but that it will one day be said: "In those times there must have been faith, for humanity then laid the foundations, by its sufferings, of the better state of things which we enjoy. But for that, our fathers would not have had the courage to endure the heat of the day. But now we have the key of the enigma, and God is justified by the greater good of the species." So long as belief in immortality will have been necessary in order to render life endurable, so long it will have been believed.

163. As a general rule, the barbarians were received with open arms. The Bishops and all the most enlightened men such as St. Augustine, Salvian, etc., opened their arms to them. Upon the other hand, the last representatives of the old society, polite, corrupt

and effete, Sidonius Apollinaris and Aurelius Victor, heap insults upon them and cling to the abuses of the old Empire, without seeing that it was inevitably condemned to perish.

164. Mendelssohn, already celebrated, already one of the first critics in Germany, was still a warehouseman in a silk shop. Lessing, who had come on purpose to see him, found him at the counter, measuring out the silk.

165. The sordid or so-called low character of certain occupations might also indicate them to persons devoted to literary labour ; for this character would be likely to correspond either to a higher rate of pay ; or, what amounts to the same thing, to fewer hours of labour. Lowness of condition, according to worldly ideas, does not exist for man, regarded from a moral point of view.

166. Gymnastics, for instance, are considered by many people as a useful diversion for indoor work. But would it not be more useful and pleasanter to follow for two or three hours the calling of carpenter or gardener, than to tire oneself out with movements which have no sort of aim ?

167. Aristotle, *Polit.*, book i. ch. ii. 5. (See Translation by M. Barthélemy St. Hilaire.)

168. I depict to myself the mind as a tree the branches of which are studded with iron hooks. Study is, as it were, a cornucopia pouring out from the top upon this tree objects of a thousand different colours and shapes. The hooks do not catch all these objects or hold them for an indefinite time. Such an object, after having hung to one of the hooks for a certain time, drops, and then comes the turn for another. So the mind, at its different epochs, is as it were provided with a various assortment of things, and that, added to the inward modifications of its being, causes the diversity of its aspects.

169. I carry the respect of the individuality so far that I should like to see women given a share in critical and scientific work, being persuaded that they would open up new views of which we have no idea. If we are better critics than the savants of the seventeenth century, it is not that we know more than they did, but that we see more clearly. Well, I am convinced that women would import into this their individuality, and would reflect the object in fresh colours. The socialists are quite mistaken as to the intellectual rôle of woman ; they would like to make a man of her. But woman will never be more than a poor sort of man. She must remain what she is, but must be pre-eminently what she is. She is different from but not inferior to man. A perfect woman is quite as good as a perfect man. But she should be perfect in her own way, and not by resembling man. She differs from him like positive from negative electricity, that is to say as regards sense and direction, not as regards essence. The negative is not inferior to the positive, but it goes in the opposite direction ; any quantity may be considered either as negative or positive at pleasure. The negative and the positive combined form the complete. Everything seeks for its complement, the positive naturally attracts the negative, the inner angle invites

the outward angle. Thus life is divided ; all have the better part, and there is room for love.

170. In his beautiful piece *le Crépuscule.*

171. "We shall have all that in Paradise" a clever answer made by kind-hearted nuns, rather out of patience, to a scientific maniac, who, having gone into a hospital, bored the worthy sisters who were nursing him with his misplaced elucubrations.

172. Wherever you go in our towns and public promenades, you will come upon barriers and notices, which are necessary, it is true, for the preservation of order, but which make anything like free action impossible. We all of us must have felt the humiliating and disagreeable effect which a prohibitive notice produces, even when one knows that it is general ; it is a limit fixed. When I walk through the gardens at Versailles, with a hedge on each side of me, I am never satisfied. I should like to go in among the shrubs, and that is forbidden. How monotonous our broad, straight roads are ! I prefer by far the crooked roads of Brittany, with the sheep feeding at the sides. What is more distasteful than a high road, what can be more charming than a pathway ?

173. One of the most beautiful deaths conceivable is that of the inquisitive man, indifferent to his end, and with his attention wholly fixed upon the rising of the curtain which is about to take place, and upon the mighty problems about to be solved for him.

174. " When he thinks he has put forth something exaggerated," says Goethe speaking of Albert, "something too general or doubtful, he keeps on limiting, modifying, adding or retrenching until nothing is left of his proposition." Many people will no doubt distort my idea, because I have not adopted that plan.

175. Augustin Thierry, *Dix Années d'Etudes Historiques*, preface.

176. Study the characters of Polus and of Kallicles in Plato's *Gorgius.*

177. See the curious conversation with Le Maistre de Sacy preserved by Fontaine.

178. *Méthode pour arriver à la vie bienheureuse*, last lesson. The whole of this lesson is an admirable one. Never did the pious wrath of honest souls against scepticism find more eloquent expression.

179. One of the characteristic traits of the men about whom I am speaking is to affect a profound contempt for ideal art, pure and noble passion. They make mock at it, and are ready to say with Byron : "Oh ! Plato, you were only a pimp." They regard idealism as a piece of stupidity, and declare that they much prefer frank epicureanism.

180. Or else the clever erudition of Barthélemy, which, though of a more elevated order, is nevertheless not the grand philosophical and scientific method.

181. Acts of the Apostles, ch v., verses 38 and 39.

182. I once saw in a wood a swarm of nasty little insects, which had surrounded with their webs and were sucking its green shoots with such a character of parasitism that one could not help feeling

disgust. I was for a moment tempted to destroy them, but then I said to myself: It is not their fault if they are ugly: it is one way of life. It shows narrowness of mind to moralize nature and impose one's judgments upon her. But now I see that I was wrong: I ought to have killed them; for the mission of man in nature is to reform what is ugly and immoral.

183. The science which is the most devoid of an object, mathematics, is the very one which excites the most ardour, not so much by its truth as by the play of faculties and the power of combination which it implies. The pleasure which mathematics procure is of the same kind as that of a game of chess. None is more despotic. When Archimedes was absorbed in his demonstration table his slaves had to tear him away from it to rub him over with oil, and even then he would trace geometrical figures upon the oily surface of his body.

184. See Note 178.

185. "Some, seeing the place of the political government invaded by incapable men, withdrew. And he who asked how long it was necessary to go on philosophizing, received this answer: 'Until there are no more donkey-drivers to lead our armies.'" (Montaigne, book i. ch. xxiv.)

186. The wars of the giants of the Revolution have made nobles of us all. We are the sons of a race of heroes. Each of our fathers was entitled to say of himself: "I am an ancestor." You are the great grandchildren of crusaders; I am the son of a soldier of the Revolution, and am as good as you.

187. I assume that one of Plato's dialogues represents a conversation in Athens really held, very different from the similar compositions of Cicero, Lucian and so many others, who merely use the dialogue as a factitious form for embodying their ideas, without seeking to render any scene in real life.

188. The presence and the essential part played by woman in our modern society is no doubt the cause of this. As nothing must be said which would go above the heads of this part of the audience, the circle of speech is somewhat limited. If the seven sages, at their banquet, had been subjected to this condition, I doubt whether their dissertation would have been as elevated as it was.

189. *Nouveau Journal Asiatique*, vol. i. p. 345.—Compare in the Saint-Brandon poem the description of that marvellous island in which the monks do not grow old and receive their bread from heaven, in which the lamps light of themselves; a life of silence, of liberty, of calm, the ideal of the monastic life amid the water-floods.

190. Chateaubriand was altogether wrong in looking for poetry in the present state of Christianity. His achievement was the revealing to criticism of a view of beauty which had been unnoticed in the Christian dogma and worship; but he ought to have confined himself to the past and not have sought for poetry in the commonplaces of Jesuitism. Christianity has lost its poesy since the sixteenth century, and this has put him quite out of tune. Admirable as he is when touching the high religious string, he lapses into

the trivialities of the preacher and the apologist when he enters into what I may call vestry details. In this respect Madame de Stael is far superior to him.

191. I would without hesitation have taken Malebranche's formula *Dieu est le lien des esprits comme l'espace est le lien des corps* if it were not conceived from the substantial point of view, this giving it a somewhat coarse and inaccurate meaning. God, the spirit, the body, are, in the sense he attaches to them, too objective words.

192. It is said, for instance, that God is a spirit, that He has all the attributes of a spirit. As spirit merely signifies everything which is not body, this reasoning is equivalent to saying that there are two classes of animals ; those which are horses and those which are not horses. The bird is a "not-horse." The fish is a not-horse. Therefore the bird and the fish are of the same species, and what is said of the one will also apply to the other.

193. Christianity only received its full development in the hands of the Greeks ; so that it did not, in its definite form, gain the sympathy of oriental peoples. If, upon the contrary, it had remained what it was for the early Judæo-Christians, for St. James let us say, it would have conquered the East and there would have been no Islam. But then, again, it would never have acquired any influence in Europe.

PRINTED BY WILLIAM CLOWES AND SONS, LIMITED, LONDON AND BECCLES.